D0788160

Horizons

Horizons

The Global Origins of Modern Science

JAMES POSKETT

MARINER BOOKS
Boston New York

HarperCollins books may be purchased for educational, business, or sales promotional use. For information, please email the Special Markets Department at SPsales@harpercollins.com.

Originally published in the UK by Penguin Random House in 2022.

FIRST MARINER BOOKS EDITION PUBLISHED 2022.

Library of Congress Cataloging-in-Publication Data has been applied for.

ISBN 978-0-358-25179-8

22 23 24 25 LSC 10 9 8 7 6 5 4 3 2 1

For Alice and Nancy

Contents

Illustrations

Plates

Section 1

Section 2

1. Skeleton of the *Megatherium* on display in nineteenth-century Madrid (Alamy)
2. Zoologist and evolutionary thinker Ilya Mechnikov (Wikipedia)
3. Physicist Jagadish Chandra Bose, 1897 (Getty)
4. Postcard from the Paris Exposition of 1900 (Alamy)
5. Theoretical physicist Zhou Peiyuan (Wikipedia)
6. Albert Einstein and his wife, Elsa, in Japan, 1922 (Wikipedia)
7. Physicist Aikitsu Tanakadate in his office at the University of Tokyo (Alamy)
8. Chandrasekhara Venkata Raman, the first Indian scientist to win a Nobel Prize (Alamy)
9. Physicist Hideki Yukawa, the first Japanese scientist to win a Nobel Prize (Getty)
10. Chinese Communist Party propaganda poster from the 1960s (Getty/Translation: Anne Gerritsen)
11. Geneticists Obaid Siddiqi and Veronica Rodrigues, 1976 (Archives at NCBS)
12. Japanese doctor examining a young patient in Hiroshima, 1949 (Getty)
13. Yemeni Jewish families arriving at an immigration camp in Israel, 1949 (Wikipedia)
14. Israeli population geneticist Elisabeth Goldschmidt (Wellcome Collection)
15. Sarah Al Amiri, Chair of the United Arab Emirates Space Agency and Deputy Project Manager for the 2020 Emirates Mars Mission (Wikipedia)
16. Moustapha Cissé, Staff Research Scientist and Director of the Google AI Center, Accra, Ghana (Getty)

Note on Spelling and Translation

This book covers a great range of languages, regions, and periods of history. Broadly, I have followed the rule of using the spelling and ordering conventions common to the particular place and period I am writing about in each chapter. In a few instances, I have made an exception in order to improve readability. Diacritical marks have been omitted from the transliteration of non-Latin scripts. All translations, unless otherwise indicated, are taken from the sources listed in the notes.

Introduction: The Origins of Modern Science

Where did modern science come from? Until very recently, most historians would tell you the following story. Sometime between 1500 and 1700, modern science was invented in Europe. This is a history which usually begins with the Polish astronomer Nicolaus Copernicus. In *On the Revolutions of the Heavenly Spheres* (1543), Copernicus argued that the Earth goes around the Sun. This was a radical idea. Since the time of the ancient Greeks, astronomers had believed that the Earth was at the centre of the universe. For the first time, scientific thinkers in sixteenth-century Europe started to challenge ancient wisdom. Copernicus was followed by other pioneers of what is often called the 'scientific revolution' – the Italian astronomer Galileo Galilei, who first observed the moons of Jupiter in 1609, and the English mathematician Isaac Newton, who set out the laws of motion in 1687. Most historians would then tell you that this pattern continued for the next 400 years. The history of modern science, as traditionally told, is a story focused almost exclusively on men like Charles Darwin, the nineteenth-century British naturalist who advanced the theory of evolution by natural selection, and Albert Einstein, the twentieth-century German physicist who proposed the theory of special relativity. From evolutionary thought in the nineteenth century to cosmic physics in the twentieth century, modern science – we are told – is a product of Europe alone.[1]

This story is a myth. In this book, I want to tell a very different story about the origins of modern science. Science was not a product of a unique European culture. Rather, modern science has always depended upon bringing together people and ideas from different cultures around the world. Copernicus is a good example of this. He was writing at a time when Europe was forging new connections with Asia, with caravans travelling along the Silk Road as well as galleons sailing across the Indian Ocean. In his scientific work, Copernicus relied upon mathematical techniques borrowed from Arabic and Persian texts, many of which had only recently been imported into Europe. Similar kinds of scientific exchange were taking place throughout Asia and Africa. This was the same period in which Ottoman astronomers journeyed across the Mediterranean, combining their knowledge of Islamic science with new ideas borrowed from Christian and Jewish thinkers. In West Africa, at the courts of Timbuktu and Kano, mathematicians studied Arabic manuscripts imported from across the Sahara. To the east, astronomers in Beijing read Chinese classics alongside Latin scientific texts. And in India, a wealthy maharaja employed Hindu, Muslim, and Christian mathematicians to compile some of the most accurate astronomical tables ever made.[2]

All this suggests a very different way of understanding the history of modern science. In this book, I argue that we need to think of the history of modern science in terms of key moments in global history. We begin with the colonization of the Americas in the fifteenth century and move all the way through to the present. Along the way we explore major developments in the history of science, from the new astronomy of the sixteenth century through to genetics in the twenty-first. In each case, I show how the development of modern science depended upon global cultural exchange. It is worth emphasizing, however, that this is not simply a story of the triumph of globalization. After all, cultural exchange came in lots of different forms, many of which were deeply exploitative. For much of the early modern period, science was shaped by the growth of slavery and empire. In the nineteenth century, science was transformed by the development of industrial capitalism. Whilst in the twentieth century, the history of science is best explained in terms of the Cold War and decolonization. Yet despite these deep imbalances of power, people from across the world made significant contributions to the development of modern science. Whatever period we look at, the history of science cannot be told as a story which focuses solely on Europe.[3]

The need for such a history has never been so great. The balance of the scientific world is shifting. China has already overtaken the United States in terms of science funding, and for the last few years researchers based in China have produced more scientific articles than anywhere else in the world. The United Arab Emirates launched an unmanned mission to Mars in the summer of 2020, whilst computer scientists in

Kenya and Ghana play an increasingly important role in the development of artificial intelligence. At the same time, European scientists face the fallout from Brexit, whilst Russian and American security services continue to wage cyberwarfare.[4]

Science itself is plagued by controversy. In November 2018, the Chinese biologist He Jiankui shocked the world by announcing that he had successfully edited the genes of two human babies. Many scientists believed that such a procedure was too risky to justify trying on human subjects. However, as the world quickly learned, it is very hard to enforce an international code of scientific ethics. Officially, the Chinese government distanced itself from He's research, serving him with a three-year prison sentence. But in 2021, researchers in Russia are already threatening to replicate his controversial experiment. Alongside issues surrounding ethics, science today, as in the past, suffers from deep inequalities. Scientists from minority ethnic backgrounds are underrepresented at the top of the profession, Jewish scientists and students continue to suffer antisemitic abuse, whilst researchers working outside of Europe and the United States are often denied visas for travel to international conferences. If we are to tackle such problems, we need a new history of science, one that better reflects the world in which we live.[5]

Scientists today are quick to acknowledge the international nature of their work. But they tend to think of this as a relatively recent phenomenon, a product of the 'big science' of the twentieth century, rather than something with a history stretching back more than 500 years. When contributions to science from outside of Europe are acknowledged, they are typically relegated to the distant past, not part of the story of the scientific revolution and the rise of modern science. We hear a lot about the 'golden age' of medieval Islamic science, the period around the ninth and tenth centuries, when scientific thinkers in Baghdad first developed algebra and many other new mathematical techniques. There is a similar emphasis on the scientific accomplishments of ancient China, such as the invention of the compass and gunpowder, both well over 1,000 years ago. But these stories only serve to reinforce the narrative that places like China and the Middle East have little to do with the history of modern science. Indeed, we often forget that the notion of a 'golden age' had originally been invented during the nineteenth century in order to justify the expansion of European empires. British and French imperialists promoted the false idea that the civilizations of Asia and the Middle East had been in decline since the medieval period, and so needed to modernize.[6]

Perhaps surprisingly, these stories are still just as popular in Asia as they are in Europe. Cast your mind back to the 2008 Beijing Olympics. The opening ceremony began with an enormous scroll unfolding, signifying the invention of paper in ancient China. Throughout the ceremony, a television audience of over one billion watched as China showcased its other ancient scientific achievements, including the compass. Fittingly, the ceremony closed with a spectacular display of another Chinese discovery. Fireworks lit up the sky above the Bird's Nest Stadium, a nod towards the invention of gunpowder during the Song dynasty. Yet throughout the ceremony, there was very little reference to the many scientific breakthroughs that China has contributed to since then, such as the development of natural history in the eighteenth century or quantum mechanics in the twentieth century. The same is true of the Middle East. In 2016, the Turkish President, Recep Tayyip Erdoğan, gave a lecture at the Turkish–Arab Congress on Higher Education in Istanbul. In his talk, Erdoğan described the 'golden age of Islamic civilization', the medieval period in which 'Islamic cities . . . acted as a science center'. Yet Erdoğan was seemingly unaware of the fact that many Muslims, including those living in what is today modern Turkey, had also contributed just as much to the development of modern science. From astronomy in sixteenth-century Istanbul to human genetics in twentieth-century Cairo, the Islamic world of scientific advance continued well beyond the medieval 'golden age'.[7]

Why are these stories so common? Like many myths, the idea that modern science was invented in Europe did not come about by accident. During the middle of the twentieth century, a group of historians in Britain and the United States started to publish books with titles like *The Origins of Modern Science*. Almost all were convinced that modern science – and with it modern civilization – originated in Europe, sometime around the sixteenth century. 'The scientific revolution we must regard . . . as a creative product of the West,' wrote the influential Cambridge historian Herbert Butterfield in 1949. Similar views were expressed on the other side of the Atlantic. Students at Yale University in the 1950s were taught that 'the West generated the natural sciences . . . the East did not', whilst readers of *Science* – one of the most prestigious scientific magazines in the world – were informed that 'a small circle of Western European nations provided the original home for modern science'.[8]

The politics of all this couldn't be clearer. These historians lived through the early decades of the Cold War, a period in which the struggle between capitalism and communism dominated world politics. They thought about the contemporary world in terms of a strict divide between East and West, and then – whether intentionally or not – projected this back onto the past. During this period, science and technology were widely seen as markers of political success, particularly after the Soviet Union launched Sputnik, the first artificial satellite, in October 1957. The idea that modern science was invented in Europe therefore served as a convenient fiction. For leaders in Western Europe and the United States, it was essential that their citizens saw themselves on the right side of history, as bearers of scientific and technological progress. This was also a history of science designed to convince post-colonial states around the world to follow the path of capitalism, and to steer clear of communism. Throughout the Cold War, the United States spent billions of dollars on foreign aid, promoting a combination of free market economics and scientific development in countries across Asia, Africa, and Latin America. This was intended to counter the foreign assistance programme run by the Soviet Union. 'Western science', when combined with 'market economies', promised nothing less than an economic 'miracle', at least according to American policymakers.[9]

Somewhat ironically, Soviet historians ended up reinforcing a very similar narrative concerning the origins of modern science. They tended to ignore the earlier achievements of Russian scientists working under the Tsars, instead promoting the spectacular rise of science under communism. 'Up to the twentieth century, there was really no physics in Russia,' wrote the President of the Soviet Academy of Sciences in 1933. As we'll see, this was not true. Peter the Great supported some of the most important astronomical observations made during the early eighteenth century, whilst Russian physicists played a key role in the development of the radio in the nineteenth century. Some later Soviet historians did try and highlight earlier Russian scientific achievements.

But at least in the early decades of the twentieth century, it was much more important to emphasize the revolutionary advances made under communism rather than anything achieved under the old regime.[10]

Things played out slightly differently in Asia and the Middle East, although ultimately with similar consequences. The Cold War was a period of decolonization, in which many countries finally gained independence from European colonial powers. Political leaders in places like India and Egypt desperately wanted to forge a new sense of national identity. Many looked to the ancient past. They celebrated the achievements of medieval and ancient scientific thinkers, ignoring much of what had happened during the period of colonialism. It was in fact in the 1950s that the very idea of an Islamic or Hindu 'golden age' started to become popular – not just in Europe, as it had been in the nineteenth century, but also in the Middle East and Asia. Indian and Egyptian historians seized on the idea of a glorious scientific past, one waiting to be rediscovered. In doing so, they unwittingly reinforced the very myth being peddled by European and American historians. Modern science was Western, ancient science was Eastern, or so people were told.[11]

The Cold War is over, but the history of science is still stuck in the past. From popular history to academic textbooks, the idea that modern science was invented in Europe remains one of the most widespread myths in modern history. Yet there is very little evidence to support it. In this book, I provide a new history of modern science, one that is both better supported by the available evidence and more suited to the times in which we live. I show how the development of modern science fundamentally relied on the exchange of ideas between different cultures across the world. That was true in the fifteenth century, just as it is true today.

From Aztec palaces and Ottoman astronomical observatories to Indian laboratories and Chinese universities, this book follows the history of modern science across the globe. However, it is important to remember that this is not an encyclopaedia. I have not tried to cover every country in the world, nor every scientific discovery. Such an approach would be foolhardy, and not particularly enjoyable to read. Rather, the aim of this book is to show how global history shaped modern science. For that reason, I have picked four key periods of world historical change, linking each of these to some of the most important developments in the history of science. By placing the history of science at the heart of world history, this book also uncovers a new perspective on the making of the modern world – from the history of empire to the history of capitalism, if we want to understand modern history, we need to pay attention to the global history of science.

Finally, I want to emphasize that I see science as very much a human activity. Modern science was undoubtedly shaped by wider world events, but it was nonetheless made through the efforts of real people. These were individuals who, whilst living in a very different time and place, were not fundamentally different from you or me. They had families and relationships. They struggled with their emotions and health. And each of them wanted more than anything else to better understand the universe in which we live. Throughout this book, I have tried to give a sense of that more human side of science: an Ottoman astronomer captured by pirates in the Mediterranean; an enslaved African collecting medicinal herbs on a plantation in South America; a Chinese physicist fleeing the Japanese assault on Beijing; and a Mexican geneticist collecting blood samples from Olympic athletes. Each of these individuals, although largely forgotten today, made important contributions to the development of modern science. This is their story – the scientists who have been written out of history.

PART ONE

Scientific Revolution, *c.* 1450–1700

1. New Worlds

Stepping out into the Mexican sun, Emperor Moctezuma II could hear the birds calling. His palace – located at the heart of the Aztec capital city of Tenochtitlan – housed an aviary, in which birds from all over the Americas were kept. Green parakeets perched on the latticework, whilst purple hummingbirds flashed through the trees. Alongside the aviary, Moctezuma's palace featured a menagerie in which larger animals lived, including a jaguar and a coyote. But of all the wonders of nature, Moctezuma most appreciated flowers. Each morning, he would take a turn around the royal botanical garden. Roses and vanilla flowers lined the paths, whilst hundreds of Aztec gardeners tended to rows of medicinal plants.[1]

Built in 1467, this Aztec botanical garden predated European examples by almost a century. And it wasn't just for show. The Aztecs developed a sophisticated understanding of the natural world. They categorized plants according to their structure as well as use, particularly distinguishing between decorative and medicinal plants. Aztec scholars also reflected on the relationship between the natural world and the heavens, arguing – much like in the Christian tradition – that plants and animals were the handiwork of the gods. Moctezuma himself took great interest in all this. He commissioned surveys of the natural history of the Aztec Empire and made vast collections of animal skins and dried flowers. An accomplished scholar in his own right, Moctezuma is described in Aztec chronicles as 'by nature wise, an astrologer, a philosopher, skilled in all the arts'. He stood at the head of a vast empire, one in which science reached new heights.[2]

Tenochtitlan was an engineering marvel. Built on an island at the centre of Lake Texcoco in 1325, the Aztec capital could only be reached by crossing one of three causeways, each stretching several miles across the water. Just like Venice, the city was criss-crossed by canals, with Aztec merchants paddling back and forth in canoes as they went about their daily business. An aqueduct provided the city with a supply of fresh water, whilst, out on the lake, farmers tended to strips of reclaimed land, growing maize, tomatoes, and chillies. At the centre of the city stood the Great Temple, an immense stone pyramid, over sixty metres tall. Aztec architects had designed the temple to align perfectly with the rising and setting of the Sun on key feast days. Moctezuma himself would attend ceremonies, praising the gods and offering tribute in the form of flowers, animal skins, and sometimes human sacrifice. By the middle of the fifteenth century, Tenochtitlan had grown to an unprecedented size. With a population of over 200,000, this Aztec megacity was much larger than most European capitals, including London and Rome. Over the following decades, the Aztec Empire continued to expand, stretching right across the Mexican plateau and incorporating over three million people.[3]

All this was made possible thanks to the advanced state of Aztec science and technology. From observing the heavens to studying the natural world, the Aztecs placed great emphasis on the cultivation of knowledge. Unlike most European kingdoms at the time, a significant proportion of Aztec children, both male and female, received some kind of formal education. There were also specialist schools for noble boys who wished to train as priests, a profession that required expert knowledge of astronomy and mathematics in order to compile the Aztec calendar. Alongside priests, there was a special class of people referred to as 'knowers of things'. These were highly trained individuals, the equivalent of a university-educated scholar in Europe. They built up great libraries, often contributing new works themselves. The Aztecs also developed one of the most advanced medical systems in the world at that time. In Tenochtitlan, you could consult a range of medical practitioners, from physicians known as *ticitl*, to surgeons, midwives, and apothecaries. The city even housed a medical market, where traders from across the empire brought herbs, roots, and ointments for sale. Today we know that many Aztec medicinal plants do have pharmacologically active properties. These include a type of daisy that can be used to induce labour, as well as a species of Mexican marigold that helps reduce inflammation.[4]

Much of what we know about Tenochtitlan comes from accounts written by the people who destroyed it. On 8 November 1519, the Spanish conquistador Hernán Cortés entered the city for the first time.

Initially, Moctezuma welcomed the Spanish, housing Cortés and his men in the royal palace. They were overwhelmed by what they saw. Bernal Díaz del Castillo, one of the soldiers who accompanied Cortés, later described Moctezuma's gardens in *The True History of the Conquest of New Spain* (1576):

> We went to the orchard and garden, which was such a wonderful thing to see and walk in, that I was never tired of looking at the diversity of the trees, and noting the scent which each one had, and the paths full of roses and flowers, and the many fruit trees and native roses, and the pond full of fresh water.

Díaz also described the aviary. He recalled seeing 'everything from the royal eagle . . . down to tiny birds of many-coloured plumage . . . feathers of five colours – green, red, white, yellow and blue'. There was also a 'great tank of fresh water and in it all other sorts of birds with long stilted legs, with body, wings, and tail all red'.[5]

The tranquillity did not last. Cortés took advantage of the situation, taking Moctezuma hostage and fighting his way through the city. And although the Spanish were initially repelled, Cortés returned with a far greater force two years later. Ships armed with cannon surrounded the city on the lake, as Spanish soldiers drove through the gates. Moctezuma was murdered and the Great Temple was destroyed. Cortés himself set fire to the palace. The aviary, the menagerie, and the gardens all burned. As Díaz noted, somewhat mournfully for a soldier, 'of all these wonders that I then beheld . . . today all is overthrown and lost, nothing left standing'. The conquest of the Aztecs marked the beginning of the Spanish Empire in the Americas. In 1533, Charles V established the Viceroyalty of New Spain. The capital, Mexico City, was built on the ashes of Moctezuma's palace.[6]

Most histories of science do not begin with the Aztecs in Mexico. Traditionally, the history of modern science begins in sixteenth-century Europe, with what is often called the 'scientific revolution'. We are told that, in the period between around 1500 and 1700, an incredible transformation in scientific thought took place. In Italy, Galileo Galilei observed the moons of Jupiter, whilst in England, Robert Boyle first described the behaviour of gases. In France, René Descartes developed a new way of doing geometry, whilst in Holland, Antonie van Leeuwenhoek first observed bacteria under a microscope. Typically, this story culminates with the work of Isaac Newton, the great English mathematician who set out the laws of motion in 1687.[7]

Historians have long argued over the nature and causes of the scientific revolution. Some see this as a period of intellectual advance, one in which a few lone geniuses made new observations and challenged medieval superstition. Others argue that this was a period of great social and religious change, one in which the English Civil War and the Protestant Reformation forced people to reassess a range of basic beliefs about the nature of the world. Then there are those who see the scientific revolution as a product of technological change. From the printing press to the telescope, this period saw the invention of an assortment of new tools, each of which allowed for the investigation of nature and the dissemination of scientific ideas on an unprecedented scale. Finally, some historians deny that this really was a period of significant change. After all, many of the great thinkers of the scientific revolution continued to rely in some ways on much older ideas, such as those found in the Bible or in ancient Greek philosophy.[8]

Until recently, however, very few historians have stopped to consider whether they are looking in the right place to begin with. Is the history of the scientific revolution really a story about Europe alone? The answer is no. From the Aztec Empire in the Americas to the Ming Empire in China, the history of the scientific revolution is a story which incorporates the entire world. And it isn't just that people in the Americas, Africa, and Asia happened to be developing advanced scientific cultures at the same time as those in Europe. Rather, it is the history of encounters between these different cultures which explains precisely why the scientific revolution occurred when it did.

With this in mind, I want to tell a new history of the scientific revolution. In this chapter, we explore how encounters between Europe and the Americas kickstarted a major reassessment of natural history, medicine, and geography. Much of what we know about the science produced in the New World during this period comes from the perspective of European explorers, a legacy of the history of colonization that this chapter examines. But if we look a little closer, using sources such as Aztec codices and Inca histories, we can also uncover another side to this story, one that highlights the hidden contributions of Indigenous peoples to the scientific revolution. In the next chapter, we move east, revealing how connections between Europe, Africa, and Asia shaped the development of mathematics and astronomy. Together, these chapters represent the beginning of a recurrent theme concerning the importance of global history for understanding the history of modern science. Ultimately, to account for the scientific revolution we need to look, not just to London and Paris, but to the ships and caravans which connected the early modern world.[9]

I. Natural History in the New World

After over two months at sea, Christopher Columbus finally sighted land. Sailing aboard the *Santa María* on behalf of the Spanish Crown, Columbus was in search of a western passage to the Indies. Instead, he encountered a whole new continent. On 12 October 1492, Columbus landed on an island he named San Salvador, part of the Bahamas. This was the beginning of a long history of European colonization in the Americas. Like many subsequent travellers to the New World, Columbus was amazed by the diversity of plant and animal life he encountered. He recorded in his diary that 'all the trees were as different from ours as day from night, and so were the fruits, the herbage, the roots, and all things'. Columbus also quickly recognized the commercial potential of the Americas, noting that there were 'many plants and many trees, which are worth a lot in Spain for dyes and for medicines'. Most alarmingly, the island was inhabited. On landing, the Spanish crew encountered a group of Indigenous people. Still believing he had reached the East Indies, Columbus named them *indios*, or 'Indians'. Encouraged by the abundance of plant, animal, and human life, Columbus continued to explore the West Indies over the following months, reaching Cuba and Hispaniola. He later returned on three separate voyages, travelling as far as Central and South America.[10]

The colonization of the Americas was one of the most important events in world history. It was also an event which profoundly shaped the development of modern science, challenging longstanding assumptions about how scientific knowledge was best acquired. Prior to the sixteenth century, scientific knowledge was thought to be found almost exclusively within ancient texts. This was especially the case in Europe, although, as we'll see in the following chapter, similar traditions existed across much of Asia and Africa. Surprising as it may sound today, the idea of making observations or performing experiments was largely unknown to medieval thinkers. Instead, students at medieval universities in Europe spent their time reading, reciting, and discussing the works of ancient Greek and Roman authors. This was a tradition known as scholasticism. Commonly read texts included Aristotle's *Physics*, written in the fourth century BCE, and Pliny the Elder's *Natural History*, written in the first century CE. The same approach was common to medicine. Studying medicine at a medieval university in Europe involved almost no contact with actual human bodies. There were certainly no dissections or experiments on the workings of particular organs. Instead, medieval medical students read and recited the works of the ancient Greek physician Galen.[11]

Why, then, sometime between 1500 and 1700, did European scholars turn away from ancient texts and start investigating the natural world for themselves? The answer has a lot to do with the colonization of the New World alongside the accompanying appropriation of Aztec and Inca knowledge, something that traditional histories of science fail to account for. As many early European explorers were quick to recognize, the plants, animals, and people they encountered in the Americas were not described in any of the ancient works. Aristotle had never seen a tomato, let alone an Aztec palace or an Inca temple. It was this revelation which brought about a fundamental shift in how Europeans understood science.[12]

The Italian explorer Amerigo Vespucci, after whom 'America' is named, was one of the first to recognize the implications of Columbus's 'discovery' for natural history. After returning from his own voyage to the New World in 1499, Vespucci wrote to a friend in Florence. He reported seeing all kinds of incredible animals, including a 'serpent' – most likely an iguana – which the Indigenous people roasted and ate. Vespucci also recalled seeing birds 'so numerous and of so many species and varieties of plumage that it is astounding to behold'. Most significantly, Vespucci made a direct connection between the natural history of the New World and what was known from ancient texts. He concluded with a damning criticism of Pliny's *Natural History*, the traditional authority on the subject. As Vespucci noted, 'Pliny did not touch upon a thousandth part of the species of parrots and other birds and animals' which were found in the Americas.[13]

Vespucci's criticism of Pliny was just the start. Over the following years, thousands of travellers returned from the New World with reports of things unknown to the ancients. One of the most influential accounts was written by a Spanish priest named José de Acosta. Born to a prosperous merchant family in 1540, Acosta was always looking to escape his comfortable but rather mundane upbringing. At the age of twelve, he ran away from home to join the Society of Jesus, a Catholic missionary organization which played a major role in the development of early modern science. The founder of the order, Ignatius of Loyola, urged his followers to 'find God in all things', whether that was in reading the Bible or studying the natural world. The Jesuits therefore placed great emphasis on the study of science, both as a way to appreciate God's wisdom, but also as a means to demonstrate the power of the Christian faith to potential converts. After joining the Jesuits, Acosta attended the University of Alcalá, where he studied the classical works of Aristotle and Pliny. On graduating, Acosta was asked to go as a missionary to the New World, setting sail in 1571. He spent the next fifteen years in the Americas, travelling across the Andes in search of converts. On returning to Spain, Acosta began to write a book describing everything he had seen, from the volcanoes of Peru to the parrots of Mexico. The finished work was titled *Natural and Moral History of the Indies* (1590).[14]

Acosta witnessed many strange things in the Americas. But perhaps the most important experience Acosta had was during his initial voyage across the Atlantic Ocean. The young priest was anxious about the journey, not least because of what ancient authorities said about the equator. According to Aristotle, the world was divided into three climatic zones. The north and south poles were characterized by extreme cold and known as the 'frigid zone'. Around the equator was the 'torrid zone', a region of burning dry heat. Finally, between these two extremes, at around the same latitudes as Europe, was the 'temperate zone'. Crucially, Aristotle argued that life, particularly human life, could only be sustained in the 'temperate zone'. Everywhere else was either too hot or too cold.[15]

Acosta therefore expected to experience incredible heat as he approached the equator. But this was not the case. 'The reality was so different that at the very time I was crossing I felt such cold that at times I went out into the sun to keep warm,' explained Acosta. The implications for ancient philosophy were clear. Acosta went on, writing:

> I must confess I laughed and jeered at Aristotle's meteorological theories and his philosophy, seeing that in the very place where, according to his rules, everything must be burning and on fire, I and all my companions were cold.

As he travelled across South and Central America, Acosta confirmed that the region around the equator was not always as hot, and certainly not as dry, as Aristotle believed. Indeed, Acosta experienced a great diversity of climates, explaining how 'in Quito, and on the plains of Peru' it was 'quite temperate', whereas in Potosí it was 'very cold'. Not only that, but, most strikingly of all, the region was full of life – not just plants and animals, but also people. As Acosta concluded, 'the Torrid Zone is habitable and very abundantly inhabited, even though the ancients said that this was impossible'.[16]

This was certainly a blow to classical authority. If Aristotle had been mistaken about the climatic zones, what else might he have been wrong about? Worried by this thought, Acosta spent much of his life trying to reconcile what he had learned from ancient texts with what he had experienced in the New World. The diversity of previously unknown animals proved particularly difficult to explain. From sloths in Peru to hummingbirds in Mexico, there were 'a thousand kinds of birds and fowl and forest animals that have never been known before either in name or shape, nor is there any memory of them in the Latins and Greeks, nor in any nations of our world', explained Acosta. Clearly, Pliny's *Natural History* was incomplete.[17]

Acosta understood the implications of his discoveries. However, he wasn't ready to completely abandon classical learning. As a Christian, Acosta still placed great value on ancient authority. The Bible after all was the ultimate classical text. Like many early travellers to the Americas, Acosta therefore mixed the old with the new. In some instances he claimed that, whilst Aristotle might have been wrong, other ancient sources were right. In the case of the torrid zone, Acosta pointed out that the ancient

Greek geographer Ptolemy took a different view, and 'believed that there were commodious habitable regions under the tropics'. Acosta also noted that some ancient texts even suggested the existence of new worlds beyond the known oceans. Plato described the mythical island of Atlantis, whilst the Bible referred to a faraway land called Ophir from which King Solomon received shipments of silver. Indeed, classical texts were full of unknown countries, each of which could easily be interpreted as the Americas. At first, then, encounters in the New World did not lead to a complete rejection of ancient learning. Instead, European scholars were forced to revisit classical texts in light of new experiences.[18]

Bernardino de Sahagún spent most of his life in the Americas. Born in Spain in 1499, Sahagún joined the Franciscan order whilst studying at the University of Salamanca. Like José de Acosta, he received an education typical of the time, studying the ancient works of Aristotle and Pliny as preparation for the priesthood. In 1529, Sahagún crossed the Atlantic and arrived in New Spain, one of the first cohort of missionaries to reach the New World. He spent the rest of his life in the Americas, dying in Mexico City aged ninety. During his time there, Sahagún helped compile one of the most comprehensive accounts of sixteenth-century Mexico. He called it the *General History of the Things of New Spain* (1578). Better known as the *Florentine Codex*, this monumental work described, not only the plants and animals of the New World, but also Aztec medicine, religion, and history. The complete work was made up of twelve books and contained over 2,000 hand-coloured drawings.[19]

The *Florentine Codex* was not the work of Sahagún alone. Rather, it was a collaborative effort with Indigenous people. Shortly after arriving in New Spain, Sahagún took up a post teaching Latin at the Royal College of Santa Cruz in Tlatelolco, on the outskirts of Mexico City. The Royal College had been established in 1534 in order to train the sons of Aztec nobles in preparation to join the clergy. Over seventy Indigenous boys lived at the college, receiving a traditional scholastic education much as Sahagún had in Spain. The boys learned Latin and read Aristotle, Plato, and Pliny. Alongside this, the Aztec students at the Royal College were taught to write their own language, Nahuatl, in the Latin alphabet. This was an important development, as traditionally the Aztecs did not use a written alphabet. Instead, Nahuatl was a pictorial language in which certain images represented different words or phrases. The Spanish often dismissed Aztec pictorial books as primitive, even idolatrous. As another missionary claimed, the Aztecs were 'a people without writing, without letters, without written chronicles, and without any kind of enlightenment'. This, as we now know, was not true. But such attitudes served the Spanish well as they tried to transform the Aztecs into Europeanized Christians. This was part of a broader European attempt to justify the conquest of the Americas under the guise of bringing Christianity to the New World.[20]

Sahagún, however, recognized the value of Aztec culture more than many of his contemporaries. He learned Nahuatl and, in 1547, began work on the *Florentine Codex*. Sahagún realized that, to really understand the natural history of the New World, he would have to learn from the people who already lived there. With this in mind, Sahagún assembled a group of students at the Royal College. We know the names of four of them: Antonio Valeriano, Alonso Vegerano, Martín Jacobita, and Pedro de San Buenaventura. (Unfortunately, their original Nahuatl names are lost.) Together, Sahagún and his party set off across New Spain in search of Aztec knowledge. On arriving in a town, Sahagún would arrange an interview with a group of Indigenous elders. Often, the elders would recite ancient Aztec histories or describe an unknown plant or animal. Sometimes the elders would even bring out a surviving Aztec codex, each page painted with a complex array of glyphs. 'They gave me all the matters we discussed in pictures, for that was the writing deployed in ancient times,' explained Sahagún. As he could not interpret these himself, Sahagún then relied on his Aztec students to translate what they saw into written Nahuatl. Later, back at the Royal College, Sahagún and his assistants translated the Nahuatl into Spanish. He also commissioned a group of Indigenous artists to paint illustrations to accompany the text. In 1578, after over two decades of work, Sahagún finally sent the complete manuscript to Philip II of Spain.[21]

Like Acosta, Sahagún fused the old with the new. The *Florentine Codex* took Pliny's *Natural History* as a model. Indeed, Sahagún's students at the Royal College would have been familiar with this ancient work. Much like Pliny, the *Florentine Codex* is composed of a series of books covering geography, medicine, anthropology, plants, animals, agriculture, and religion. The main book that covers natural history is titled 'Earthly Things'. On opening this volume, however, we discover a world of plants and animals unknown to the ancients. Appropriately enough, this volume is also the most heavily illustrated, including paintings of 39 mammals, 120 birds, and over 600 plants. The vibrancy of the images is striking, depicting not only the natural world, but also animal behaviour, the uses of plants, and associated Aztec beliefs.[22]

The *Florentine Codex* listed hundreds of New World plants, all divided according to an Aztec system of taxonomy. The Aztecs typically arranged plants into four broad groups: edible, decorative, economic, and medicinal. These divisions were reflected in the naming of plants: for example, plants ending with the suffix *-patli* were medicinal, whereas those ending with the suffix *-xochitl* were decorative. This organization was then reproduced in the *Florentine Codex*. All the medicinal plants are listed together, with names such as *iztac patli* (a herb that could be used as a cure for fever). These are then followed by all the flowering plants, with names such as *cacaloxochitl* (known in Europe as the frangipani, after a sixteenth-century Italian noble who imported it).[23]

1. An illustration of hummingbirds from the *Florentine Codex* (1578). Note the hummingbird hanging from the tree in a state of 'torpor'.

Animals also feature heavily in the *Florentine Codex*. There is an image of a rattlesnake catching a rabbit and another of ants building a mound. Hummingbirds in particular appear in a number of the illustrations. One depicts a hummingbird extracting nectar from a flower, whilst another shows a group of hummingbirds migrating south for the winter. This focus on the hummingbird in fact reflected an important Aztec belief. Huitzilopochtli, or the Hummingbird God, was the patron deity of Tenochtitlan. The Great Temple in the city was dedicated to Huitzilopochtli, and warriors who died in battle were said to transform into hummingbirds. The Aztecs therefore studied the hummingbird closely. They were fascinated by its ability to enter a state of hibernation known as torpor. No European had ever seen this before, and so Sahagún was relying on the word of his Aztec informants, some of whom had actually worked in Moctezuma's aviary:

> In the winter, it hibernates. It inserts its bill in a tree; there it shrinks, shrivels, molts . . . when the sun warms, the tree sprouts, when it leafs out, at this time [the hummingbird] also grows feathers once again. And when it thunders for rain, at that time it awakens, moves, comes to life.[24]

The hummingbird's behaviour fitted perfectly with an Aztec view of the world, one regulated by a constant cycle of life and death. Warriors, much like the hummingbird, might be reborn. Death was never the end.[25]

II. Aztec Medicine

For Bernardino de Sahagún, the *Florentine Codex* was primarily a religious work. By compiling a comprehensive account of Aztec wisdom, he sought to show 'the degree of the perfection of this Mexican people'. This, Sahagún hoped, would help convince Christians back in Europe that the Aztecs were a 'civilized' race capable of receiving the word of God. Others, however, saw the New World in more commercial terms. In 1580, Ferdinando de' Medici, Grand Duke of Tuscany and head of the famous Italian Medici family, purchased the *Florentine Codex*. He put it on display in the famous Uffizi Gallery in Florence, hence the name by which it is known today. In the Uffizi Gallery, the *Florentine Codex* sat alongside the Medici family's incredible collection of art, sculptures, and curiosities from across the globe. These included a green feather headdress as well as a turquoise Aztec mask. At this time, the Medicis were developing a strong commercial interest in the New World. Ferdinando de' Medici began importing cochineal (used in the manufacture of crimson red dyes) from Mexico and Peru, whilst maize and tomatoes – both native to the Americas – were grown in the gardens of the Medici palazzo in Florence. For Ferdinando de' Medici, the *Florentine Codex* was essentially a commercial catalogue: a list of the most valuable natural resources that the New World had to offer.[26]

It was this commercial attitude towards the New World that really transformed the study of natural history. Merchants and doctors tended to place much greater emphasis on collecting and experimentation over classical authority. American plants represented a potentially lucrative source of revenue, and there was a clear commercial advantage in promoting these discoveries as novel. Tobacco, avocadoes, and chillies were all marketed as incredible new cures, whilst the earliest record of a potato being sold in Europe comes from the account books of a sixteenth-century Spanish hospital. At the same time, universities across Europe started establishing their own botanical gardens. These were not dissimilar from the Aztec botanical gardens the Spanish saw in Mexico, specialist sites for the study and cultivation of medicinal herbs. In 1545, the University of Padua established the first botanical garden in Europe. This was soon followed by gardens at Pisa and Florence. By the middle of the seventeenth century, botanical gardens – all growing New World plants – could be found at every major European university. Some wealthy physicians even started to establish their own private botanical gardens, marketing new medical cures derived from American plants.[27]

Much of what Europeans knew about the medical uses of New World plants was derived from Aztec sources. The Spanish Crown in particular invested enormous effort, not only in collecting and cataloguing specimens from the New World, but also in recording what the Aztecs knew about them. In 1570, Philip II of Spain ordered a major survey of the natural history of the New World. At the head of the survey, Philip appointed his personal physician, Francisco Hernández. Over the next seven years, Hernández travelled across New Spain, collecting herbs and learning about Aztec medical practices.[28]

Born in 1514, Hernández studied at the University of Alcalá before setting up a successful medical practice in Seville. Like most sixteenth-century physicians, and as noted earlier, Hernández's medical training involved little more than reading ancient texts. He read the works of Galen and Dioscorides, both ancient Greek physicians. Dioscorides's *On Medical Material* provided a list of herbal treatments for various ailments, whilst Galen's vast corpus described the basic theory underlying ancient Greek medicine. This theory centred on achieving a balance between the four humours: blood, phlegm, black bile, and yellow bile. Bloodletting was commonly recommended to cure fever, whilst laurel leaves could be taken to purge excess yellow bile.[29]

Hernández, however, lived through a time of great medical change. Many physicians began to turn away from ancient authority, instead placing much greater emphasis on dissection and experimentation. Many were inspired by the work of Andreas Vesalius, whose *On the Fabric of the Human Body* (1543) provided a new account of human anatomy based on dissection. Others followed the work of Paracelsus, a controversial Swiss alchemist who promoted all kinds of new herbal and mineral cures. Hernández himself was a great champion of these medical reforms, undertaking dissections and establishing a botanical garden whilst working at a hospital in western Spain. It would be wrong, however, to assume that this new way of thinking about medicine can be explained by looking at Europe alone. Rather, knowledge originating in the New World, produced by the Indigenous peoples of the Americas, helped shape a vision of medicine as an experimental and practical science.[30]

Francisco Hernández arrived in Mexico City in February 1571, accompanied by his son, Juan, and a team of scribes, painters, and interpreters. The city was in the midst of an epidemic, known as *cocoliztli* by the Indigenous people and 'the great pestilence' by the Spanish. Victims died within days of contracting the disease, suffering horrific pain and bleeding from the eyes and nose. Hernández, who had been appointed Chief Medical Officer of the Indies, spent the first few weeks undertaking dissections of the recently deceased. When the outbreak calmed down, Hernández and his party set off on a tour of New Spain, spending seven years scouring the land for new plants, animals, and minerals, anything that might be medically useful. He even visited an abandoned Aztec botanical garden at Texcoco, copying some of the paintings of flowers from the ruined walls. Altogether, Hernández identified over 3,000 plants previously unknown to Europeans. For comparison, the ancient Greek physician Dioscorides only listed 500 plants in his *On Medical Material*. This then really was a complete challenge to the idea that ancient authors knew everything.[31]

In undertaking this survey, Hernández was absolutely reliant on Indigenous people and their medical knowledge. In fact, Philip II had explicitly recommended that Hernández quiz the local population. The official instructions for the expedition ordered Hernández to 'consult, wherever you go, all the doctors, medicine men, herbalists, Indians, and other people with knowledge of such matters'. Hernández took these orders seriously and began to learn Nahuatl. He then set about interviewing Indigenous medical practitioners, carefully recording the names of the plants and animals they described and making sure to use the native terms. Hernández described the properties of *zacanélhuatl*, a kind of root that, when crushed and mixed with water by Indigenous doctors, could help cure kidney stones. Hernández noted that this concoction 'provokes urination and cleans out its tract'. He also learned of a herb called *zocobut* with 'leaves like a peach, but wider and thicker'. It could be used to cure migraines, dampen swelling, and 'combat poisons and poisonous stings and bites'. Indeed, this particular herb was 'highly esteemed by the natives', so much so that 'it is not easy to get them to tell you its properties'. Hernández also investigated the medical uses of New World animals. After describing the possum, he noted that 'the tail of this animal is an excellent medicament'. Ground and mixed with water, 'it cleans the urinary tract . . . cures fractures and colic . . . comforts the belly'. Most intriguingly, Indigenous medical practitioners reported the possum tail acted as an aphrodisiac, Hernández writing that 'it excites sexual activity'. Whilst we can't be sure about every plant listed by Hernández, scientists today have shown that some do indeed have medicinal properties. The leaves of the thorn apple, for example, contain an analgesic. Others, such as the seeds of the Mexican apple, have been shown to help prevent certain forms of cancer.[32]

Describing the appearance and properties of plants and animals was all very well. But when everything was so new, at least for Europeans,

2. An engraving of an armadillo, copied from a drawing made by an Indigenous artist in sixteenth-century Mexico, from Francisco Hernández, *The Treasury of Medical Matters of New Spain* (1628).

only a picture could really communicate the diversity of American natural history. Like Sahagún, Hernández therefore decided to employ a group of Indigenous artists to paint pictures of everything he saw. Over six years these artists – named Pedro Vázguez, Baltazar Elías, and Antón Elías – made hundreds of paintings, all in situ, including one of a sunflower and another of an armadillo. Many of these images were later reproduced in European works of natural history, including Hernández's own publications. In 1577, Hernández returned to Spain with sixteen handwritten volumes along with the paintings. Later published as *The Treasury of Medical Matters of New Spain* (1628), Hernández's manuscript was deposited at the Escorial Palace Library, located just outside Madrid. The royal librarian, José de Sigüenza, was impressed, particularly by the artwork. 'This is a history of all the animals and plants that could be seen in the West Indies, painted in their native colours,' he explained, adding, 'it is something that offers great delight and variety to those that look at it; and no small profit to those whose task it is to consider nature'.[33]

Francisco Hernández's *Treasury* was typical of a new genre of natural histories, ones that repackaged Aztec medical knowledge for European audiences. Nonetheless, it was still ultimately the work of a conquistador.

Hernández was a man who had been sent by the King of Spain on an expedition which, at its core, was about the extraction of knowledge and wealth. Indeed, the choice of title is telling – this really was a 'treasury' for the Spanish. Nonetheless, Europeans were not the only authors of significant works of natural history during this period. At just the same time as Hernández was writing, an Aztec scholar compiled his own natural history of the New World, which later reached Europe and influenced a number of early modern medical texts.

Martín de la Cruz was born in Mexico before the Spanish conquest. Unfortunately, we know little of his early life. We don't even know his Nahuatl name. Cruz later simply described himself as 'an Indian doctor', and was probably a middle-ranking Aztec physician. What we do know is that Cruz converted to Christianity and taught medicine at the Royal College of Santa Cruz in Tlatelolco, the same institution at which Bernardino de Sahagún had begun work on the *Florentine Codex*. On 22 May 1552, he presented a manuscript titled *The Little Book of the Medicinal Herbs of the Indians* to the master of the college. Cruz had originally written the book in Nahuatl, but had it translated into Latin by another Indigenous tutor at the college, Juan Badiano. More than any other work of the period, *The Little Book of the Medicinal Herbs of the Indians* represents a fusion of European and Aztec knowledge. At first glance, it looks much like a typical classical compendium of herbs, not dissimilar from Dioscorides's *On Medical Material*. Cruz divided his book into thirteen chapters, beginning with the head, moving down the body, all the way to the feet. Each page identifies a particular condition, such as 'toothache' or 'difficulty in passing urine', and then describes the preparation of herbs used to treat it. Most pages also feature an illustration of the individual herbs, sketched and painted by Cruz himself.[34]

Look a little closer, however, and it is clear that Cruz was drawing heavily on Aztec medical knowledge. All the plants' names are given in Nahuatl and, like the *Florentine Codex*, reflect Aztec classificatory schemes. In this case, the names not only indicate the use of the plant but also where it might be found: for example, plants with the prefix *a-* (meaning 'water') could be found near lakes or rivers, whereas those with the prefix *xal-* (meaning 'sand') could be found in deserts. Throughout the book, Cruz also draws on traditional Aztec understandings of the body. The

Aztecs typically believed that the body contained three forces, located in the head, the liver, and the heart respectively. Disease resulted from an imbalance of these forces, something that was often caused by an excess of heat or cold in a particular part of the body. (This was not so different from the ancient Greek theory of the four humours.)[35]

Reading Cruz's description of the herbs carefully, we can see that his focus is on restoring this balance. Pain and swelling in the eyes, for example, was understood to be the result of excessive heat in the head. The cure involved preparing a concoction of cooling herbs. Flowers of *matlal-xochitl* (known in Europe as spiderwort) and leaves from the mesquite tree were to be ground up and mixed with breastmilk and 'limpid water'. This ointment would then be applied to the face. Cruz also advised avoiding 'sexual acts' or eating chilli sauce until the condition improved, as both of these could cause an excess of heat as well.[36]

The final clue to Aztec influence is the most important, but also the most difficult to spot. Earlier historians often read Cruz's illustrations as imitations of typical European botanical drawings, with each plant pictured in isolation, roots and leaves visible for easy identification. More

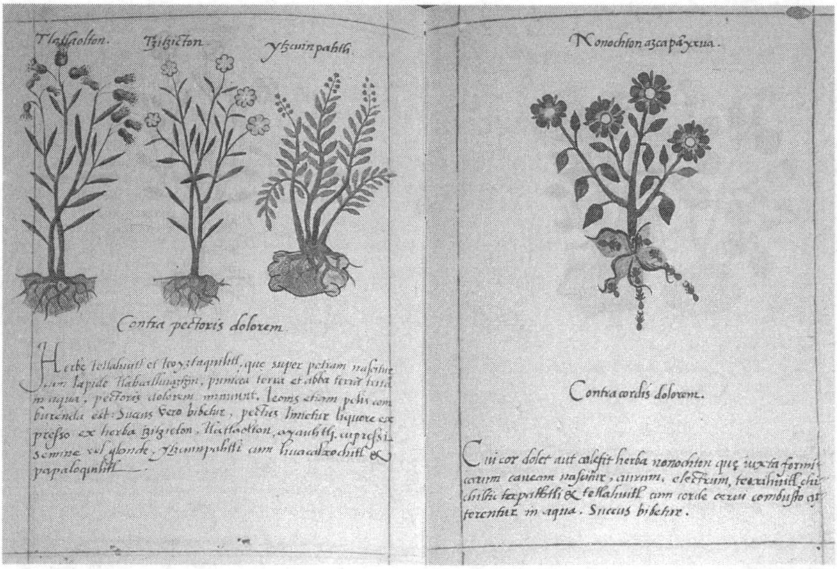

3. An illustration from Martín de la Cruz, *The Little Book of the Medicinal Herbs of the Indians* (1552). The roots on the plant named *itzquin-patli* (*third from left*) feature the Nahuatl glyph for 'stone'.

recently, however, experts in Aztec culture have re-examined the images and noticed that they incorporate Nahuatl glyphs. Cruz was in fact trying to combine the style of European botanical illustrations with a traditional Aztec pictorial codex. He used glyphs throughout to indicate the place at which a plant may be found, reinforcing the naming system described earlier. The specific Aztec glyph for 'stone' appears around the roots of a number of plants in Cruz's drawings, as does the glyph for 'water'. Cruz was ultimately combining European and Aztec traditions, both medical and artistic, in order to create a completely new kind of natural history. In doing so, he was typical of the way in which science was being practised in the sixteenth century, a product of cultural exchange and encounter.[37]

By the end of the sixteenth century, New World plants could be found in gardens across Europe. Sunflowers bloomed in Bologna, whilst a yucca even flowered in London. These plants soon came to feature in new works of natural history and medicine, many of which promoted the value of experience over ancient texts. In London, the apothecary John Gerard described the medicinal uses of tobacco in his bestselling *Herball* (1597), whilst, in Seville, the physician Nicolás Monardes advised patients to buy cacao in his *Medical Study of the Products Imported from Our West Indian Possessions* (1565). (Monardes also ran a successful business growing American plants in his private botanical garden.) Even Andreas Vesalius, probably the most famous anatomist of the sixteenth century, showed an interest in the New World, discussing the possibility that the gum of guaiacum (a flowering plant native to Mexico) could be used to treat syphilis. This idea stemmed from a widespread – although today widely disputed – belief that syphilis itself had originated in the Americas, and hence a cure for it was most likely to be found there.[38]

European naturalists and apothecaries soon amassed vast collections of exotic plants and animals. They were supported by wealthy patrons, like the Medicis in Florence and the Spanish king in Madrid, filling the museums of Europe with objects and specimens from the New World. This new approach to natural history was also reflected in the increasing use of images. Whereas ancient texts on natural history tended not to be illustrated, the new natural histories of the sixteenth and seventeenth centuries were full of drawings and engravings, many of which were hand-coloured. This was partly a reaction to the novelty of what had been discovered. How else would those in Europe know what a vanilla plant or a hummingbird looked like? But it was also a way to incorporate existing Aztec traditions of codifying knowledge through pictograms.

Crucially, this whole enterprise relied, not only on specimens coming from the New World, but also Indigenous knowledge. Aztec understandings of the body and of nature filtered subtly into European texts of this period. In Naples, the botanist Carolus Clusius consulted Hernández's manuscripts when writing his influential *History of Rare Plants* (1601). Similarly, in Padua, Pietro Mattioli incorporated Cruz's *Little Book of the Medicinal Herbs* into his commentary on ancient Greek medicine. Today, the influence of Aztec natural history is still with us. The words 'tomato' and 'chocolate' are both derived from Nahuatl. The same is true of many other New World plants and animals. From the 'coyote' to the 'chilli', the way we speak about the natural world is ultimately the legacy of an encounter between the Old World and the New, something that is often forgotten when we focus on the achievements of European naturalists alone. And, as we will see, encounters between Europe and the Americas in the sixteenth century shaped more than just medicine and natural history – these encounters also shaped scientific understandings of the origins of humanity.[39]

III. The Discovery of Humankind

Antonio Pigafetta could hardly believe his eyes. In June 1520, on the southern tip of the American continent, the Italian explorer came face to face with a 'giant'. Nine months earlier, Pigafetta had joined a Spanish voyage to circumnavigate the globe. Led by Ferdinand Magellan, the first challenge was to sail across the Atlantic and around the South American coast. Winter had set in and the crew pulled into a bay they named Port San Julián, located in modern-day Argentina. 'We passed two months there without seeing any people,' Pigafetta recalled. But then, 'one day we suddenly saw a naked man of giant stature on the shore of the port, dancing, singing, and throwing dust on his head'. Somewhat improbably, Pigafetta estimated the man was over eight feet tall. 'He was so tall that the tallest of us only came up to his waist,'

Pigafetta wrote in his diary. This 'giant' had a face 'painted red all over, while about his eyes he was painted yellow'. Initially, the European explorers tried to make gestures of peace. They invited the 'giant' onto the ship, offering food and drink. But it didn't take long for this friendly encounter to turn violent. A few days later, Magellan ordered his crew to capture two of the 'giants' as trophies to present to the Spanish king. Fighting broke out, a Spanish sailor was killed, and the 'giants' ran away, apparently 'faster than horses'.[40]

Encounters in the Americas brought Europeans into contact with new animal and plant life. But for many, the most striking thing about the New World was its people. Pigafetta's diary was just one of the many reports of previously unknown people to reach Europe from the Americas in the sixteenth century. Descriptions of cannibalism and human sacrifice sparked the popular imagination, with New World peoples featuring in plays and poems of the era, including in Shakespeare's *The Tempest*. And although Magellan was unable to take any prisoners, a number of other explorers did bring Indigenous people, usually by force, from the Americas to Europe. Columbus himself captured six Caribbean islanders, presenting them at the court of Queen Isabella and King Ferdinand of Spain in 1493. Cortés also rounded up seventy defeated Aztecs in Tenochtitlan, transporting them in chains across the Atlantic in 1528. These included three of the sons of Moctezuma, presented alongside parrots and a jaguar at the court of Charles V in Madrid.[41]

For Europeans, the existence of Indigenous people in the Americas raised serious questions about the nature of humanity. Were these people human? Or were they monsters? If they were human, were they descended from Adam, as the Bible taught? Or were they created separately? And if they had originated in Europe, then how had they reached the Americas? Answering these questions demanded an entirely new way of thinking about humankind. Once again, there was a limit to how useful ancient texts could be. After all, Pliny did not contemplate the existence of unknown people, whilst Aristotle denied the possibility that somewhere like the Americas could be inhabited. For the first time, European scholars started to study humans in the same way as they were beginning to study natural history, by collecting evidence and testing ideas against experience. As they did so, people increasingly came to be seen as part of the natural world, rather than separate from it. The sixteenth century therefore witnessed the development of the first human sciences, not as a reaction to religious or intellectual change in Europe, but rather as a response to encounters in the Americas. The discovery of the New World was also the discovery of humankind.[42]

Antonio Pigafetta's description of the 'giants' of South America was typical of many early encounters with New World peoples. Europeans were remarkably willing to believe that the Americas were inhabited by monstrous beings. On landing on Cuba, Columbus described seeing 'one-eyed people and others with dog muzzles who ate human beings'. Similarly, Amerigo Vespucci reported that the people of Brazil were 'lightly covered with feathers' and could 'live to be a hundred and fifty years old'. These beliefs actually stemmed from an ancient tradition. Pliny had described the world beyond the Mediterranean as teeming with marvels – giants, pygmies, and troglodytes. This was later incorporated into a Christian idea that the further one got from Jerusalem, the more monstrous humans became. But despite these early fantastical descriptions, it didn't take long for European explorers to realize the truth: the people of the Americas really were human. In 1537, Pope Paul III settled the matter, declaring that 'the Indians are truly men, and that they are not only capable of understanding the Catholic faith but, according to our information, they desire exceedingly to receive it'. For Europeans, this was in some ways more disturbing, as it implied once again that ancient philosophy was lacking. Even the Bible was worryingly silent on this matter. José de Acosta, the Jesuit missionary we met earlier, noted this, writing, 'many of the ancients believed that there were neither men nor land nor even sky in these parts'.[43]

A different approach was clearly needed. Acosta in particular emphasized the importance of experience in studying the origins of American people. He complained that some authors 'assert without evidence that everything about the Indians is the work of superstition'. In contrast, Acosta proposed to study people much as he did plants and animals. Indeed, the clue to Acosta's approach is in the title of his book: *Natural and Moral History of the Indies*. This was a 'natural' history, but it was also a 'moral' history, meaning it was concerned with humans. The two went together. Acosta therefore set about exploring the history of humankind alongside the history of the natural world. Once again, he tried to mix the old and the new. As a Jesuit missionary, Acosta's starting point was still the Bible. 'Holy Writ clearly teaches, that all men were preceded by a first man,' Acosta explained. The Aztecs, Incas, and other Indigenous peoples he encountered on his travels must have been descended from Adam.[44]

This, however, raised a serious question: how did they get there? Acosta rejected any kind of miraculous explanation. 'We are not to think that there was a second Noah's Ark . . . nor much less that some angel brought the first inhabitants of this world holding them by the hair,' he wrote. Acosta also rejected the idea that the people of the Americas had travelled from Europe across the Atlantic in the ancient past. 'I find no traces in all of antiquity of anything as important and celebrated as this would have been,' he explained. Instead, Acosta proposed that 'the land of the Indies are connected to other lands in the world, or at least lie very close to them'. In short, Acosta argued that there must be some kind of land bridge between the Old World and the New, probably somewhere to the north. (As we now know, he was right. Humans first reached the Americas across a land bridge between Siberia and Alaska approximately 15,000 years ago.) This explanation, Acosta noted, also had the advantage that it could be used to account for the plants and animals found in the New World – they too must have crossed the same land bridge as humans.[45]

The question of the origins of American people wasn't just a scientific matter, it was also a political one. Throughout sixteenth-century Europe, there was great debate on the morality of the Spanish conquest. Some argued, based on the misguided beliefs outlined earlier, that the Aztecs were little more than barbarians and needed to be driven out by force. The typical comparison at the time was with the Catholic conquest of Muslim Spain, an event which coincided with the colonization of the New World. Others, however, suggested that the Aztecs were clearly an advanced civilization. The Indigenous people of the Americas possessed sophisticated medical theories, had built impressive cities, and developed complex legal and political systems. The Spanish destruction of Tenochtitlan and enslavement of its people was therefore immoral. And whilst few Europeans argued that the Spanish should withdraw entirely from the Americas, many did make the case that Indigenous people should be granted more rights. One of those to make this argument most forcefully was a Spanish priest named Bartolomé de las Casas.[46]

Las Casas was aged just nine when he first saw an Aztec. His father had travelled to the Americas on Columbus's second voyage, returning in 1499 with an 'Indian' alongside 'many green and red parrots', all of which were kept at the family home in Seville. Initially, Las Casas looked set to follow his father as a conquistador. In 1501, he travelled to the Spanish colony of Santo Domingo, in what is today the Dominican Republic, and managed a small plantation worked by enslaved Caribbean people. However, the realities of Spanish colonialism soon wore Las Casas down. In 1523, he joined the Dominican order, emerging as one of the great champions of Indigenous rights.[47]

Over the following years, Las Casas travelled back and forth between Europe and the Americas, trekking through Peru and New Spain, trying to understand the cultures of the people he encountered. In 1550, Las Casas returned to Spain to take part in a major debate, held at the College of San Gregorio in Valladolid. On one side of the debate was the conservative theologian Juan Ginés de Sepúlveda, who argued that the Indigenous people of the Americas were irrational beings who did not deserve freedom. 'How can we doubt that these people, so uncivilized, so barbaric, so contaminated with so many sins and obscenities . . . have been justly conquered,' thundered Sepúlveda. Las Casas took the opposite position. According to Las Casas, 'the native people of these Indies' were 'naturally of good reason and good understanding'. The key word here was 'naturally'. Las Casas, like Acosta, was starting to think about humans as a product of the natural world. During the debate, Las Casas listed 'the natural causes of rationality in the Indians'. These included the 'condition of the lands', 'the makeup of the parts and organs of the exterior and interior senses', 'the climate', and 'the excellence and wholesomeness of the food'. In short, Las Casas provided an entirely natural explanation of both similarity and difference between human populations.[48]

Clearly, the people of the Americas were similar in many ways to Europeans. They were intelligent, they built great cities, and – as the Bible made clear – they must have been descended from Adam. But at the same time, people like the Aztecs and the Incas clearly looked and behaved very differently from Europeans. Their skin was typically darker, they were taller, and they rarely had any facial hair. They also engaged in human sacrifice and worshipped the Sun. Rather than looking for answers in ancient texts, Las Casas suggested that the climate, the landscape, and the food explained these differences. The Aztec diet was chiefly composed of 'roots and herbs and things from the earth', Las Casas noted, whereas the Spanish mainly ate bread and meat. Similarly, the hot climate, Las Casas argued, was the best explanation of why people in the Americas had darker skin.[49]

The same arguments could equally apply to Europeans. After all, if the climate explained why the Aztecs were so different, then what might happen to the Spanish who made the New World their home? Francisco Hernández, the Spanish doctor we met earlier, worried that Europeans might 'degenerate to the point of adopting the customs of the Indians'. There were similar debates surrounding diet. Although many marketed New World foods as fantastic cures, others argued that eating maize or potatoes might be dangerous for Europeans, that this could lead to degeneration or even death. In some ways, there was a classical precedent for these ideas. The ancient Greek physician Hippocrates had argued that climate could influence disease and the balance of the four humours. But in the sixteenth century, a new generation of thinkers took things a step further. They developed an environmental theory, not just of disease, but of human nature itself. And in doing so, they brought together the study of natural history, medicine, and humankind.[50]

For one group of people in particular, these debates were especially personal. In the years following the colonization of the New World, a number of conquistadors fathered children by Indigenous women. Known as *mestizos* by the Spanish, there was much at stake for these mixed-race individuals in the debate over human nature. Did diet or lineage matter most? Were the Aztecs civilized or barbarians? The answers to these questions would dictate all aspects of *mestizo* life, from who they could marry to whether they could inherit. Many mounted passionate defences of Indigenous culture, countering European charges of barbarism and irrationality. Some *mestizos* wrote detailed accounts of the Indigenous people of the Americas, many of which were later used by European writers. Brought up in the Americas, far from European centres of learning, *mestizos* were also less committed to the idea of ancient Greek and Roman authority. They understood, much as with natural history, that the best source of information on the history of the Americas was the very people who lived there. All you had to do was ask.[51]

Garcilaso de la Vega was born in 1539 in the Inca capital of Cusco, Peru. His father was a conquistador, descended from Spanish nobility. His mother was an Inca princess, the niece of the last Inca ruler. Garcilaso entered the world amidst ongoing conflict, as the Spanish did not completely defeat the Incas until 1572. Nonetheless, in the relative safety of Cusco, Garcilaso spent his early years navigating between two worlds. At his father's house, he was taught to read and write Spanish. At his mother's house, he was taught the Inca language of Quechua. Significantly, however, Garcilaso never attended university. Whilst he later learned about the works of Aristotle and Pliny, he didn't hold these ancient authors in particularly high regard. Instead, his knowledge of human history and culture came mainly from his mother, who told him about the long and proud traditions of the Incas.[52]

In 1560, Garcilaso left Peru for Spain, where he was known as 'El Inca'. His father had recently passed away and he needed to petition the Spanish court to retain his noble title. Garcilaso arrived in Spain at the height of the debates about the nature of American peoples. He met Las Casas – the Dominican friar who defended Indigenous rights – and also learned about the debate with Sepúlveda – the Spanish theologian who believed that Indigenous people were little more than barbarians. Remembering the tales of his mother, Garcilaso decided it was time to set the record straight, writing *The Royal Commentaries of the Incas* (1609). In this book, he criticized European scholars for failing to ground their accounts in evidence and experience. 'Though there have been learned Spaniards who have written accounts of the States of the New World . . . they have not described these realms so fully as they might have done,' explained Garcilaso. 'I have fuller and more accurate information than that provided by previous writers,' he added. The Incas, unlike the Aztecs, did not have a system of writing. The memorization and repetition of Inca history was therefore an important part of the education of a young Inca noble, something Garcilaso and his family clearly took seriously. Most of the *Royal Commentaries* was written from memory. 'It seemed to me the best scheme . . . was to recount what I often heard as a child from the lips of my mother and her brothers and uncles,' explained Garcilaso. Drawing on this oral history, Garcilaso promised to uncover 'the origin of the Inca'.[53]

The *Royal Commentaries* opens well before the Spanish conquest, with the founding of the Inca Empire in the twelfth century. Garcilaso set out the traditional origin myth in which the first Inca ruler, Manco Cápac, was created by the Sun god, rising from a great lake. Manco Cápac then led his people into the Andes, establishing the capital city of Cusco and with it the Inca Empire. Garcilaso, much like his contemporaries in Europe, also discussed the importance of climate in shaping human history. Cusco was presented as a kind of earthly paradise. The city sat in a 'beautiful vale . . . surrounded on all sides with lofty heights, with four streams of water irrigating the land'. High in the Andes, it was neither too hot nor too cold. 'The climate is most agreeable, fresh, and soft, with constant fine weather, and free from heat and cold,' explained Garcilaso. Unlike the lowlands, there were 'very few flies' and 'no stinging mosquitos'. In this idyllic setting, Manco Cápac transformed Garcilaso's wandering ancestors into an advanced civilization. Before long, the Incas were tilling the soil, growing crops, and building temples – all practices that at the time were understood by Europeans as markers of civilization. The Incas began, as Garcilaso explained, to 'use the fruits of the earth like rational beings'. The message was clear. Sepúlveda was wrong. The people of the Americas were no barbarians.[54]

IV. Mapping America

In May 1493, Pope Alexander VI divided the world in two. Spain and Portugal had been squabbling over the New World since its 'discovery', each claiming the islands of the Caribbean and the coastline around Brazil as its own. To resolve this conflict, Pope Alexander issued a decree. A line would be drawn right down the middle of the New World. All the land to the west of this line would be claimed by the Spanish. All the land to the east of this line would be claimed by the Portuguese. Spain and

Portugal agreed, signing the Treaty of Tordesillas a year later in 1494. They settled on a line just over 1,000 miles west of the Cape Verde Islands. Portugal got Brazil. Spain got Mexico and Peru. There was only one problem. No one had a good map of the New World.[55]

The majority of European maps produced before the sixteenth century were based on the work of the ancient Greek geographer Claudius Ptolemy. Originally written in the second century, Ptolemy's *Geography* was still widely read over 1,000 years later in fifteenth-century Europe. It was usually accompanied by a map of the world which ran from the coast of West Africa all the way to the Gulf of Thailand in the east. Ptolemy knew about India and China, and he also knew that the Earth was round. Ptolemy, however, had no knowledge of the American continent. He simply assumed that the Atlantic Ocean extended all the way to the East Indies. In fact, it was this idea that inspired Christopher Columbus in the first place. Setting sail in August 1492, he hoped to discover, not a new continent, but a westward route to China.[56]

Columbus himself never quite gave up on this idea. He died in 1506 still believing he had reached the East Indies. But others were quick to point out the implications of the 'discovery' of the New World for geography. 'The opinion of the ancients was, that the greater part of the world beyond the equinoctial line to the south was not land, but only sea,' explained Amerigo Vespucci after returning from Brazil in 1503. 'But this opinion is false, and entirely opposed to the truth,' he concluded. As with natural history and medicine, the encounters in the Americas brought about a transformation in the study of geography. Many began to question the authority of ancient texts, instead emphasizing the importance of collecting evidence and testing ideas against experience.[57]

Initially, cartographers in Europe found it hard to reconcile the many conflicting accounts of American geography. The earliest surviving map of the New World, dating from 1500, depicts the American continent as a series of islands. It was based largely on Columbus's account of his first and second voyages, and his claim to have reached 'the Indian Islands beyond the Ganges'. Other early sixteenth-century maps depicted the North and South American land masses as separate, suggesting that it might be possible to sail between them. Cartographers also had to grapple with the challenges of working at far greater scales

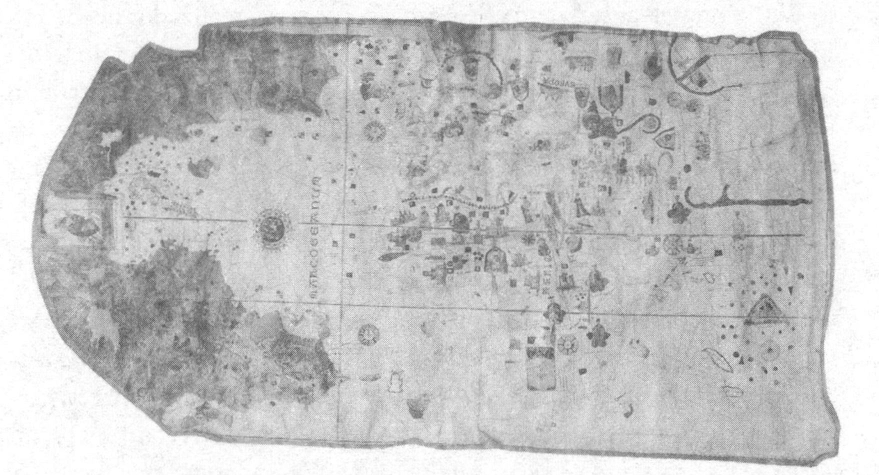

4. The oldest surviving European map to include the Americas, produced by Juan de la Cosa, captain of the *Santa María*, in 1500.

than before. Mapping the Mediterranean was one thing, but the entire world, and a new continent, quite another.[58]

The basic problem, which was now much more pressing, stemmed from the fact the world is round, but a map is flat. What then was the best way to represent three-dimensional space on a two-dimensional plane? Ptolemy had used what is known as a 'conic' projection, in which the world is divided into arcs radiating outwards from the north pole, rather like a fan. This worked well for depicting one hemisphere, but not both. It also made it difficult for navigators to follow compass bearings, as the lines spread outwards the further one got from the north pole. In the sixteenth century, European cartographers started experimenting with new projections. In 1569, the Flemish cartographer Gerardus Mercator produced an influential map he titled 'New and More Complete Representation of the Terrestrial Globe Properly Adapted for Use in Navigation'. Mercator effectively stretched the Earth at the poles, and shrunk it in the middle. This allowed him to produce a map of the world in which the lines of latitude and longitude are always at right angles to one another. This was particularly useful for sailors, as it allowed them to follow compass bearings on the map as straight lines. Today, Mercator's projection, originally designed to aid navigation to the Americas, is used as the basis of all modern world maps.[59]

With so much at stake, the Spanish Crown soon realized it needed to adopt a more systematic approach to studying the Americas. In 1503, Queen Isabella and King Ferdinand established the House of Trade in Seville. This acted as a central hub for all information arriving from the New World. Every report of a new island, new animal, or a new plant was sent to Seville, to be recorded and catalogued. The House of Trade worked closely alongside the Council of the Indies, established in 1524 to centralize the administration of the Spanish Empire. Together, these two organizations provided some of the first salaried positions for studying science outside of universities in Europe. Geographers, astronomers, natural historians, and navigators were all employed directly by the Spanish Crown. Together, they produced new charts and maps, all with the ultimate aim of securing Spanish territory under the Treaty of Tordesillas. Every captain returning from the New World was also expected to report to the House of Trade, noting any discrepancies between his voyage and the map he had been provided with. This then was the first time that modern science became fully institutionalized in Europe. Not in universities or academic societies, but as part of a Spanish project to know, and to conquer, the Americas.[60]

Juan López de Velasco was the ultimate polymath. He occupied the position of Chief Cosmographer at the Council of the Indies, one of the new roles supported by the Spanish Crown. The discipline of cosmography combined aspects of geography, natural history, anthropology, and cartography. Indeed, it brought together many of the different sciences we've uncovered in this chapter. Velasco's job was essentially to take all this knowledge and provide the most complete account of the Spanish Empire in the Americas, something which would then aid in its administration. Mapping was top of the agenda. However, Velasco soon realized that, to produce a really accurate map of the Americas, he would need to mobilize the entire Spanish Empire. This is exactly what he tried to do in 1577.

Through his position at the Council of the Indies, Velasco organized a questionnaire to be sent out to every Spanish American province. There were fifty questions, ranging from queries about the natural produce of the region to the exact latitude and longitude of major towns. 'What are the ports and landings along the coast?' Velasco asked, continuing, 'Give the names of the mountains, valleys, and districts, and for each, tell what the name means in the indigenous language.' A number of questions also directly asked for the respondent to draw a map. The local governor or mayor of the province was then expected to compose a response. This, often accompanied by a hand-drawn map of the local area, would be sent back to Velasco in Spain. The responses are known as the 'Geographical Reports'. All in all, Velasco received 208 reports, ranging from Peru to Hispaniola. The majority of the responses, however, came from the largest colony: New Spain.[61]

A questionnaire might seem like an obvious way to collect geographical information, but in the sixteenth century this idea was entirely novel. It represented a new way of doing geography, one that – like science more generally in this period – relied less and less on ancient Greek and Roman authority. It also represented a particularly centralized and institutionalized approach to science, one that had not been tried before in Europe. Nonetheless, what is most fascinating about the Geographical Reports is the way that Indigenous people contributed to the project. As with natural history and medicine, the only way to truly know the geography of the Americas was to ask the people who lived there.

Europeans were often impressed by the geographical knowledge possessed by Indigenous people. Columbus himself described how the Arawak people of the Caribbean 'sail all these seas, and it is wonderful how good an account they give of everything'. Columbus even reported that he had found one who 'would draw a sort of chart of the coast'. Similarly, in the 1540s, the Spanish explorer Francisco Vázquez de Coronado acquired a map of New Mexico from the local Native American tribe, the Zuni. As was common amongst various Indigenous groups, the map had been painted on a deer hide. Other Indigenous people simply memorized maps, scratching them into the sand or arranging sticks at a camp site when needed. But of all the American peoples, it was once again the Aztecs who possessed a particularly advanced method of mapping. This in part stemmed from the status of the Aztec Empire as a large, centralized, tributary state.[62]

Like the Spanish, the Aztecs recognized the importance of maps as tools of government. Moctezuma himself had commissioned an enormous map of the Aztec Empire in the 1510s. Painted on cloth, and incorporating the entire Gulf of Mexico, the map featured all the roads, rivers, and cities surrounding the capital of Tenochtitlan. This had been compiled following a major geographical and historical survey of the empire, all recorded in a series of Nahuatl pictorial codices. This Aztec tradition of mapping turned out to be an important source for the Geographical Reports sent to the Council of the Indies. In fact, of the sixty-nine maps Velasco received from New Spain, forty-five were produced by Indigenous artists. This makes sense. After all, most local Spanish governors had not travelled far beyond the cities they worked in. Velasco himself subtly recognized this. In the instructions accompanying the questionnaire, Velasco noted that, if a governor couldn't answer the questions, he was to 'encharge them to intelligent persons with knowledge of matters of the area'. More often than not, this meant an Aztec elder.[63]

The process of mapping the Americas was therefore not dissimilar from Sahagún's research into natural history. The local Spanish governor would consult a group of Indigenous elders. They would be supplied with the questionnaire, sometimes translated orally into Nahuatl, and asked to respond. The elders would then call a 'native painter' to produce a map – often copied directly from an existing Aztec codex. As with works of natural history and medicine, these tended to incorporate Nahuatl glyphs or traditional Aztec imagery. In 1582, for example, Velasco received a stunning map of the region known as Minas de Zumpango. At first glance, it doesn't look so different from a European map of the period. But look a little closer, and the Nahuatl glyphs are there again. A series of pictograms at the top of the map record the Nahuatl names for surrounding towns. These are then separated by a string of tiny footprints, a traditional Aztec symbol used to represent a border.[64]

Other maps received by Velasco followed a similar pattern. A map of the region known as Misquiahuala, painted by an Indigenous artist on animal hide, features a series of Nahuatl glyphs. Like the map of Minas de Zumpango, pictograms surround the edge indicating the names of nearby towns. A great river runs through the middle, whilst a large hill – again in the form of a Nahuatl glyph – sits to the west. The artist of the map then provided a pictorial answer to many of Velasco's questions concerning natural history, covering the hill with pictograms representing cacti and animals. Aware that Velasco might have trouble interpreting the Nahuatl glyphs, a Spanish missionary then annotated the map, writing, 'this is a hill of Misquiahuala, where there are many

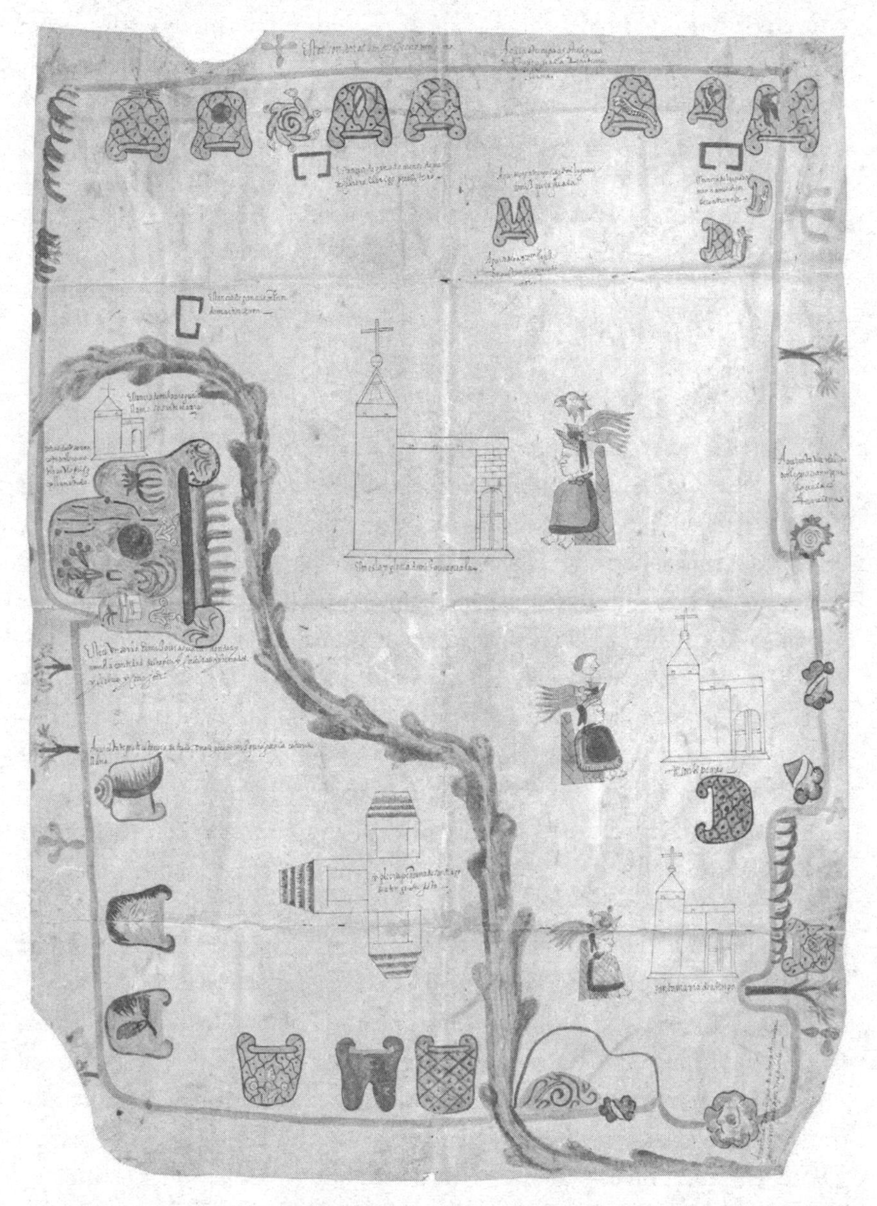

5. An Aztec map of Misquiahuala, New Spain, sent to the Council of the Indies as part of the Geographical Reports, *c.*1579.

lions, serpents, deers, hare, and rabbits'. This map is also one of the few to feature Indigenous people. Right at the centre, next to the main church of Misquiahuala, there is an Aztec elder, complete with feather headdress, sitting on a throne. The image served as a reminder of the ambivalent position the Spanish found themselves in. On the one hand, the Spanish wanted to map the Americas so that they could more easily claim and administer colonial territory. But on the other hand, such a project was clearly impossible without the help of the very people the Spanish hoped to displace.[65]

V. Conclusion

'The further one travels, the more one learns.' Christopher Columbus wrote these words shortly after returning from his third voyage to the New World in 1500. He was right. From the early sixteenth century onwards, the sciences were transformed by conquistadors, missionaries, and *mestizos* travelling to and from the Americas. In this chapter, we began to uncover the vital importance of global history for understanding the history of modern science. Opening with the colonization of the Americas in 1492, we've seen that to really explain the scientific revolution we need to examine connections between Europe and the wider world. We saw how the development of natural history, medicine, and geography was closely tied to the political and commercial goals of the Spanish Empire in the Americas. Maps were used to claim territory, whilst explorers searched for valuable plants and minerals. These efforts to conquer and colonize the Americas sparked a transformation, not just in what was known, but in how science was actually done.[66]

Prior to the sixteenth century, European scholars relied almost exclusively on ancient Greek and Roman texts. For natural history they read Pliny, for geography they read Ptolemy. However, following the colonization of the Americas, a new generation of thinkers started to place a greater emphasis on experience as the main source of scientific knowledge. They conducted experiments, collected specimens, and organized geographical surveys. This might seem an obvious way to do science to us today, but at the time it was a revelation. This new emphasis on experience was in part a response to the fact that the Americas were completely unknown to the ancients. Pliny had never seen a potato, and Ptolemy believed the Atlantic Ocean extended all the way to Asia. Today, we still talk of scientists making 'discoveries'. This is a metaphor which has its origins in the sixteenth century, when scientific discoveries and geographical discoveries went hand in hand. Nonetheless, the scientific revolution didn't simply follow from new evidence that contradicted ancient texts. It was also a product of encounters between different cultures.[67]

Often forgotten today, the Indigenous people of the Americas possessed an advanced scientific culture of their own. Europeans were fascinated by Aztec and Inca ideas and what they might offer. Drawing on this knowledge, explorers and missionaries – as well as Indigenous people – produced new works of natural history, medicine, and geography. There was an irony here. Scholars back in Europe increasingly promoted themselves as rejecting ancient texts, replacing what Pliny or Ptolemy had written with first-hand experience. But in reality, many were simply exchanging one text for another. Missionaries such as Bernardino de Sahagún sought out Aztec codices, translating them from Nahuatl into Latin and Spanish. These codices, many of which were destroyed in the sixteenth century by Catholic missionaries who saw them as a threat to Christian doctrine, then formed the basis of some of the most important works of early modern science produced in Europe between 1500 and 1700.

It wasn't just in the Americas that Europeans encountered new ways of thinking about science. In 1497, just five years after Columbus first arrived in the Americas, the Portuguese navigator Vasco da Gama sailed around the Cape of Good Hope, reaching the Indian Ocean for the first time. In doing so, he initiated a new era of contact between Europe and Asia, one that had an equally profound effect on the development of science. It is also important to recognize that Europeans weren't the only ones encountering new cultures in this period. As we'll see in the following chapter, scientific thinkers from across Asia and Africa were also travelling the world and exchanging ideas. With the expansion of religious and trading networks in the sixteenth and seventeenth centuries, the scientific revolution soon transformed into a global movement.

2. Heaven and Earth

Standing atop the observatory, Ulugh Beg gazed at the heavens. Every night, the young Muslim prince would walk to the astronomical observatory on the outskirts of Samarkand, located in modern-day Uzbekistan. The Samarkand observatory was at the centre of an Islamic world of scientific advance, one that would profoundly influence the development of astronomy and mathematics in Christian Europe. Built in 1420 on a hill overlooking the city, this was the perfect place to watch the stars. From the rooftop, Ulugh Beg could identify the constellations and spot comets. Like many fifteenth-century rulers, whether in Europe, Asia, or Africa, he placed great trust in astrology. An unfavourable alignment of the stars, perhaps if Cancer was too low in the sky, could signal disaster. Plague and crop failure might follow. And whilst today we might associate astrology with superstition, in the early modern period it was an important aspect of religious and political life. Rulers used astrological predictions to help make important political decisions, such as when to go to war, or whom to form an alliance with, whilst most world religions tied key festivals – whether that was Ramadan or Easter – to astronomical phenomena.

For over twenty-five years, between 1420 and 1447, astronomers at Samarkand undertook a programme of meticulous observation, measuring and predicting the movements of the stars and planets. The main building of the Samarkand observatory consisted of a large tower, three storeys high. The outer walls were covered in sparkling turquoise tiles, inlaid with geometrical designs typical of the Islamic architecture of the day. At the centre of the observatory stood the great 'Fakhri Sextant'. Over forty metres tall, this was one of the most accurate scientific instruments in the world at the time. Built of brick and limestone, the Fakhri Sextant was used to measure the exact position of the stars and planets in the sky. Today, if you visit the Samarkand observatory, you can still see the bottom section of this immense stone structure. Only a few metres remain, but you quickly get a sense of the

6. The Fakhri Sextant, built in 1420 in Samarkand, in modern-day Uzbekistan.

scale. The Fakhri Sextant sinks deep into the ground, built right into the bedrock.[1]

Born in 1394, Ulugh Beg was the grandson of Tamerlane, founder of the Timurid Empire. Throughout the fourteenth century, Tamerlane conquered much of Central Asia, hoping to unite the region under a single Islamic ruler. Ulugh Beg spent his early years following his grandfather on military campaigns, and it was during this time that he first became interested in astronomy. In the course of his travels, the young prince visited the remains of the Maragheh astronomical observatory, built in northern Persia during the thirteenth century. Inspired by this observatory, which featured its own enormous stone quadrant, Ulugh Beg ordered the construction of a similar institution back in Samarkand. This was part of a broader programme of works initiated by Ulugh Beg when he became governor of the city. From colleges and public baths to mosques and ornamental gardens, Ulugh Beg transformed Samarkand into a vibrant cultural hub, right at the heart of the Silk Road – a long-distance trading route which stretched from Africa through Europe and Central Asia all the way to China.[2]

For Ulugh Beg, the astronomical observatory was a site of religious devotion as much as scientific enquiry. In the Islamic world, science and faith had always gone hand in hand. From the times of the five daily prayers to the start and end of Ramadan, Islam is a religion which, perhaps more than any other, relies on accurate astronomical information. For this reason, most large mosques employed a timekeeper, and most Islamic courts employed an astronomer. Today, we often think of the job of an astronomer (someone who tracks the movements of the stars and planets) as completely separate from the work of an astrologer (someone who makes predictions about the future based on the movement of the heavens), but in the early modern period these roles overlapped. In Islamic courts, an astronomer would also double as an astrologer (in fact, the Arabic term *munajjim* refers to both), casting horoscopes and offering religious and political guidance. In establishing the observatory at Samarkand, Ulugh Beg was therefore fulfilling what he saw as a religious obligation. 'It is the duty of every true Muslim . . . to strive after knowledge,' Ulugh Beg would say, quoting directly from the words of the Prophet Muhammad.[3]

Patronage of the sciences, particularly astronomy, was part of a long tradition amongst Muslim rulers, dating back to the medieval period. In ninth-century Baghdad, the ruler of the Abbasid caliphate established the 'House of Wisdom'. It was here that a number of Islamic scientific thinkers made important contributions to fields ranging from mathematics to chemistry. These contributions included the invention of algebra and the development of the laws of optics. Many of the scientific terms we use today, including algebra, alchemy, and algorithm, have their origins in Arabic, or are named after Muslim thinkers. For this reason, historians of science often refer to the period between the ninth and fourteenth centuries as the medieval Islamic 'golden age'.[4]

There is, however, a major problem with the idea of an Islamic 'golden age'. It relies on the false notion that Islamic science – along with Islamic civilization in general – went into a period of decline immediately after the medieval period. This serves to separate out the Muslim world from the story of the scientific revolution, which took place between the fifteenth and seventeenth centuries. In fact, as we learned in the introduction to this book, the idea of an Islamic 'golden age' had been invented during the nineteenth century in order to justify the expansion of European empires into the Middle East. It was then later reinforced during the Cold War by historians of science in Western Europe and the United States, as well as postcolonial nationalists, all of whom were keen to relegate Muslim achievements to the distant past. And so, whilst it is certainly true that Islamic scholars played a key role in the development of medieval science, their contributions did not suddenly end in the fourteenth century. Ulugh Beg and his observatory are an important reminder of this. He followed in the tradition of patronage established by earlier Muslim rulers, but pushed this well beyond the medieval 'golden age' typically associated with Islamic science.[5]

Unlike many Muslim rulers, who simply acted as patrons, Ulugh Beg was himself an accomplished mathematician and astronomer. Records from the time describe him as the 'sahib of the observatory', indicating that he took an active role in directing the programme of astronomical work. According to Ulugh Beg's own proclamation, the Sun and Moon were to be observed every day, Mercury every five days, and the remaining planets every ten days. We also know that he carefully studied the astronomical works of those who went before him. He owned a copy of a medieval Arabic astronomical catalogue titled *Book of the Fixed Stars* (964), scribbling notes in Persian in the margin. Contemporary astronomers also praised Ulugh Beg's mathematical ability. One even described an occasion in which Ulugh Beg had apparently 'derived the longitude of the Sun to a fraction of two minutes by mental calculation while riding on horseback'. Before long, scholars from all over Central Asia came to Samarkand to work with this great 'king and astronomer'.[6]

Ali Qushji was the star of the show. Born in Samarkand in 1403, he enjoyed a comfortable upbringing in the royal court. The son of the royal falconer, he studied at one of the new colleges established by Ulugh Beg in the city. Ali Qushji soon learned how to operate an astrolabe, a classic Islamic scientific instrument used to make astronomical observations and aid in mathematical calculations, as well as how to read Persian manuscripts describing the laws of planetary motion. Before long, Ali Qushji was ready to put what he had learned into practice. He trekked across the desert to the Gulf of Oman, studying the relationship between the Moon and the tides. This resulted in his first astronomical work, a short manuscript on the phases of the Moon. Any astronomer who could better predict the motion of the Moon was bound to find favour with Muslim patrons, particularly as the Islamic calendar is based around the lunar month. Ulugh Beg was impressed, and quickly invited Ali Qushji to return to Samarkand and join the observatory. It was here that he helped compile one of the most influential works in the history of astronomy: *The Tables of the Sultan* (1473).[7]

Written in Persian, *The Tables of the Sultan* was the most accurate compilation of astronomical measurements ever produced, and would remain so for the next 150 years. Ali Qushji undertook much of the astronomical work himself. Ulugh Beg also helped, running up and down the stairs at the centre of the observatory, tracing the path of a particular star or planet along the Fakhri Sextant. The finished result, a product of over fifteen years of daily observations, included a catalogue listing the coordinates of 1,018 stars as well as data for the orbits of each of the five known planets (Mercury, Venus, Mars, Jupiter, and Saturn). Alongside this, *The Tables of the Sultan* contained a calculation for the length of the solar year, a crucial figure for the compilation of the annual calendar. At 365 days, 5 hours, 49 minutes, and 15 seconds, the final result was within 25 seconds of the value known today, over 500 years later.[8]

Whilst his grandfather Tamerlane had sought to unite the Islamic world through conquest, Ulugh Beg turned to science. *The Tables of the Sultan* structured the daily lives of Muslims across the Timurid Empire. Whether in Baghdad or Bukhara, Ulugh Beg's tables determined the times of prayer as well as major religious festivals. *The Tables of the Sultan* also allowed astronomers to accurately establish the direction of Mecca, another fundamental aspect of Islamic prayer. Through astronomy, Ulugh Beg hoped to bring the people of Central Asia together, under one religion, and one ruler. Before long, *The Tables of the Sultan* spread beyond the Timurid Empire, travelling east and west along the Silk Road. In Egypt, the Mamluk sultan ordered a copy of the tables. Local astronomers then translated them from Persian into Arabic, recalculating many of the coordinates relative to Cairo as they went. As we'll see, copies of *The Tables of the Sultan* later reached as far as Istanbul and Delhi, helping to standardize Islamic religious practice right across the world.[9]

However, just as Ulugh Beg was pressing for unity, the Timurid Empire began to crumble. Perhaps it was foretold in the stars. After the death of Ulugh Beg's father in 1447, civil war ensued. Rival factions vied for control and Ulugh Beg ended up fighting wars against his uncles and cousins, as each tried to lay claim to the throne. Even his own children turned against him. Abdul Latif, Ulugh Beg's eldest son, had come under the influence of a group of religious fanatics. They stoked his jealousy, convincing him that he had been unfairly treated and should take the throne for himself. Angry and resentful, Abdul Latif ordered the assassination of his own father. On 27 October 1449, Ulugh Beg, the great Samarkand astronomer, was dragged from his horse and murdered.[10]

The death of Ulugh Beg marked the end of astronomy in Samarkand, but it was just the beginning of a wider transformation in how people understood the heavens. As we saw in the previous chapter, the scientific revolution is best interpreted as a product of global cultural exchange. In this chapter, we build on this story, moving east rather than west. We follow how connections between Europe, Asia, and Africa shaped the development of astronomy and mathematics from around 1450 to 1700. This was a period in which religious and trading networks expanded significantly, bringing different people into contact with a range of new scientific ideas. Caravans travelling along the Silk Road, as well as missionaries sailing across the Indian Ocean, returned with copies of Arabic manuscripts, Chinese star catalogues, and Indian astronomical tables.

At almost exactly the same time as Ulugh Beg was building his observatory in Samarkand, stargazers in Europe were entering the Renaissance. This was a period of major intellectual advance in the arts and sciences, stretching from the fifteenth century to the seventeenth.

During the Renaissance, a term which literally means 'rebirth', European scientific thinkers reinterpreted the works of ancient Greek and Roman writers. Astronomers like the famous Nicolaus Copernicus, who first suggested that the Sun was at the centre of the universe, ultimately rejected the wisdom of the ancients, proposing radical new theories of planetary motion.

That story is at the heart of most traditional histories of science. But as we'll see, it is impossible to properly account for the scientific revolution in Europe without paying attention to what was going on elsewhere. Copernicus himself relied on ideas found in Arabic and Persian manuscripts imported from places like Samarkand and Istanbul. And during the same period, Chinese, Indian, and African astronomers fused their own ideas with those originating from Europe and the Islamic world. When we look across Europe, Africa, and Asia we in fact see remarkable similarities with how scientific thinkers were combining new and old ideas, as well as drawing from different cultures. This then was a global Renaissance, one that stretched all the way from Rome to Beijing. As ideas travelled back and forth across oceans and along the Silk Road, the empires of Europe, Africa, and Asia witnessed a major transformation in the sciences. In order to understand the history of astronomy and mathematics during the scientific revolution we therefore need to begin, not with the traditional story of Copernicus in Europe, but with the Islamic world of science that inspired him.[11]

I. Translating the Ancients

European astronomers had long relied on Arabic sources. After all, Muslim scholars were the first to show a serious interest in ancient Greek science, which later underpinned much of the curriculum at medieval European universities. In ninth-century Baghdad, a group of Muslim scholars first translated the works of Claudius Ptolemy from ancient Greek into Arabic. Ptolemy's *Almagest*, originally written in second-century Egypt, proved incredibly influential in both medieval Europe and the Islamic world. Ptolemy described a classical model of the universe, with the Earth rather than the Sun at the centre. Ptolemy's astronomy, however, was not without its problems. For a start, it was exceptionally complicated. Many of these complications arose from Ptolemy's commitment to Aristotle's philosophy of the cosmos. In his *Physics*, written in the fourth century BCE, Aristotle described a fundamental separation between the Earth and the heavens. The heavens were perfect, unchanging, and eternal. The Sun, the stars, and the planets therefore moved at a constant speed in perfect circles around the Earth. In contrast, the Earth was 'corruptible'. Motion on the Earth was therefore discontinuous and linear. Things moved in straight lines, at changing speeds, and could come to rest.[12]

Now, even Ptolemy realized that the planets do not move in perfect circles. Rather, the planets appeared to wobble, moving closer and further away from the Earth throughout the year. They also seemed to speed up and slow down as they went, at least from the perspective of a stationary Earth. (Today, we know this is because they are orbiting around the Sun, in ellipses rather than circles.) In order to reconcile this, Ptolemy introduced all kinds of mathematical tricks. He had planets orbit a point slightly offset from the Earth, known as the 'eccentric'. Ptolemy also introduced the concept of an 'epicycle' – literally 'upon the circle' – in which the planets did a kind of double rotation. Each planet moved in an orbit around a smaller circle which itself moved in a larger orbit around the Earth. Finally, Ptolemy introduced another imaginary point: the 'equant'. From this point, which was also distinct from the Earth, planets appeared to move at a constant speed. All this allowed Ptolemy, through some mental gymnastics, to preserve Aristotle's claim that everything in the heavens moved in perfect circles at a constant speed, whilst also providing a reasonable model of planetary motion.[13]

Arabic translators were well aware of the deficiencies of Ptolemy's model. Ibn al-Haytham, an eleventh-century astronomer based in Cairo, wrote a damning critique entitled *Doubts on Ptolemy* (1028). Al-Haytham wasn't fooled by Ptolemy's tricks. He argued that the introduction of all these imaginary points, like the equant and the eccentric, made a mockery of the ideal of uniform circular motion. Clearly, al-Haytham reasoned, the planets did not move in perfect circles. 'Ptolemy assumed an arrangement that cannot exist,' he concluded. This was the beginning of a long tradition in the Islamic world, which later reached Christian Europe, whereby translators of ancient Greek

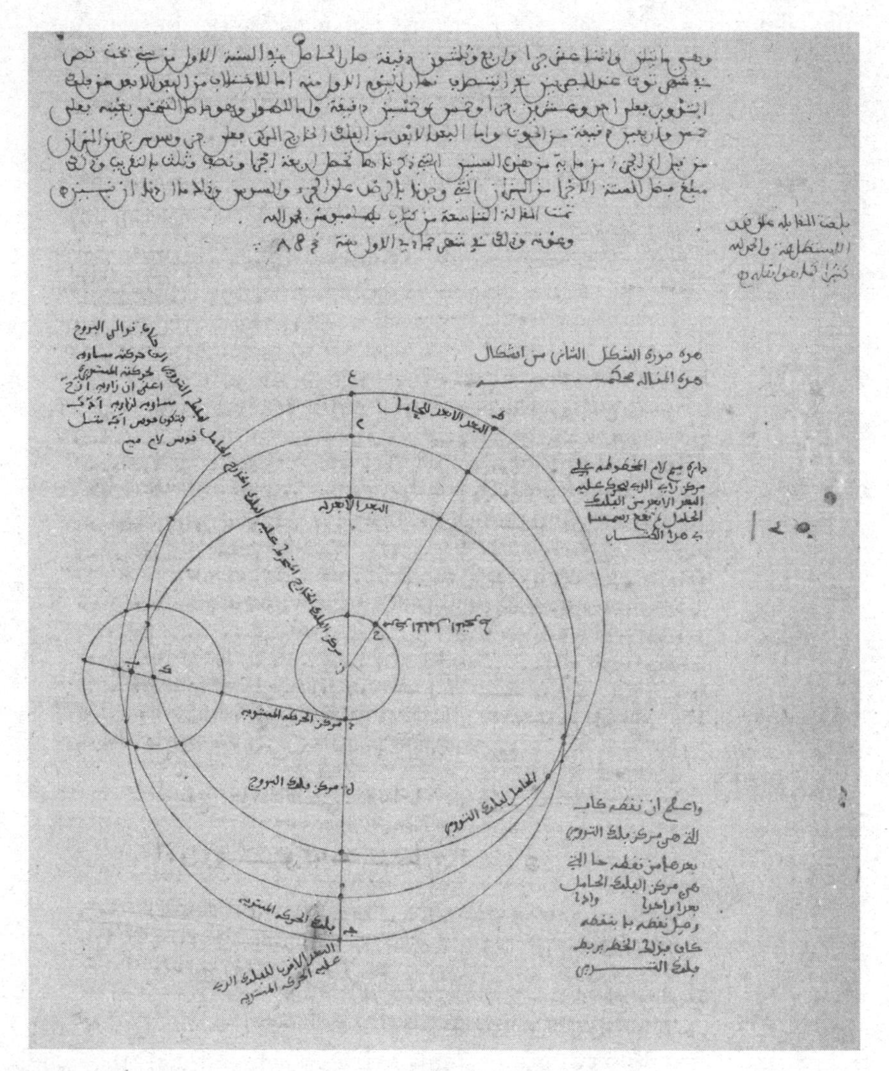

7. An Arabic manuscript translation of Claudius Ptolemy's *Almagest*, copied in Spain in 1381. The diagram illustrates Ptolemy's model of the universe, with the Earth at the centre, along with the use of the epicycle and eccentric.

science also offered commentaries and critiques. The most influential of these was penned by Nasir al-Din al-Tusi. Born in 1201, al-Tusi was the leading astronomer at the Maragheh observatory in northern Persia, then part of the Mongol Empire. This was the very observatory that Ulugh Beg later visited as a young man, inspiring him to establish a similar enterprise in Samarkand. At Maragheh, al-Tusi conducted daily observations of the heavens, compiling astronomical tables, or *zij*. He also had access to a wealth of ancient Greek and Arabic manuscripts, particularly after Mongol forces raided Baghdad in 1258.[14]

Al-Tusi immediately saw the flaws in Ptolemy's system. In his *Memoir on Astronomy* (1261), al-Tusi followed al-Haytham in pointing out the contradictions between Ptolemy's model of the universe and Aristotle's physics. But al-Tusi went one better. Rather than just critiquing Ptolemy, al-Tusi offered a solution. He invented a geometric tool known as the 'Tusi couple'. This was a combination of two circles: a smaller one rotating around the circumference of a larger one that was exactly twice the size. Al-Tusi realized that this movement almost perfectly modelled the characteristic wobble of planets, without the need for the epicycle or equant invented by Ptolemy. The Tusi couple also suggested that Aristotle's distinction between linear and circular motion made no sense. If you take a point on the smaller circle and follow it, it seems to oscillate up and down along a straight line. Al-Tusi therefore showed that it was possible to generate linear motion – that is, movement in a straight line – simply by combining rotating circles. Later, as we'll see, the Tusi couple had a profound influence on the development of new astronomical ideas in Europe.[15]

By the twelfth century, most of the ancient Greek corpus had been translated into Arabic, ranging from Pythagoras's mathematics to Plato's philosophy. It was through these Arabic editions, as well as commentaries by al-Haytham and al-Tusi, that scholars in medieval Europe first encountered the ancients. Gerard of Cremona, an Italian living in the Kingdom of Castile, completed a Latin translation of Ptolemy in 1175. He did so by piecing together the text from Arabic manuscripts collected in Muslim Spain. Gerard even decided to keep the Arabic title: the *Almagest*, simply meaning 'the greatest'. Latin editions of other ancient Greek works soon followed, all translated from medieval Arabic manuscripts. By the 1400s, European astronomers had long accepted the

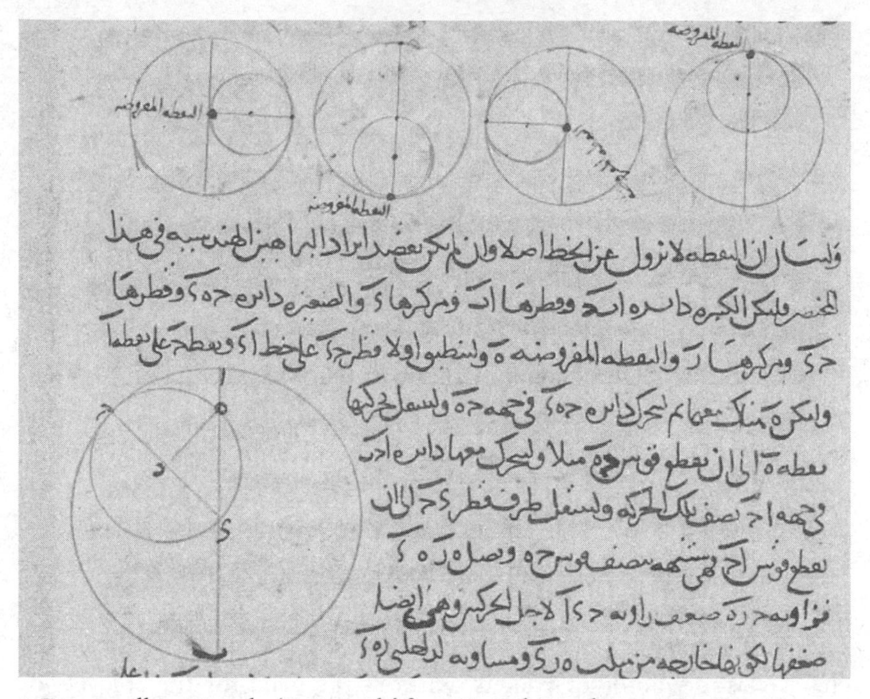

8. Diagram illustrating the 'Tusi couple' from Nasir al-Din al-Tusi, *Memoir on Astronomy* (1261).

need to work with Arabic sources, usually translated into Latin. Many believed that the ancient Greek originals had been lost forever. They were wrong.[16]

II. Islamic Science in Renaissance Europe

The people of Istanbul were preparing for the worst. For nearly two months, the capital of the Byzantine Empire had been under siege. Following the collapse of the Timurid Empire, a new Muslim power began to dominate Central and Western Asia: the Ottomans. Sultan Mehmed II lay siege to the city, raining fire from his galleys stationed on the Bosporus and blasting the Roman walls with immense iron cannon. On 29 May 1453, the city fell. Many Christians fled, whilst the Greek Orthodox basilica, the Hagia Sophia, was converted into a mosque. This signalled the beginning of over 400 years of Ottoman rule in the region, with an empire stretching from Istanbul to Cairo. It also marked a period of renewed European engagement with the Islamic world, one that would transform the sciences.

By the end of 1453, Istanbul lay in ruins, the air thick with smoke following weeks of bombardment. As Ottoman troops looted the city, many Byzantine Christians decided it would be safest to leave. The majority fled across the Adriatic, settling in the Italian city states of Venice and Padua. They brought with them a treasure trove of books and manuscripts, many of which had been locked away in church vaults for centuries. These included copies of ancient Greek editions of Aristotle and Ptolemy. Very few people in Europe had ever seen or read these before. Suddenly, many began to question the wisdom of relying solely on Arabic translations of ancient works, particularly as these editions were often extensively edited. There was also a worry that errors might have been introduced through multiple translations. Perhaps it would be better to return to the originals? It was this idea which underpinned the Renaissance movement known as 'humanism'. Humanists believed that the only way to revitalize European civilization was to return to the ancient past, an idea which soon spread to the sciences. In 1456, George of Trebizond, a Byzantine born on Crete, completed a new Latin translation of Ptolemy's *Almagest*. This edition was based entirely on ancient Greek manuscripts, bypassing the Arabic translations.[17]

The Renaissance, however, was never simply about the rejection of Arabic knowledge. Rather, this was a time in which a whole range of traditions collided. Alongside Byzantine refugees, Italian city states hosted Ottoman envoys seeking to establish trading networks or negotiate military treaties. At the same time, Europeans sent trading and diplomatic missions to the east. Venetian traders and Vatican diplomats could be found in the streets of Damascus and Istanbul. It was through these exchanges that new Arabic manuscripts, as well as Byzantine translations of Islamic sources, reached Europe. Today, many of the most valuable collections of Arabic and Byzantine manuscripts are housed in the libraries of Venice and the Vatican. And it was by combining these sources, from East and West, that Renaissance astronomers transformed our understanding of the heavens.[18]

*

Johannes von Königsberg, better known as Regiomontanus, was something of a child prodigy. In 1448, aged just twelve, he enrolled at the University of Leipzig. Finding the mathematics course there too easy, Regiomontanus decided to transfer to the University of Vienna, a more prestigious institution at the time. Arriving in 1450, the young mathematician and astronomer spent his free time compiling almanacs and casting horoscopes for wealthy patrons. It was also at the University of Vienna that Regiomontanus first met Georg von Peurbach, his great mentor. Peurbach was a typical Renaissance man. He could be found lecturing on everything from Roman poetry to Aristotle's physics. And together, Peurbach and Regiomontanus undertook a major reassessment of the astronomical sciences, starting with Ptolemy's *Almagest*.[19]

The duo were supported by Basilios Bessarion, a Byzantine Greek who had fled Istanbul following the Ottoman conquest. Bessarion arrived in Vienna in 1460, seeking an audience with the Holy Roman Emperor, Frederick III. Pope Pius II had recently declared a new crusade against the Ottomans, and Bessarion was sent to Vienna to secure the support of the Holy Roman Empire. Whilst there, he met Peurbach, by that time Frederick III's court astronomer. Bessarion, an accomplished scholar in his own right, had read George of Trebizond's new translation of Ptolemy's *Almagest*, and wasn't impressed. Peurbach shared these concerns. On close inspection, George of Trebizond's edition was full of errors and failed to accurately convey the ancient Greek. With this in mind, Bessarion invited Peurbach to complete a new translation of the *Almagest*, promising him unfettered access to all the latest manuscripts arriving from Istanbul at the time, both Greek and Arabic. It was an opportunity too good to miss, and Peurbach set to work.[20]

In 1461, after just a year working on the new translation, Peurbach fell gravely ill. He was only half-way through his new translation. Worried that all his hard work might go to waste, Peurbach made the young Regiomontanus promise to finish what he had started. Regiomontanus was as good as his word. He spent the next ten years travelling across Italy, collecting all the manuscripts he could get his hands on. The finished result was the most up-to-date work on astronomy for generations. Titled *Epitome of the Almagest* (1496), this was a quintessential work of Renaissance science. As its title suggests, the *Epitome* was much more than a new translation. Regiomontanus had instead combined the best bits from all the sources he could find – ancient Greek, Arabic, and Latin – producing a much improved version of Ptolemy's astronomy in the process. True, the Earth was still at the centre of the universe, but Regiomontanus had been able to solve a number of technical problems that had confounded European astronomers for centuries.[21]

One of Regiomontanus's major innovations was actually borrowed directly from Ali Qushji, the leading astronomer at the Samarkand observatory. Following Ulugh Beg's death in 1449, Ali Qushji fled the Timurid Empire. He wandered the desert for many years, seeking patronage at princely courts across Central Asia. In 1471, he arrived in Istanbul, recently conquered by the Ottomans. Sultan Mehmed II had received word of the great Samarkand astronomer and sent for Ali Qushji. He was to work as a professor of mathematics at one of the new *medreses*, or colleges, established in the city. It was this connection to Istanbul and the Ottomans which brought Ali Qushji's work to the notice of astronomers in Europe. In the *Epitome of the Almagest*, Regiomontanus copied out a diagram from a manuscript Ali Qushji had originally composed back in Samarkand in the 1420s. The diagram, a complex array of circles, proved that Ptolemy's epicycles could be dispensed with. All that was needed, Ali Qushji argued, was the eccentric. In short, Ali Qushji argued that the motion of all the planets could be modelled simply by imagining that the centre of their orbits was at a point other than the Earth. Neither he nor Regiomontanus went as far as to suggest this point might in fact be the Sun. But by dispensing with Ptolemy's notion of the epicycle, Ali Qushji opened the door for a much more radical vision of the structure of the cosmos.[22]

Nicolaus Copernicus was born in Poland in 1473. His family hoped he might become a Catholic priest, and sent him to the University of Bologna in 1497 to take a higher degree in canon law. But Copernicus soon found that Renaissance Italy had much more to offer. In Bologna, Copernicus attended lectures by Domenico Maria Novara, a controversial astrologer who had studied under Regiomontanus. Influenced by the growing criticism of Ptolemy, Novara argued that it was possible to detect a subtle shift in the Earth's axis. This would explain why the fixed stars appear to gradually move over long periods of time. (A phenomenon known as the 'precession of the equinoxes'.) Once again, this contradicted the classical teachings of Ptolemy, who held that the Earth was completely stationary. Novara also introduced Copernicus to Regiomontanus's *Epitome of the Almagest*, a copy of which Copernicus purchased whilst in Bologna. From then on, Copernicus was hooked on astronomy. He spent the next few years travelling across Italy, studying briefly in Padua, where Regiomontanus had lectured on Persian astronomy, before graduating in 1503 and returning to Poland. He settled in Frombork, where he became a canon of the city's cathedral. It was here that Copernicus developed one of the most famous theories in the history of science.[23]

Written in Latin, Nicolaus Copernicus's *On the Revolutions of the Heavenly Spheres* (1543) set out a heliocentric model of the universe: the Sun, rather than the Earth, was now at the centre. It proved contentious to say the least, challenging both religious and scientific understandings of the heavens. What Copernicus did was to draw together all the existing work, pushing the debate over Ptolemy, which had been rumbling on for centuries, to its logical conclusion. He borrowed philosophical ideas from Persia, astronomical tables from Muslim Spain, and planetary models from Egyptian mathematicians. In this respect *On the Revolutions of the Heavenly Spheres* was a classic Renaissance work of synthesis, drawing on both European and Islamic learning. Copernicus opened his book with a criticism that by this time was well known: Ptolemy's astronomy was inconsistent. It failed to preserve Aristotle's ideal of uniform circular motion and it introduced all kinds of mathematical tricks that made it needlessly complicated.

As we've seen, these ideas had been floating around the Islamic world since the ninth century, and were starting to infiltrate European astronomy. Copernicus cited no fewer than five Islamic authors in *On the Revolutions of the Heavenly Spheres*, many of whom were critical of Ptolemy. These included Thabit ibn Quarra, a ninth-century Syrian mathematician, and Nur ad-Din al-Bitruji, a twelfth-century astronomer from Muslim Spain. Copernicus himself couldn't read Arabic. But he didn't need to. Latin and Greek editions of major works of Arabic astronomy were widely available in sixteenth-century Europe. And studying in Italy, Copernicus had plenty of opportunities to learn about Islamic science from those who could read Arabic, men like Andrea Alpago at the University of Padua, who had spent over a decade in Damascus.[24]

Next, Copernicus argued that Ptolemy's model didn't correspond to the actual movement of the planets. In making this argument, Copernicus largely relied on existing astronomical tables, making few original observations himself. He based most of his work on the *Alfonsine Tables*, a collection of earlier Islamic tables compiled at the request of Alfonso X of Castile in the 1250s. These are a prime example of cultural exchange in action: a group of Jewish mathematicians collated a series of Arabic tables, before translating them into Spanish and Latin. Finally, Copernicus suggested that all these problems could be solved if we imagined that the Sun was at the centre of the universe. In making this move he was directly inspired by the *Epitome of the Almagest*. Regiomontanus, drawing on Ali Qushji, had shown it was possible to imagine that the centre of all the orbits of the planets was somewhere other than the Earth. Copernicus took the final step, arguing that this point was in fact the Sun. Projecting an image of divine order, Copernicus concluded that 'the Sun, as if seated on a kingly throne, governs the family of planets that wheel around it'.[25]

Having made this claim, Copernicus still had a lot of work to do. By itself, placing the Sun at the centre didn't produce a completely accurate model of the universe. For a start, like Aristotle and Ptolemy, Copernicus was still committed to the idea that heavenly bodies moved in perfect circles. But even with the Sun at the centre, the planets still seemed to wobble. To solve this problem, Copernicus turned to the work of one of the Islamic astronomers we met earlier: Nasir al-Din al-Tusi. *On the Revolutions of the Heavenly Spheres* contains a diagram which is identical to the one found in al-Tusi's Arabic writings. The similarity is striking, right down to the equivalent choice of letters, in both Latin and Arabic, used to label many of the elements. Copernicus most likely learned about al-Tusi from a Byzantine Greek translation of the Arabic original. Copies of such manuscripts, brought from Istanbul following the Ottoman conquest, could be found in a number of Italian libraries at the time. The diagram in *On the Revolutions of the Heavenly Spheres* shows the Tusi couple in action. Copernicus used this idea to solve exactly the same problem as al-Tusi. He wanted a way to generate an oscillating movement without sacrificing a commitment to uniform circular motion. Copernicus then took things a step further. He used the Tusi couple to model planetary motion around the Sun rather than the Earth. This mathematical tool, invented in thirteenth-century Persia, now found its way into the most important work in the history of European

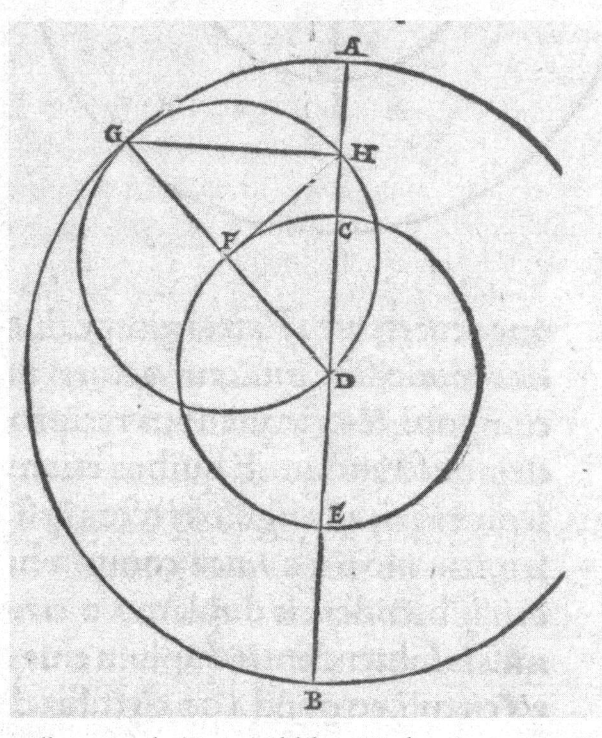

9. A diagram illustrating the 'Tusi couple' from Nicolaus Copernicus, *On the Revolutions of the Heavenly Spheres* (1543).

astronomy. Without it, Copernicus would not have been able to place the Sun at the centre of the universe.[26]

The publication of *On the Revolutions of the Heavenly Spheres* in 1543 has long been considered the starting point for the scientific revolution. However, what is less often recognized is that Nicolaus Copernicus was in fact building on a much longer Islamic tradition. Ibn al-Haytham, writing in eleventh-century Egypt, had long ago pointed out the contradictions in Ptolemy's model of the universe, particularly the idea that the planets moved in perfect circles. Nasir al-Din al-Tusi, writing in thirteenth-century Persia, had later suggested a way to solve this problem, essentially by imagining that the planets are revolving around two circles. And Ali Qushji, writing in fifteenth-century Samarkand, had also provided another solution, arguing that it was much easier to model planetary motion by assuming that the Earth was not at the centre of the orbits. Even the idea that the Sun might be at the centre of the universe was not entirely new. A number of Muslim astronomers, dating back to the ninth century, had discussed this possibility, although the idea never gained widespread acceptance in the medieval Islamic world.[27]

Rather than thinking of Copernicus as some kind of lone genius who single-handedly initiated the scientific revolution, we should instead see him as part of a much broader story of global cultural exchange. The key event here was the rise of the Ottoman Empire in the eastern Mediterranean, particularly following the conquest of Istanbul in 1453. Byzantine refugees and Venetian traders returned from the Ottoman lands with hundreds of new scientific manuscripts. Some of these were ancient Greek originals, others were more recent Arabic and Persian commentaries. It was this exposure to all these new texts and ideas that really kickstarted the scientific revolution in Europe. Copernicus is a perfect example of this. *On the Revolutions of the Heavenly Spheres* combined ideas found in Arabic, Persian, Latin, and Byzantine Greek sources to produce a radical new model of the universe.

Cultural exchange had a profound effect on the development of science in Renaissance Europe. But what about the rest of the world? In what follows, we explore the global history of the scientific revolution, travelling across Asia and Africa. From Istanbul and Timbuktu to Beijing and Delhi, scientific thinkers in cities across the world were beginning to reassess the ancient past, making new observations, and developing new astronomical and mathematical theories. This was all made possible thanks to the significant expansion in trading and religious networks from the fifteenth century onwards. These networks brought people into contact with new ideas and cultures, transforming the scientific revolution into a global movement. As we'll see, there are in fact a number of remarkable parallels between the story of the scientific revolution in Europe and the story of the scientific revolution elsewhere. With this in mind, we begin with an Ottoman astronomer at sea.

III. Ottoman Renaissance

Taqi al-Din was sailing across the Mediterranean, on his way from Alexandria to Istanbul. Having spent years mastering the art of astronomy in

Egypt, he was hoping to win favour at the court of the new Ottoman sultan, Murad III. Born in Damascus in 1526, and educated in Cairo, Taqi al-Din planned to offer his services as an astronomer, perhaps determining the times of the five daily prayers or the direction of Mecca. He might even provide the sultan with a horoscope, a service that fetched a high price. That was the plan, but, as Taqi al-Din soon found out, getting to Istanbul was no easy task. The Mediterranean was a dangerous place to travel in the sixteenth century, as European and North African pirates roamed the waters, taking captives that would either be sold as slaves or be ransomed. Suddenly, a galley pulled up alongside the ship that Taqi al-Din was sailing on. Fighting quickly broke out, as the pirates launched themselves onto the deck. Most of the crew were killed, their bodies tossed overboard. But Taqi al-Din was spared. As a learned man, the pirates knew, he would be worth a lot of money.[28]

A few months later, Taqi al-Din was sold as a slave to a Renaissance scholar in Rome. Educated Muslims were greatly prized for their ability to translate new astronomical manuscripts arriving from the east. Whilst in Rome, Taqi al-Din was asked to translate Arabic works on Euclid and Ptolemy. At the same time, this experience brought Muslims like Taqi al-Din into contact with Renaissance scientific culture. By the time he finally bought his freedom, Taqi al-Din was aware of the latest European astronomical theories. He had even learned a little Italian. Once he left Rome, Taqi al-Din was at last able to complete his journey to Istanbul. He arrived in 1571, over a decade after leaving Egypt, and was appointed as the chief astronomer to the Ottoman sultan. Taqi al-Din then convinced Murad III that science in Christian Europe was fast catching up with the Islamic world. In order to advance in science, and obtain good astrological predictions, the Ottoman sultan would need to establish a new astronomical observatory.[29]

Murad III agreed to the plan and ordered the construction of a new observatory in Istanbul in 1577. It was built on a hill overlooking the Bosporus, providing a breathtaking view of the city by day and the heavens by night. Although nothing of the original structure remains, we know a great deal about how the observatory operated thanks to a series of beautifully painted Persian miniatures. These accompanied an epic poem entitled *Book of the King of Kings*, written by Ala al-Din al-Mansur in 1580. The poem, as the title suggests, charted the reign of Murad III, a record

10. Taqi al-Din (*top row, third from right*) working in the Istanbul observatory.

of all his great deeds. 'When he issues orders for making observations and compiling astronomical tables; the stars will descend and prostrate themselves before him,' al-Mansur wrote. The observatory itself was coated in ornamental brass and copper, a golden dome on the Istanbul skyline. It also featured an enormous Fakhri Sextant. At fifty metres tall, this was even bigger than the one at Samarkand. There was also a separate well, twenty-five metres deep, which allowed astronomers to observe the stars during the day by blocking out the surrounding sunlight.

Inside the observatory, Taqi al-Din spent his time taking astronomical readings and compiling new tables. He owned a copy of Ulugh Beg's

Tables of the Sultan, writing corrections in the manuscript as he went. In one of the Persian miniatures, Taqi al-Din sits alongside fifteen other astronomers, mathematicians, and scribes. Each is dressed in typical Ottoman clothing – red and green robes complete with pristine white turbans. Some are observing the heavens, holding up an astrolabe. Others are measuring the passage of time. There is an hourglass in the centre, a terrestrial globe at the bottom and, intriguingly, a mechanical clock in one corner.[30]

At first glance, the clock may seem insignificant. But it is actually an indicator of just how closely connected Ottoman and European science had become. Mechanical, spring-driven clocks were invented in Europe at the end of the fourteenth century. They were primarily installed on church towers, or used as showpieces in princely courts. Taqi al-Din, however, saw the potential for this new invention for astronomy. After all, measuring the time it took a star or planet to cross the night sky, right down to the nearest second, was essential for compiling accurate astronomical tables. And so whilst earlier observatories, like those at Maragheh and Samarkand, used water clocks and sundials, Taqi al-Din built a mechanical device. In doing so, he became one of the first astronomers, in Europe or Asia, to install a dedicated mechanical clock in an observatory. The clock we see in the painting was most likely built by a European artisan. Throughout the sixteenth century, clockmakers in Holland and France catered for the growing Ottoman market in mechanical devices, constructing clocks with Turkish numerals. Some even featured the different phases of the Moon, corresponding to the Islamic lunar calendar. These clocks were often presented as gifts by European ambassadors, eager to curry favour with the Ottoman court. One clock kept at the observatory was constructed especially for Murad III. According to an official, it was 'in the shape of a castle, the gate of which opens on the stroke of the hour and out comes a figure of the Sultan on horseback'.[31]

Taqi al-Din was obsessed with these new devices. He inspected all the clocks in the sultan's collection, quickly understanding how they were constructed. He had also likely encountered mechanical clocks whilst imprisoned in Rome. Worried about reliance on European artisans, Taqi al-Din then set about designing and building his own. His manuscripts include incredibly detailed diagrams setting out exactly how to build a clock with an accurate second hand, crucial for astronomical work. One manuscript even described a clock-like machine used for cooking a shish kebab. Clearly, Taqi al-Din had mechanics on the mind. In fact, he was starting to think that the universe itself might be rather like a giant clock. (The same idea proved incredibly influential in Europe, particularly during the seventeenth century when it was taken up by René Descartes.) In a work that mixed theology, philosophy, and mathematics, Taqi al-Din set out his vision of a clockwork universe. He explained that he hoped 'to build a machine and a clock that would reflect the spiritual structure of the heavens'. Indeed, the Istanbul observatory featured just such a machine. In another one of the Persian miniatures we see an enormous spherical device, made of metal, supported by a wooden frame. Known as an 'armillary sphere', this was essentially a mechanical model of the heavens, one that astronomers could use to quickly perform complex geometrical calculations. Armillary spheres had been used in antiquity, but few people thought of them as anything more than a helpful tool, a bit like a calculator. Taqi al-Din in contrast was one of the first to push the philosophical implications of such a device. The universe really was like a machine.[32]

Stocked with the latest mechanical marvels, the Istanbul observatory emerged as a new centre for scientific advance in the eastern Mediterranean. But this wasn't just a site of Islamic science. Jews and Christians also worked at the observatory, reflecting the ethnic and religious diversity of the expanding Ottoman Empire. Some were brought as slaves, with one report stating that there were 'twelve captured Christians' at the Istanbul observatory. Others were fleeing religious persecution elsewhere. Amongst these was a Jewish man known as 'Dawud al-Riyadi', or David the Mathematician.[33]

In 1577, just as the observatory was being built, Taqi al-Din had been trying to observe a solar eclipse. However, it was far too cloudy in Istanbul that day, and Taqi al-Din could not get the measurements he needed. However, he had recently heard of a great astronomer and mathematician living in Salonika, 300 miles to the west. That man, whose real name was David Ben-Shushan, had been living in the Ottoman lands since the 1550s. He was an Italian Jew who, like many others, fled Europe during a period of rising antisemitism. In 1492, Spain expelled its Jewish population, followed by Portugal in 1497. Many initially travelled to Italy, but the Roman Inquisition of 1542 marked another wave of persecution. Jewish refugees were forced to flee once again. Many travelled further east towards the territory occupied by the Ottoman Empire. In Salonika, Ben-Shushan joined a Jewish population of around 20,000. He taught mathematics to the sons of local Ottoman governors, hence the name by which he was known in Arabic and Turkish. And it was through these court networks that Taqi al-Din learned of Ben-Shushan back in Istanbul.[34]

The two excitedly swapped astronomical data, discussing the latest scientific theories. Much to Taqi al-Din's delight, Ben-Shushan had been able to observe the 1577 eclipse and take detailed measurements. Taqi al-Din was so impressed that he invited Ben-Shushan to join the staff of the Istanbul observatory. Ben-Shushan then travelled to the heart of the Ottoman Empire. An Italian Jew who could read Latin, Hebrew, and Turkish, he represents – perhaps more than any other individual – the importance of cultural exchange for the development of science during the sixteenth century. Ben-Shushan introduced Taqi al-Din to all the latest works of Renaissance science, including the new translations of Ptolemy. He was also familiar with the kinds of mechanical sciences which fascinated Taqi al-Din, particularly the operation of European clocks. As a consequence, Ben-Shushan soon rose to become assistant astronomer at the Istanbul observatory. He can be seen in the Persian miniatures sitting right next to Taqi al-Din himself.[35]

Ben-Shushan arrived in Istanbul just in time to assist with a particularly important observation. In November 1577, a burning white light appeared in the night sky. It was a comet, one that was subsequently observed all over the world, from Peru to Japan. Taqi al-Din and Ben-Shushan watched the comet as it soared over Istanbul. One Persian miniature depicts it directly above the Hagia Sophia. As the chief astronomer to the sultan, Taqi al-Din immediately reported to the Ottoman court. Murad III wanted to know the meaning of this sudden change in the heavens. The Islamic millennium was approaching, due to occur in the year 1591 of the Christian calendar, and the Ottoman sultan wanted to be sure that all was well. Taqi al-Din assured his master that the comet signalled good tidings. The comet had apparently appeared on the first day of Ramadan, an auspicious sign. It was 'like a Turban Sash over the Ursa Minor stars', suggesting that Murad III ruled over both heaven and

Earth. Finally, Taqi al-Din argued that the Ottoman sultan would ultimately prevail in his struggle against Christian Europe. The comet, according to Taqi al-Din, 'sent a gush of light from the east to the west . . . its arrow promptly fell upon the enemies of Religion'.[36]

At the same time that Nicolaus Copernicus was causing a stir in Europe, astronomers and mathematicians in the Ottoman Empire were entering their own Renaissance. Between the fifteenth and sixteenth centuries, Ottoman scientific thinkers produced over 200 original works of astronomy, once again challenging the idea that Islamic science ended with the medieval 'golden age'. Taqi al-Din was just one amongst a number of Muslim scholars who travelled to work under the patronage of the Ottoman sultan following the conquest of Istanbul in 1453. After the death of Ulugh Beg, Ali Qushji, the leading astronomer at the Samarkand observatory, journeyed to Istanbul, where he was employed to work at one of the hundreds of *medreses*, or colleges, established by the Ottomans in the city. Other scholars made their way to Istanbul from across the Islamic world, including from Persia and Mughal India. At the same time, it is worth remembering that Istanbul was never an exclusively Muslim city. Jews and Christians also found patronage in the Ottoman court. The Jewish astronomer David Ben-Shushan worked in the Istanbul observatory alongside Taqi al-Din, whilst Mehmed II's private physician was also Jewish, a refugee from Renaissance Italy. Sitting at the crossroads between Europe and Asia, early modern Istanbul was a cosmopolitan city, in which – just as we've seen elsewhere – the expansion of religious and trading networks during the fifteenth and sixteenth centuries led to a transformation in the sciences.[37]

There are in fact many similarities between this Ottoman story and the history of the scientific revolution in Europe. Much like in Renaissance Europe, Ottoman scientific thinkers took a deep interest in the writings of ancient Greek authors. Mehmed II owned a vast collection of ancient Greek manuscripts, seized during the conquest of Istanbul. Building on a long Islamic tradition, the sultan ordered a series of new translations of these ancient Greek works into Arabic. In keeping with the cosmopolitan nature of the Ottoman court, these translations were completed by a Byzantine Greek. Like in Europe, Ottoman scientific thinkers also began to read and translate the works of earlier Islamic thinkers during this period. Ali Qushji's astronomical manuscripts were translated into Ottoman Turkish, as were those of Nasir al-Din al-Tusi, the thirteenth-century astronomer whose ideas had such an impact on Copernicus. By the middle of the seventeenth century, Ottoman scientific thinkers were also starting to read European works of astronomy. 'Copernicus laid a new foundation and compiled a small *zij* supposing that the Earth is in motion,' explained an Ottoman astronomer named Tezkireci Köse Ibrahim in 1662. Ibrahim even sketched out a diagram of Copernicus's famous heliocentric model of the universe.[38]

We can therefore start to see a number of parallels with the traditional European story of the scientific revolution. Ottoman scientific thinkers were reading and translating ancient Greek texts, and they were also learning to critique these older ideas drawing on the work of more recent Islamic writers. Istanbul, after all, was a city in which – thanks to its position on the Silk Road – you could easily access scientific manuscripts written in languages ranging from Latin and Greek to Persian and Arabic. Not only this, but the very idea at the heart of the European Renaissance also had its parallel in the Islamic world. In Arabic, this was referred to as *tajdid* (literally, 'renewal'). Traditionally, this was a term applied by religious scholars to the reform of Islam. But from the fifteenth century onwards, the idea of *tajdid* started to be used much more broadly, as part of a movement to revitalize, not only religion, but also the Islamic sciences. This movement was not confined to Istanbul. As we'll see in the following section, the connection between astronomy, mathematics, and Islam spread west along the Silk Road, across the Sahara, to Africa.[39]

IV. African Astronomers

In November 1577, a spectacular meteor shower appeared above the city of Timbuktu, in what is today modern Mali. Reports of astronomical phenomena in West Africa continued throughout the sixteenth and seventeenth centuries. 'A comet was seen to appear. First of all it rose on the horizon at dawn, then, rising little by little, it reached the middle of the sky between sunset and night. Finally it disappeared,' reported Abd al-Sadi, a West African chronicler writing in the early seventeenth century. In this chapter, we've already seen how rulers across the Islamic world, from Samarkand to Istanbul, took a great interest in astronomy during this period. The same was true in sub-Saharan Africa. A number of astronomers were employed at the court of Askia Muhammad, ruler of the Songhay Empire – an Islamic sultanate which controlled much of West Africa during the sixteenth century. These astronomers contributed to the running of the Songhay Empire by helping to compile the annual calendar and offering religious guidance. A devout Muslim himself, Askia Muhammad rewarded his astronomers handsomely, paying them to help calculate the times of prayer and dates of Ramadan. Others were asked to determine the direction of Mecca.[40]

The presence of astronomers in sixteenth-century Timbuktu is an important reminder of the place of sub-Saharan Africa in the history of modern science. This is a region that, more than any other, has been excluded from histories of the scientific revolution. Even in histories of science which acknowledge the importance of the wider world, sub-Saharan Africa is conspicuous by its absence. However, the idea that there was no science in Africa before the era of European colonization is a myth, one that is in urgent need of correction. Africa, like the rest of the world, had a rich scientific tradition, one that underwent a major transformation during the fifteenth and sixteenth centuries with the expansion of both religious and trading networks. Rather than seeing sub-Saharan Africa as separate from the rest of the world, we should therefore see it as part of the same story that we've been following throughout this chapter – a story of global cultural exchange.[41]

The city of Timbuktu was founded in the twelfth century. It then underwent a significant expansion during the fifteenth and sixteenth centuries, particularly following the rise of the Songhay Empire, which took control of the city in 1468. This expansion was driven primarily by increased trade across the Sahara, as caravans travelling from Timbuktu transported gold, salt, and slaves to Egypt and beyond, connecting West Africa through the Silk Road to Asia. During the same period, other West African kingdoms began trading with Europeans on the coast. This marked the beginning of the transatlantic slave trade, the impact of which we will explore in further detail in the next two chapters. Timbuktu soon grew rich, allowing the ruler of the Songhay Empire to support 'a magnificent and well-furnished court' complete with 'numerous doctors, judges, scholars, [and] priests'. Alongside trade, religion was another key factor connecting Africa to the wider world. Following the Muslim conquest of North Africa in the seventh century, Islam began to spread across the Sahara to West Africa during the tenth century. It then started to be taken up much more widely from the fourteenth century onwards, particularly in the countryside. This was also a period in which West African Islamic scholars began to produce more and more original manuscripts locally, in cities like Timbuktu, rather than just importing them. African rulers had long recognized the importance of Islam for consolidating political power. Askia Muhammad even completed a pilgrimage to Mecca in 1496, accompanied by many of the scholars from Timbuktu.[42]

With trade and pilgrimage came knowledge. Askia Muhammad arrived back from Mecca with hundreds of Arabic manuscripts, detailing everything from new astronomical ideas to the principles of Islamic law. Traders also returned to West Africa from across the Sahara with collections of Arabic manuscripts purchased in Istanbul and Cairo. 'Here are brought manuscript books from Barbary [North Africa], which are sold at greater profit than any other merchandise,' noted the famous sixteenth-century traveller Leo Africanus during his visit to Timbuktu. Other manuscripts arrived with the many Islamic scholars who fled the Catholic conquest of Muslim Spain, a campaign which culminated in the defeat of the Emirate of Granada at the end of the fifteenth century. As we'll see, the spread of Arabic manuscripts in West Africa ultimately ushered in a transformation in the sciences, a story with remarkable parallels to that of Renaissance Europe.[43]

Even before the spread of Islam, African peoples watched the heavens. The Dogon people of ancient Mali gave names to all the different stars, whilst the Xhosa of southern Africa used Jupiter as a guide to travel at night. The medieval ruler of the Kingdom of Benin, in what is today modern Nigeria, even employed a special group of astronomers, referred to as the *Iwo-Uki* ('Society of the Rising Moon'), to keep track of the movement of the Sun, Moon, and stars throughout the year. This was particularly important for planning the agricultural calendar. Medieval astronomers in the capital of the Kingdom of Benin watched the passage of Orion's Belt closely, declaring that 'when this star disappears from the sky, the people know it is time to plant their yams'. The medieval ruler of the Kingdom of Ife, also in modern-day Nigeria, similarly recognized the importance of astronomy for the agricultural and religious life of the city. A centre of Yoruba culture, the city of Ife contained many temples. Close to these, the king built large granite pillars which were used to track the movement of the Sun and determine the times of religious festivals as well as the annual harvest.[44]

These existing astronomical traditions underwent a significant transformation from the fifteenth century onwards. Much like in Europe, African scholars began to learn about the work of ancient Greek thinkers, such as Aristotle and Ptolemy, through Arabic translations. At night, groups of students would huddle around a campfire, watching the passage of the stars, and comparing what they measured with the astronomical tables found in various Arabic manuscripts. One manuscript, most likely used for teaching astronomy in sixteenth-century Timbuktu, was titled 'Knowledge of the Movement of the Stars'. It began by explaining the astronomical theories of ancient Greek and Roman authors, before moving on to more recent Islamic thinkers, such as Ibn al-Haytham, the eleventh-century author of an influential critique of Ptolemy's astronomy. The manuscript then explained how to determine the location of particular stars, as well as their astrological significance.[45]

Another manuscript, written by a Timbuktu scholar named Muhammad Baghayogho, explained how to calculate the times of prayer during the day (using a sundial) as well as at night (using the location of the Moon). Baghayogho, who had completed a pilgrimage to Mecca in the early sixteenth century, owned one of the largest collections of Arabic manuscripts in Timbuktu, and also authored a commentary on the work of a sixteenth-century Ottoman astronomer named Muhammed al-Tajuri. Indeed, in Timbuktu you could find manuscripts written, not only in Arabic, but also in Ottoman Turkish, indicating the close connection between the development of Ottoman and West African science during this period.[46]

Timbuktu was undoubtedly one of the most important sites for the advancement of science in early modern West Africa. But it was by no means unique. A number of other African cities, particularly those connected to the wider world of trade and religion, experienced a similar expansion in scientific knowledge during this period. In the Sultanate of Borno, an Islamic kingdom in what is today modern Nigeria, scholars at the Great Mosque studied 'several scientific works', according to one later account. Similarly, the ruler of the Sultanate of Kano, another Islamic kingdom in what later became Nigeria, invited scholars from across the Muslim world to teach at his court. At the beginning of the fifteenth century, one travelled all the way from Medina, bringing with him a vast collection of Arabic manuscripts, many of which covered scientific subjects, such as astronomy and mathematics. Just like in Timbuktu, African scholars in fifteenth-century Kano were reading Arabic summaries of ancient Greek texts, as well as those of influential Muslim scientific thinkers such as Ibn al-Haytham.[47]

Much as we've seen elsewhere, astronomers working at the court in Kano helped compile the annual calendar. One scholar, named Abdullah bin Muhammad, even wrote a detailed manuscript describing the traditional Islamic astrological calendar, in which the Moon moves through different constellations throughout the year. Alongside this, Abdullah bin Muhammad described the 'revolutions of the planets' along with their various astrological meanings. Most significantly, this manuscript was written in Hausa, the language of the Hausa ethnic group which made up the majority of the population in Kano. Abdullah bin Muhammad even noted the Hausa names for individual stars and planets alongside the traditional Arabic ones. Mercury, for example, was listed in Hausa as 'Magatakard' (meaning 'scribe'), whilst the Sun was named 'Sarki' (meaning 'king'). This again is an important reminder of the existence of a pre-Islamic astronomical tradition in Africa, one that was transformed following the arrival of new Arabic manuscripts during the fifteenth and sixteenth centuries.[48]

The development of new scientific ideas continued in West Africa right through to the early eighteenth century. In 1732, a mathematician working in Katsina (also in modern-day Nigeria) wrote a manuscript titled 'A Treatise on the Magical Use of the Letters of the Alphabet'. The author, Muhammad ibn Muhammad, had been taught astronomy, astrology, and mathematics by a leading Muslim scholar in the Sultanate of Borno, almost 800 miles to the east. He had also, like many of the African scientific thinkers we've encountered in this chapter, recently completed a pilgrimage to Mecca. Despite the somewhat arcane title, Muhammad ibn Muhammad's manuscript was in fact a work of mathematics. It described in detail the principles behind what are known as

11. Two 'magic squares' from an early modern Arabic mathematical manuscript. Similar manuscripts were produced in seventeenth-century Timbuktu and Kano.

'magic squares'. These are the sort of thing you might have come across at school. The simplest magic square is a 3 × 3 grid, filled in with the numbers 1–9. By putting the numbers in the right places, you can make it so that all the columns, rows, and diagonals add up to the same number. And whilst there are multiple ways to arrange the numbers, there is only one 'magic number' that they will always add up to. (In the case of a 3 × 3 grid, that number is 15.) Once you've mastered this, you can then start to ask more complex mathematical questions – for example, what is the 'magic number' of a bigger square, say a 9 × 9, or even an arbitrarily large square, n × n? You can also start to work out how many different permutations of the solution there are for different sized squares, as well as what the best algorithm is for solving them.[49]

Magic squares were widely discussed by medieval Islamic mathematicians, and Muhammad ibn Muhammad almost certainly learned about them whilst reading the Arabic manuscripts that were being traded in Katsina. He was clearly fascinated, covering the pages of his manuscript with them, and offering up a formula for constructing squares of various sizes. He also showed that, for a 3 × 3 square, you can obtain all the different solutions simply by rotation and reflection. However, alongside his mathematical interest, Muhammad ibn Muhammad saw magic squares as part of his religious duty. The magic square was considered a gift from Allah. 'The letters are in God's safekeeping,' he wrote. In fact, these magic squares were considered so special that Muhammad ibn Muhammad recommended that mathematicians 'work in secret . . . you should not spread God's secrets indiscriminately'. This was also an allusion to the mystical properties that many people associated with magic squares. Like lots of scientific thinkers, whether in Africa, Asia, or Europe, Muhammad ibn Muhammad believed that magic squares functioned as a kind of talisman, something that could protect against bad omens. This is why the title of his manuscript referred to the 'magical use' of mathematics. Magic squares were also widely used to try and predict the future. Muhammad ibn Muhammad would have offered his services in early modern Katsina, producing 'readings', typically by swapping out particular numbers for words or letters. Some people even had magic squares sewn into their clothes, in order to ward off evil spirits.[50]

*

For too long, sub-Saharan Africa has been left out of the history of the scientific revolution. But once we start to explore the rich scientific culture of the region, we can in fact see many parallels with what was happening in Europe during the same period. Like in Europe, people in Africa learned about ancient Greek and Roman scientific thinkers, such as Aristotle and Ptolemy, through translations and summaries written in Arabic. As in Europe, people in Africa also learned to critique these ancient thinkers, drawing on the work of more recent Islamic astronomers and mathematicians, such as Ibn al-Haytham. And like in Europe, the scientific revolution in Africa did not completely displace older ideas: astronomy, astrology, and divination were still often indistinguishable from one another. Rather than thinking of Africa as separate from the scientific revolution, we should therefore see it as part of a shared history – a history in which the growth of trade and pilgrimage along the Silk Road led to a transformation in the sciences during the fifteenth and sixteenth centuries.

In Timbuktu and Kano, just like in Samarkand and Istanbul, Islamic scholars were supported by wealthy African patrons who recognized the religious value of astronomy and mathematics. 'One of the uses of this science is knowing prayer times,' noted an astronomer at the court of the Songhay Empire. At the same time, astronomers helped guide caravans across the Sahara, contributing further to the growth of trade in the region. They travelled across the vast desert 'as it were upon the sea, having guides to pilot them by the stars', explained one writer. Africa, sitting at the far western end of the Silk Road, ultimately experienced its own scientific revolution during the fifteenth and sixteenth centuries. Now we move east along the Silk Road, uncovering how similar kinds of commercial, religious, and intellectual exchanges helped bring about a scientific revolution in China and India.[51]

V. Astronomy in Beijing

Dressed in a red silk robe, Matteo Ricci entered the Forbidden City. In doing so, he became the first European to gain access to the Chinese emperor's inner sanctum, right at the heart of Beijing. Ricci chose to adopt the clothing of a Confucian scholar, hoping to impress the emperor.

He even grew a long beard for the occasion, typical of the Chinese literati. Walking into the great marble courtyard in February 1601, Ricci fulfilled an ambition which stretched back nearly two decades. He had come to China in 1582 as a member of the Society of Jesus. As we saw in the previous chapter, the Jesuits' missionary activity was closely tied to the development of early modern science. They saw the study of the heavens as a way to appreciate God's wisdom, as well as a means to demonstrate the power of the Christian faith to potential converts. This is exactly how Ricci approached his missionary work in China.

Born in 1552 in the Papal State of Macerata, Ricci studied mathematics and astronomy under the leading Jesuit scholar Christopher Clavius at the Roman College in the early 1570s. It was an exciting time to be an astronomer. Copernicus's heliocentric model was causing a stir, whilst in November 1572 a 'new star' appeared in the sky, further challenging the idea that the heavens were completely unchanging. (This 'new star' was in fact a supernova.) On completing his training, Ricci was asked to join the Jesuit mission in the Far East. Leaving Rome in 1577, Ricci travelled to Lisbon, where he boarded a ship to China. The journey took nearly four years, including a stopover in India. Ricci finally arrived in the Portuguese trading port of Macao in August 1582. He spent the rest of his life in China, playing a major role in the development of both Christianity and science in Asia.[52]

Ricci was certain that astronomy and mathematics would help the Jesuits gain a foothold in China. The Ming dynasty, which came to power in the middle of the fourteenth century, had long been wary of European visitors. The Wanli Emperor, who ascended to the throne in 1572, tolerated the presence of the Portuguese at Macao, but only allowed a couple of vessels a year to travel further inland. Like the Portuguese traders, the Jesuits initially struggled to establish a presence. These 'foreign devils', as they were often called, were not generally welcome. During his travels, Ricci was locked up multiple times and had his house pelted with stones. In the end, he did manage to set up a small mission station in the city of Zhaoqing in the south. But even this proved temporary, as the Jesuits were expelled in 1589 following the appointment of a new governor. Ricci decided that in order to secure the future of the Jesuits in China, he would have to petition the emperor himself. It was with this in mind that Ricci travelled to Beijing in 1601. He brought with him an assortment of gifts. These included a painting of the Virgin Mary and a crucifix embellished with pearls and glass beads. Ricci also brought two mechanical clocks – a large one driven by iron weights, and a smaller one driven by springs.

The Wanli Emperor was not particularly impressed by the painting or the crucifix, but he did like the clocks. These 'bells that ring by themselves', as he called them, fascinated the emperor. He ordered the larger one to be installed in his private garden and the smaller one in his apartment. As the cogs turned and the springs compressed, Wanli tried to understand how the mechanism worked. Before long, however, the clocks ground to a halt and stopped chiming. Dismayed, the emperor requested Ricci return to the court and fix them. The choice of gift had been shrewd. The mechanical clocks, brought all the way from Italy, had clearly impressed. But they also needed an intimate knowledge of European mathematics to operate, requiring daily adjustment and regular winding. The emperor soon realized that, if he wanted the clocks to keep chiming, he would need to allow Ricci into the Forbidden City. Wanli requested that the Jesuits return four times a year to service the clocks and, as a reward, Ricci was granted permission to reside in Beijing and establish a mission.[53]

Ricci's faith in science had come good. He wrote to Rome in 1605, arguing that astronomy and mathematics proved the best means to win favour with the Chinese elite. 'Because of my world-maps, clocks, spheres, astrolabes, and the other things I do and teach, I have gained the reputation of being the greatest mathematician in the world,' explained Ricci. He then suggested that this strategy should be expanded, arguing that 'nothing could be more advantageous than to send some father or brother who is a good astrologer to this court'. According to Ricci, this would 'enhance our reputation, give us freer entry into China, and assure us of greater security and liberty'. Ricci got what he wanted, and over the next fifty years the Jesuits sent a string of brilliant astronomers and mathematicians to China. This marked the beginning of a much broader exchange of scientific knowledge between Europe and East Asia. Many of the debates about the nature of the heavens and the role of ancient knowledge now played out in a different setting, as European and Chinese approaches to astronomy and mathematics came into contact, transforming one another in the process.[54]

Before long, the Jesuits had their first high-profile convert. Xu Guangqi converted to Christianity shortly after the establishment of the Beijing mission in 1601. He was a *jinshi*, a high-ranking official within the Chinese imperial bureaucracy. Xu, who was known to the Jesuits as 'Dr Paul', was exactly the kind of convert that Matteo Ricci hoped to attract – an influential scholar who could help promote the Jesuits' cause at court. Xu also showed a great appreciation for the sciences, working with Ricci and other Jesuits to help translate many of the most important works of ancient Greek and Renaissance science into Chinese. Born into a humble agricultural family, Xu studied at a small Buddhist monastery before climbing the government ranks. Now, working with Ricci in Beijing, he helped produce the first Chinese translation of Euclid's *Elements*, the ancient Greek text which underpinned much of European mathematics.

Ricci thought that a translation of Euclid would further the influence of the Jesuits as 'among the Chinese the mathematical disciplines are held perhaps in higher esteem than among any other nation'. Xu and Ricci worked, not from the ancient Greek original, but from a Latin edition written by Christopher Clavius, Ricci's tutor back in Rome. By this time, Ricci was proficient in spoken Chinese. But he wasn't confident in writing. The duo therefore worked as a team. Ricci translated from the Latin into Chinese orally, whilst Xu wrote down Ricci's translation, reworking it in the style of classical Chinese expected of a Confucian scholar. Other translations soon followed, including of Clavius's major work, *The Astrolabe* (1593). By the time of Ricci's death in 1610, many of the major works of ancient Greek science, alongside a number of medieval and Renaissance works, had been translated into Chinese.[55]

It would be easy to see these translations simply as an episode in the transfer of European science to China. But the real story is more complex. As we saw in the case of the Islamic world, the Renaissance ideal of rediscovering ancient knowledge was not unique to Europe. Chinese scholars also saw what they were doing as part of a very similar tradition. Xu believed that by working with Ricci he could recover a lost world of Chinese science. Just as Europe had relied on the Islamic world to access the past, so China might need to rely on Europe. Xu set out his vision for the recovery of ancient knowledge in the preface of his Chinese translation of Euclid. 'Before the Three Dynasties, mathematics was flourishing and teachers handed down complete knowledge,' Xu explained. He described the period up to the third century BCE in which Chinese philosophy and mathematics was at its height. This was the age in which the *Four Books* and *Five Classics* had been composed, the Confucian works which formed the basis of Chinese bureaucracy. It was also the period in which classical mathematical works, such as *The Nine Chapters on the Mathematical Arts* and *The Book on Number and Computation*, had been written. However, as with the science and mathematics of ancient Greece, this knowledge had been lost, 'totally destroyed in the flames of the Ancestral Dragon'. Nonetheless, Xu argued that this knowledge could be recovered by working with Europeans like Ricci. 'Why shouldn't we, if the rites are lost, retrieve them from the Barbarians?' Xu asked rhetorically.[56]

Chinese scholars such as Xu therefore acted rather like humanist scholars back in Europe. They translated ancient Greek science, but they did so with a view to rediscovering a lost world. And, much like the humanists, Chinese translators offered commentaries and critiques, hoping to not only recover, but also improve upon the originals. Xu even wrote a work entitled *Similarities and Differences of Measurement* (1608), comparing Chinese and European mathematical methods. He complained that earlier Chinese mathematics had 'only been able to state its methods, without being able to state its principles'. Xu correctly identified that many existing works of Chinese mathematics were concerned with practical solutions to particular problems, rather than generalizable theories. And without a general mathematical theory, it was hard to produce new knowledge, as there was no easy way to apply what you had already learned to a new situation. As one of Xu's contemporaries put it, 'Chinese mathematical texts contain only examples, but no proofs.'[57]

Ancient Greek works appealed to Xu because they seemed to provide a theoretical basis for existing Chinese mathematics. Euclid's *Elements*, for example, contained a proof of Pythagoras's theorem (the lengths of a right-angled triangle are given by the formula $a^2 + b^2 = c^2$). Xu took this idea, but then showed that ancient Chinese mathematical texts, including the *Nine Chapters*, contained examples of such a theorem in action, without an explicit proof. By reading Euclid, Xu argued that Chinese mathematicians could recover knowledge that had been lost, perhaps even improve upon it. 'Through Western learning one returns to the *Nine Chapters*', as another of Xu's contemporaries put it. This then was a Chinese Renaissance.[58]

All Xu Guangqi's hard work paid off. In 1629, he was appointed Vice-Minister of Rites, one of the highest appointments within the Chinese bureaucracy. The Jesuits finally had a man on the inside. Xu's department, the Ministry of Rites, was responsible for managing court ceremonies, religious practices, and imperial examinations. It also oversaw one of the most significant scientific institutions in early modern China – the Astronomical Bureau. Writing in his diary in 1601, Matteo Ricci gave a vivid description of the site, which can still be visited in Beijing today:

> There is a high hill at one side of the city, but still within the walls. On the top there is an ample terrace, capitally adapted for astronomical observation, and surrounded by magnificent buildings erected of old. Here, some of the astronomers take their stand every night to observe what may appear in the heavens, whether meteoric fires or comets, and to report them in detail to the emperor.

As Ricci's account suggests, the Astronomical Bureau was an institution of great political as well as scientific importance. In China, the emperor was understood as the 'son of heaven'. His job was to mediate between the celestial and the terrestrial realms, ensuring harmony between man, nature, and the universe. In practical terms, this meant that the emperor was expected to issue an annual calendar, establishing the dates of key religious festivals alongside the agricultural seasons. The calendar was therefore an instrument of political power. Adopting the calendar provided a means to display loyalty to the emperor, particularly amongst Chinese vassal states like Korea. But equally, if the emperor failed to predict a celestial event, such as an eclipse, he would be forced to issue an apology, weakening his position. With this in mind, a new emperor would almost always reform the calendar, in order to cement his claim to the throne.[59]

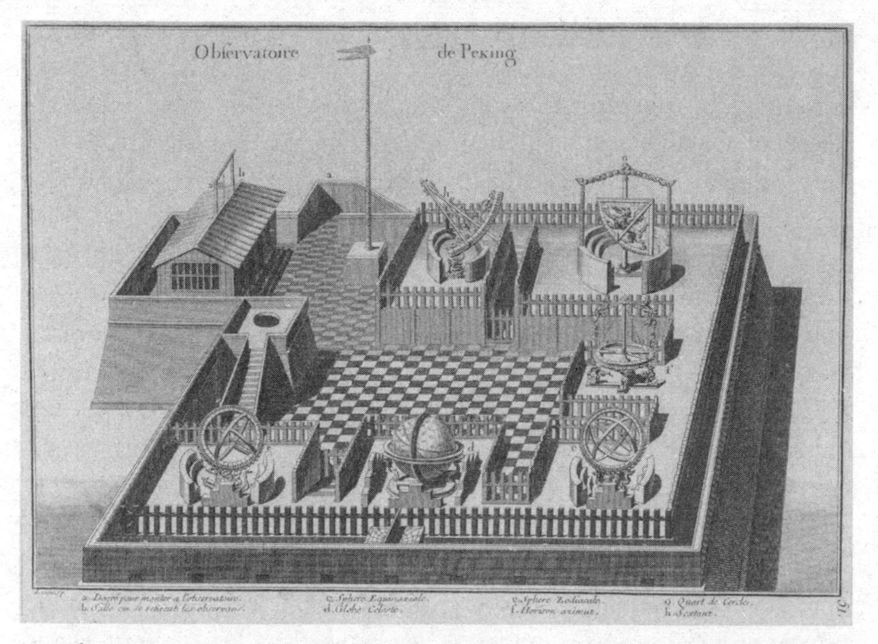

12. The Astronomical Bureau in seventeenth-century Beijing. Many of the scientific instruments feature traditional Chinese motifs, including dragons. Other instruments, such as the sextant (*top row, far left*), incorporate both Chinese and Islamic elements of design.

This is exactly what the Chongzhen Emperor did on ascending to the throne in 1627. He was concerned about previous failures to predict major celestial events, including a series of eclipses. In 1610, the Bureau had incorrectly predicted the appearance of a solar eclipse, getting the time wrong by half an hour. (That might sound inconsequential, but on a good day both European and Chinese astronomers could predict the timing of an eclipse down to the nearest minute.) Since then, the Bureau had failed to predict a further ten eclipses. In keeping with Confucian philosophy, troubles in the heavens were reflected on Earth. The previous two emperors had not lasted long, one dying less than a month after coming to the throne. Invaders also threatened from the north, with the Manchus approaching the Great Wall. Worried about all this, Chongzhen ordered the Astronomical Bureau to reform the calendar.[60]

Xu jumped at the opportunity. Making the most of his new position, he petitioned the emperor to appoint him to lead the calendar reform. Xu also argued that astronomers in Beijing needed to learn from the

Jesuits, rather than solely relying on existing traditions. By this time it was clear that there was a fundamental problem with the Chinese calendar. It was a 'lunisolar' calendar, meaning it needed to reconcile the length of the solar year with the length of the lunar month. It takes the Earth around 365 days to orbit the Sun, giving the solar year. And it takes the Moon about twenty-nine days to orbit the Earth, giving the lunar month. Unfortunately, there is no perfect combination of lunar months that make up a solar year. Twelve lunar months gets you close, but even then you still need to add extra days here and there to make the two line up. Any calendar based on combining the two was always going to diverge over time, and this is why the Ming dynasty had increasing trouble predicting the exact timing of celestial events such as eclipses.[61]

This problem was not unique to China. In 1582, Pope Gregory XIII had asked the Jesuits to help reform the Christian calendar back in Europe. As both leading astronomers and Catholic servants, the Jesuits proved an ideal group to undertake such a task. Christopher Clavius, Ricci's tutor at the Roman College, led the reforms. He integrated the latest mathematical methods, alongside data taken from Copernicus's astronomical tables. The result was the Gregorian calendar, still in use today throughout many parts of the world. Much like in China, the adoption of the Gregorian calendar was a way to signal allegiance to the Catholic Church, with many Protestant countries refusing to adopt Clavius's reforms until the eighteenth century. The Jesuits ultimately hoped that the Chongzhen Emperor might also adopt the Gregorian calendar, indicating a commitment to Catholicism.[62]

In the end, they were disappointed. As with the translation of Euclid, Xu promoted the idea that Jesuit astronomy could help reform the calendar, but that the result would still be fundamentally Chinese in character. 'Melting the material and substance of Western knowledge, we will cast them into the model of the Chinese system', is how Xu put it. He also pointed out that the Chinese had long relied on outsiders. Much like those in Renaissance Europe, Chinese astronomers owed a lot to the Islamic world. Many of the astronomical instruments at the Beijing observatory had been built by Persian astronomers during the thirteenth century. These included giant masonry instruments, similar to the Fakhri Sextant at Samarkand. Even in the seventeenth century, there was still a Muslim Section of the Astronomical Bureau, which worked with Islamic astronomical tables. Xu simply suggested that such a strategy should be extended, and that Chinese astronomers would benefit from a similar engagement with Jesuit science.[63]

In the end, the Chongzhen Emperor agreed that Xu and the Jesuits were the best people to lead the calendar reform. In 1629, Xu was appointed as head of the new Calendar Department of the Astronomical Bureau. He was joined by two German Jesuits, both of whom had studied under Clavius in Rome. Together, they produced a new star catalogue alongside a monumental collection of scientific works entitled *Astronomical Treatises of the Chongzhen Reign* (1645). True to his word, Xu ensured that the new calendar mixed Chinese and European ideas. The calendar itself was still organized around a combination of the solar year and lunar months. However, the data was all new, based on mathematical methods and tables imported from Europe.[64]

The Chongzhen Star Catalogue (1634), for example, mixed existing Chinese constellations with new ones from European works. Prior to the arrival of the Jesuits, Chinese catalogues did not contain any stars found in the southern sky, as these were only visible south of the equator. However, Xu consulted a number of works by European astronomers held at the Jesuit library in Beijing in order to fill in the gaps. When Xu came to naming the southern constellations, he adopted Chinese equivalents to those listed by European astronomers. The 'Phoenix' became *Huoniao* ('Firebird'), whilst the 'Musca' became *Feng* ('Bee'). Xu also listed the coordinates for each star according to both European and Chinese systems. Whilst Chinese astronomers adopted what are known as 'equatorial' coordinates, Europeans tended to use 'ecliptic' coordinates. Both had their advantages, depending on what you wanted to measure. The Chinese system was better for tracking stars, whereas the European system was better for tracking planets and moons. By listing both coordinates in his star catalogue, Xu ensured that Chinese astronomers benefited from the best of both worlds. And by the eighteenth century, most European astronomers had in fact adopted the Chinese 'equatorial' system.[65]

The development of astronomy and mathematics in early modern China followed a similar pattern to that which we saw in Europe and the Islamic world. With the expansion of long-distance trading and religious networks during the fifteenth and sixteenth centuries, Chinese astronomers came into contact with new scientific ideas. From the sixteenth century onwards, Jesuit missionaries and Chinese scholars helped to introduce ancient Greek science into the Astronomical Bureau in Beijing. By the early decades of the seventeenth century, all the major works of ancient Greek science, such as Euclid's *Elements*, had been translated into Chinese.

Much as we've seen elsewhere, this translation movement was part of a broader trend to recover the ancient past through engaging with other cultures. The scholars who worked at the Astronomical Bureau believed that they could better understand classical Chinese mathematics by studying ancient Greek texts. 'They can shed light on each other,' argued one seventeenth-century Chinese astronomer. The Chinese also recognized what many historians today have forgotten – that much of European science was in fact Islamic, noting that 'the Westerners who came to China all call themselves European, but their calendrical science is similar to that of Muslims'. The scientific revolution, however, was never simply about the recovery of ancient knowledge. It was also about making new observations, a trend that we can also see in China. 'Truth must be sought not only in books, but in making actual experiments with instruments . . . All the new astronomy is then found to be exact,' argued a group of Chinese mathematicians working at the Astronomical Bureau in the seventeenth century. Like in Europe and the Islamic world, it was this combination of the old and the new, text and experiment, that really characterized the scientific revolution in China. And, as we'll see, the story was very similar in Mughal India.[66]

VI. Indian Observatories

The maharaja watched as the funeral pyres burned. In 1737, Jai Singh II travelled hundreds of miles across northern India to reach the sacred city of Benares. Situated on the banks of the Ganges, it is where many in India come to cremate their dead. Mourners chant 'the name of Rama is truth', before scattering the ashes of their loved ones into the river. For Hindus, the Ganges is capable of purifying the soul, allowing one to obtain salvation, or *moksha*. Arriving in Benares, Jai Singh joined thousands of other Hindu pilgrims. He even bathed in the sacred waters. But Jai Singh was not simply a pilgrim. He was also an astronomer and a mathematician. And it was in Benares, in this most sacred of Hindu cities, that he chose to build one of the first dedicated astronomical observatories in India.

Overlooking the Ganges, just south of the main cremation site, Jai Singh's observatory in Benares was part of a larger network. Between 1721 and 1737, Jai Singh ordered the construction of five observatories, known as the Jantar Mantar, stretching across India. Like those in Samarkand and Beijing, these were institutions which combined scientific, political, and religious functions. As well as Benares, Jai Singh built observatories in Ujjain and Mathura, both sites of Hindu pilgrimage. Jai Singh also ordered the construction of observatories in Jaipur and Delhi, cities of great political significance. The Mughal Empire, which had ruled India since the middle of the sixteenth century, held court in Delhi, whilst Jaipur was the capital of Jai Singh's own province, the Kingdom of Amer. By establishing this network of observatories, Jai Singh hoped to make advances in astronomy, compiling more accurate tables than ever before. The idea was that by taking measurements from different sites, errors could be identified and corrected. At the same time, these observatories allowed Jai Singh to spread his influence across India, transforming him into one of the most powerful rulers on the subcontinent. In India, as elsewhere, to control the Earth, you needed to master the heavens.[67]

Just like in China, India was transformed during this period by the rise of a great empire, something which had a profound effect on the sciences. The Mughal Empire was founded by Babur in 1526. Born in Central Asia, Babur was a descendant of Ulugh Beg's grandfather, Tamerlane. Advancing on Delhi in the early sixteenth century, Babur's conquest brought Islamic learning to India. Persian and Arabic manuscripts, including copies of al-Tusi's astronomy and Ulugh Beg's tables, found their way into libraries in Delhi and Agra. At the same time, the Mughals were confronted with existing Hindu scientific ideas. Some of these were remarkably modern. As early as the fifth century, the Hindu astronomer Aryabhata suggested that the cause of the cycle between day and night was the Earth rotating on its axis. This idea, which proved to be correct, had been rejected by Ptolemy and most medieval European astronomers, who believed the Earth must remain completely stationary.[68]

Akbar the Great, the Mughal emperor who reigned between 1556 and 1605, worked hard to bring Muslims and Hindus together, including when it came to the sciences. He ordered the translation of the works of Ulugh Beg into Sanskrit, the traditional language of Hindu scripture. At the same time, Akbar appointed a Hindu mathematician, named Nilakantha, as his court astronomer. Akbar, although a Muslim himself, knew that he needed someone like Nilakantha in order to meet the demands of his Hindu subjects. Nilakantha was charged with issuing the annual Hindu calendar. Europeans were also making inroads into India during this period. Jesuit astronomers, hoping their success in China could be replicated, presented themselves at Akbar's court. Travellers and traders arrived too. Between 1658 and 1670, the French doctor François Bernier worked as court physician to the Mughal emperor Aurangzeb. Bernier reported that the Mughal elite showed great interest in the sciences. The governor of Delhi apparently read Persian translations of recent works by René Descartes and Pierre Gassendi, both leading French proponents of a more empirical approach to the study of the universe. Ultimately, it was this combination of Islamic, Hindu, and Christian cultures which led to a blossoming of scientific study in India, another example of the global Renaissance which swept across the world from the fifteenth century onwards.[69]

Jai Singh's observatories represented the pinnacle of this movement. At Jaipur, the largest of the five sites, Jai Singh brought together astronomers, instruments, and books from around the world to create one of the most advanced scientific institutions of the age. Completed in 1734, the Jaipur observatory still stands to this day. It consists of nineteen immense stone instruments. Some of these were based on traditional Islamic designs. Jai Singh had read about Ulugh Beg's observatory in Samarkand, and one of the instruments at Jaipur is an almost exact copy of the Fakhri Sextant. Even then, however, Hindu astronomy still played an important role. The stone instruments at Jaipur are marked with both Islamic and Hindu divisions of time. In the Islamic world, as in Europe, the day is divided into twenty-four hours, each consisting of sixty minutes. Hindu astronomers, however, divided the day into sixty parts (known as *ghatikas*), each of which was divided into a further sixty parts (known as *palas*). This system actually made a lot of sense. It is much easier to do calculations quickly if everything is a multiple of the same number, in this case sixty. With this in mind, Jai Singh ordered each of his stone instruments to be engraved with divisions of *ghatikas* and *palas* alongside hours and minutes.[70]

Not all of the instruments at the Jantar Mantar observatories were copies of earlier Islamic designs. A number were invented by Jai Singh himself. The most impressive of Jai Singh's designs was the Samrat Yantra, or 'Supreme Instrument'. At Jaipur, the Samrat Yantra measures over twenty-seven metres tall. In essence, it is a massive sundial, the largest surviving example in the world. But such a description does a disservice to the ingenuity of the design. On each side of the central stone column, Jai Singh built a curved structure onto which the shadow of the Sun is cast. This helped make the instrument much more accurate than a traditional sundial where the shadow is cast on a flat surface. As a

13. The Samrat Yantra, or 'Supreme Instrument', at the Jantar Mantar observatory in Jaipur, India.

consequence, the Samrat Yantra at Jaipur gives the local time to the nearest two seconds, achieving an accuracy greater than most mechanical clocks of the period. The other major instrument invented by Jai Singh is known as the Jai Prakash Yantra, or 'Light of Jai'. The design is much more complex, consisting of an enormous marble bowl, over eight metres wide, sunk into the ground. It acts as a reflection of the sky above, with the stars and constellations engraved into the marble. A small metal ring is then suspended on wires above the bowl, casting a shadow, allowing astronomers to track the movement of a particular celestial object throughout the day.[71]

Alongside building instruments, Jai Singh collected books. His library, held at the Royal Palace in Jaipur, included works in Latin, Portuguese, Arabic, Persian, and Sanskrit. Here, more than anywhere else, scientific knowledge from East and West came together. Jai Singh owned an Arabic translation of Ptolmey's *Almagest* as well as later commentaries by many of the astronomers we met earlier, including al-Tusi and al-Haytham. These Islamic works sat alongside over a hundred Sanskrit astronomical manuscripts, including a copy of Aryabhata's classic fifth-century text in which he describes the rotation of the Earth. Jai Singh also showed an increasing interest in new astronomical ideas originating from Europe. In 1727, he sent a scientific mission to Portugal, hoping to learn more about astronomy beyond India. The mission, which arrived in Lisbon in 1730, included a Muslim astronomer, Sheik Abdu'llah, as well as a Portuguese Jesuit named Manuel de Figueredo. Figueredo and Abdu'llah were granted an audience with King John V of Portugal. They returned to Jaipur in 1731, bringing with them copies of the latest European astronomical works, including Philippe de La Hire's *Astronomical Tables* (1687) as well as John Napier's *A Description of the Wonderful Law of Logarithms* (1614), both of which Jai Singh had translated into Sanskrit. Finally, Jai Singh's library included works arriving from further east. The Jesuits in India were in constant correspondence with those in China. A French Jesuit at the Beijing observatory even forwarded a copy of a book titled *A History of Chinese Astronomy* (1732), which explained recent developments in Chinese understandings of the heavens.[72]

All this knowledge, drawn from East and West, came together in the production of Jai Singh's new astronomical tables. Written in Persian, they were known as *The Tables of Muhammad Shah* (1732), in honour of the Mughal emperor. By choosing to dedicate this work to Muhammad Shah, Jai Singh hoped to secure his status at the Mughal court during a period of great political upheaval. Following the death of the emperor Aurangzeb in 1707, the Mughal Empire had been plunged into chaos. A number of Mughal emperors were murdered, sometimes by their own family members, and war raged over northern India. Jai Singh himself had been caught up in the conflict, fighting against the armies of one of the short-lived emperors. In the end, the reign of Muhammad Shah, from 1719 to 1748, brought about a period of relative stability. Jai Singh, keen to consolidate his position following the conflict, presented *The Tables of Muhammad Shah* to the new emperor, writing, 'Praise be to God . . . let us devote ourselves at the altar of the King of Kings.'[73]

Jai Singh modelled his tables directly on *The Tables of the Sultan*, produced almost exactly 300 years earlier in Samarkand by the Muslim prince Ulugh Beg, who we learned about at the start of this chapter. The star catalogue featured the same 1,018 stars. However, Jai Singh revised the coordinates to take into account the difference in longitude between Samarkand and Jaipur. Additionally, Jai Singh's tables listed the Hindu constellations alongside those used in Islamic and ancient Greek astronomy. This was important given that Jai Singh hoped his tables would win him favour, not just at the Mughal court, but also in Hindu religious centres such as Benares. Jai Singh even produced a Sanskrit translation of the tables for this purpose. Unlike the Persian edition, which was dedicated to the Mughal emperor, Jai Singh's Sanskrit work begins with a dedication 'to Holy Ganesh'.[74]

Alongside Islamic and Hindu astronomy, Jai Singh used what he had learned from Europe. *The Tables of Muhammad Shah* incorporated copies of tables taken from the French astronomer La Hire. In each case these tables were updated to reflect the Hindu divisions of the day. *The Tables of Muhammad Shah* also described Jai Singh's experience using a telescope, the first of which was brought to India by a French Jesuit in 1689. 'The telescope enables one to see bright stars in broad daylight,' explained Jai Singh. He also noted how 'with the telescope we have noted certain facts in contradiction to the well-known texts', describing the moons of Jupiter and the rings of Saturn.[75]

Writing in *The Tables of Muhammad Shah*, Jai Singh presented an argument which by this time was being made by astronomers stretching from Rome to Beijing. The ancients, Jai Singh claimed, had been mistaken to believe the heavens were unchanging. Instead, astronomers needed to make new observations, and re-read ancient texts, in order to arrive at a more perfect understanding of the universe. For Jai Singh, the problem with ancient astronomy was less philosophical and more practical. It was the accuracy of the instruments used. This was 'the reason why the determinations of the ancients, such as Hipparchus and Ptolemy, proved inaccurate', Jai Singh explained. In response, Jai Singh built his immense network of observatories across northern India, the Jantar Mantar, from which to take new measurements, and compare them to one another. And, most importantly, Jai Singh drew together scientific knowledge from East and West. As the Mughal emperor himself noted approvingly, 'the astronomers and geometricians of the faith of Islam, and the Brahmins and Pandits, and the astronomers of Europe' all came together in Jai Singh's observatories. This was a global scientific revolution.[76]

VII. Conclusion

By the early decades of the eighteenth century, the sciences of astronomy and mathematics had been transformed by the rise of four great empires – the Ottoman, the Songhay, the Ming, and the Mughal. These empires were connected to Europe and each other by a network of trading and pilgrimage routes stretching all the way from Timbuktu to Beijing. Merchants, missionaries, and envoys travelled along the Silk Road or boarded galleys bound for the Indian Ocean. They brought with them new ideas, new texts, and new scientific instruments. In doing so, they transformed the Renaissance into a global intellectual movement. At the core of this movement was the idea that ancient science needed to be reformed, particularly in the case of astronomy. From Christian Europe to Ming China, ancient works were no longer treated as gospel. Instead, astronomers started to find contradictions and offer alternatives. Rulers also understood astronomy as a science of great political and religious significance. As a consequence, the courts of Istanbul, Timbuktu, Delhi, and Beijing emerged as important sites for scientific and cultural exchange. It was this experience of contact and conflict with other religions and cultures which brought about a revolution in the study of astronomy and mathematics, not just in Europe, but across Asia and Africa.

This movement began in the Islamic world, with the first translations of ancient Greek science into Arabic. It then spread to Europe, particularly following the Ottoman conquest of Istanbul in 1453. From Regiomontanus to Nicolaus Copernicus, all the great European astronomers of this era were in some way influenced by ideas arriving from the Islamic world. This movement could also be found further to the east. In Beijing, the Ming emperor marvelled at mechanical clocks and telescopes brought by the Jesuits from Rome. Astronomers in Beijing believed that European and Islamic science could help recover and improve upon lost Chinese traditions. And finally, with the Mughal conquest of India, European and Islamic science fused with Hinduism, culminating in the construction of the Jantar Mantar astronomical observatories by Jai Singh.

This world, however, was about to change. The balance of power was already beginning to tip. Over the next 200 years, European empires began to expand more and more aggressively, particularly in Asia and Africa. This, coupled with the gradual weakening of the Ottomans, the Songhay, the Ming, and the Mughals, brought about the next great transformation in the history of science. The Silk Road couldn't go on forever.

PART TWO

Empire and Enlightenment, *c.* 1650–1800

3. Newton's Slaves

Isaac Newton invested in the slave trade. In the early eighteenth century, Newton – the famous English mathematician – purchased over £20,000 of shares in the South Sea Company. This was an incredible sum of money, equivalent to well over £2 million today. The South Sea Company had been established in 1711 to help raise funds to pay off the British national debt, which was soaring following years of expensive wars with France and Spain. Investors were lured with promises of immense profits, as the South Sea Company was granted a monopoly on British trade with South America. Much of this trade was in human beings. Between 1713 and 1737, the South Sea Company transported over 60,000 enslaved Africans across the Atlantic Ocean to Spanish colonies, including New Granada and Santo Domingo.[1]

In the eighteenth century, the slave trade was at its height. Between 1701 and 1800, over six million enslaved Africans crossed the Atlantic. Subject to extreme physical violence, these men and women were forced to work on plantations in the Caribbean and mines in South America. Newton, like most British investors in the slave trade, probably thought little about where his money was going. Sitting in London, he was far removed from the brutal realities of slavery. For Newton, the South Sea Company was just another financial investment. (One that in fact turned out rather badly, when the stock price plummeted in 1720.) Newton also managed shares in the British East India Company, which held a monopoly on trade with Asia, as well as the Company of the Bank of England. He even spent the last thirty years of his life working as Master of the Royal Mint in London, supervising overseas trade in gold and silver.[2]

Newton's financial dealings hint at a world of eighteenth-century science that is often overlooked: a world of slavery, colonial trade, and war. Typically, Newton, like most scientific figures of the eighteenth century, is presented as an isolated genius. We are told that, working away in seclusion at the University of Cambridge, Newton was able to make a series of major breakthroughs. He is credited with discovering gravity, inventing calculus, and setting out the laws of motion. In 1687, Newton published his monumental *Mathematical Principles of Natural Philosophy*, better known as the *Principia*. This work set out Newton's theories in precise mathematical detail, building on many of the ideas we explored in the previous chapter. With this, Newton did away completely with the philosophy of the ancients, providing a thoroughly mathematical explanation of the workings of the universe. As such, Newton and his *Principia* are often understood as marking the beginning of the Enlightenment. This was the age of Carl Linnaeus, the Swedish naturalist who invented a new way of classifying plants and animals, as well as Antoine Lavoisier, the French chemist who transformed the study of matter. This was the age of the great philosophers – John Locke on the workings of the mind and Thomas Paine on the 'rights of man'. This was the age of reason and rationality above all else.[3]

But the Enlightenment was also the age of empire. Over the course of the eighteenth century, European powers competed with one another, fighting their way across the Atlantic, Asia, and the Pacific. The old empires and dynasties – the Songhay, the Ming, the Mughals, and the Ottomans – either collapsed or were severely weakened. At the same time, the slave trade expanded exponentially. What began as a relatively small-scale operation in the sixteenth century, quickly transformed into an industrial system of exploitation. By the 1750s, over 50,000 enslaved Africans were being transported across the Atlantic every year. The rise of European empires in the eighteenth century was therefore a key moment in global history. As in previous chapters, by thinking in terms of global history we can better understand the history of science during this period.[4]

In 1660, Charles II of England granted two royal charters. The first was to a new national scientific academy, the Royal Society in London, of which Newton later became president. The second charter was granted to the Company of Royal Adventurers Trading into Africa, later known as the Royal African Company. This was another commercial vehicle for trade with West Africa, principally in slaves. There was considerable overlap between the memberships of both institutions. One third of the founding members of the Royal African Company became Fellows of the Royal Society. Not only that, but the Royal Society invested over £1,000 of its own funds in the Royal African Company. Individual fellows of the Royal Society also developed close links to similar commercial and colonial institutions. Newton was in fact pretty typical in his financial dealings. John Locke, elected a Fellow of the Royal Society in 1668, owned shares in the Royal African Company, whilst Robert Boyle, famous for his experiments with the air pump, worked as a director for the East India Company.[5]

These links weren't just institutional and financial. They were also intellectual. The idea of empire was at the very heart of how many European thinkers understood science in this period. The English philosopher Francis Bacon, often described as the 'father of empiricism', made this explicit in an influential work entitled *The New Organon* (1620). (The title was a reference to Aristotle's *Organon*, which Bacon sought to displace, along with the rest of ancient philosophy.) 'The growth of the sciences', Bacon claimed, depended upon 'the exploration of the world'. Drawing a direct comparison between scientific and geographic discovery, Bacon went on, arguing:

> Surely it would be disgraceful if, while the regions of the material globe . . . have been in our times laid widely open and revealed, the intellectual globe should remain shut up within the narrow limits of old discoveries.

Bacon, writing in the seventeenth century, was in fact drawing on an earlier colonial example, one we encountered in chapter 1. What Bacon had in mind was the Spanish Empire of the fifteenth and sixteenth centuries. His vision of science drew directly from the House of Trade in Seville. That is exactly how the Royal Society was imagined: an English equivalent to earlier Spanish institutions designed to centralize the collection of information. Bacon was reading Spanish accounts of travels to the New World, and transferring that idea wholesale to the study and organization of science. He even borrowed an illustration from an earlier Spanish book on navigation, written by a cosmographer in Seville, for the frontispiece to *The New Organon*. The engraving depicted a ship sailing between the mythical Pillars of Hercules, representing the limits of the known world in ancient times. In the caption below, Bacon quoted the Bible, but echoed the words of Columbus a century earlier: 'Many shall go to and fro and knowledge shall be increased.'[6]

By the beginning of the eighteenth century, this connection between science and empire was firmly entrenched. In this chapter, we examine the way in which state-sponsored voyages of exploration underpinned the growth of the physical sciences. Without these voyages, Newton and his followers would not have been able to tackle some of the most fundamental questions concerning the nature of the universe. At the same time, the development of the physical sciences provided a range of practical benefits, particularly in the fields of surveying and navigation, allowing European empires to expand further and further into new territory. This, then, is a new history of Enlightenment science. A history, not of isolated individuals applying the principles of reason, but rather a history that connects eighteenth-century science to the destructive world of empire, slavery, and war. Newton and his *Principia* are a good place to start.[7]

14. The frontispiece from Francis Bacon, *The New Organon* (1620) (*left*), copied from Andrés García de Céspedes, *Rules of Navigation* (1606) (*right*), depicting a ship sailing between the Pillars of Hercules.

I. Gravity in Gorée

Isaac Newton was born in Lincolnshire on Christmas Day 1642. He never left Britain, spending about half his adult life in Cambridge, where he was appointed Lucasian Professor of Mathematics in 1669, and the rest in London, where he worked as Master of the Royal Mint. But Newton was by no means isolated. When we examine Newton's writings closely, it soon becomes clear that he was absolutely reliant on information arriving from across the world. This was in fact the same world in which Newton's own money circulated through investments in the slave trade and the East India Company. Much of the information Newton used in his *Principia* came from explorers and astronomers travelling aboard slave ships and trading vessels.[8]

At the centre of the *Principia* was Newton's theory of universal gravitation. Today, we are so familiar with the idea of gravity that it can be hard to appreciate exactly what it is. People had always known that heavy bodies fall towards the ground. Newton's theory was more complex than this. He argued that all matter, whether you were dealing with an apple or the Earth, exerts an invisible force, drawing other matter towards it. Therefore, when an apple falls towards the Earth, both the Earth and the apple are actually being attracted to one another. What's more, Newton was able to express this idea with mathematical precision: you just needed to multiply the masses of the two objects together, and divide by the square of the distance between them. This explains why objects with a large mass (like the Earth) exert a greater gravitational pull than those with a small mass (like an apple). It also explains why objects that are far away from each other exert less of a gravitational pull than those that are close together.

Where did Newton get this idea from? Contrary to popular belief, Newton did not make his great discovery after an apple fell on his head. Instead, in a key passage in the *Principia*, Newton cited the experiments of a French astronomer named Jean Richer. In 1672, Richer had travelled to the French colony of Cayenne in South America. The voyage was sponsored by King Louis XIV through the Royal Academy of Sciences in Paris. The expedition also received support from the French West India Company, which provided the ship on which Richer crossed the Atlantic. Once in Cayenne, Richer made a series of astronomical observations, focusing on the movements of the planets and cataloguing the stars close to the equator. This new astronomical data could then be used by navigators to calculate their position at sea, enhancing the ability of the French Navy to project itself around the world. The Royal Academy of Sciences had been founded in 1666 to support exactly these kinds of scientific voyages. Jean-Baptiste Colbert, finance minister at the court of Louis XIV, had convinced the French king to establish a national scientific academy, one that would aid in the growth of the French Empire. The choice of Cayenne for one of these early voyages is telling – the colony had only just been returned to French hands following the Second Anglo-Dutch War of 1665–7. By travelling to Cayenne, Richer was making both a scientific and territorial claim on behalf of the French state.[9]

Whilst in Cayenne, Richer also undertook a number of experiments with a pendulum clock. This was a relatively new invention, pioneered by the Dutch mathematician Christiaan Huygens in 1653. Huygens had realized that a pendulum swings at a constant rate, proportional to its length, making it an ideal measure of time. In particular, a pendulum with a length of just under one metre makes a complete swing, left to right, every second. This became known as a 'seconds pendulum' and proved exceptionally useful for astronomers who wanted to keep track of the passage of stars and planets. But there was a problem. In Cayenne, Richer noticed that his carefully calibrated pendulum was running slow, taking longer than a second to complete each swing. Over the course of a day, the pendulum clock had lost over two minutes. This was odd. Richer had checked and double-checked that the pendulum was the correct length back in Paris. But now, in South America, he found that it needed to be shortened to keep to time.[10]

Intrigued, Richer repeated the experiments a few years later. In 1681, the Royal Academy of Sciences sponsored a second voyage, this time to West Africa. Once again, Richer's expedition was born out of a world of slavery and colonial expansion. He travelled aboard a French Senegal Company ship, spending two months at sea before arriving on the island of Gorée, just off the coast of Senegambia, in modern-day Senegal. Like Cayenne, Gorée was a French colony, recently seized from the Dutch. The small island provided a convenient base for French slave traders.

Thousands of African men, women, and children were crammed into airless cellars, awaiting transportation to the Americas. Richer himself worked above one of these rooms, his assistant complaining, in the racist language typical of the period, that 'we would have to live with the negroes'. After four months experimenting with the pendulum on Gorée, Richer made a final voyage across the Atlantic. He travelled aboard another French Senegal Company ship, this time loaded with over 250 enslaved Africans, bound for Guadeloupe in the Caribbean. Here, at the heart of the French slave trade, Richer confirmed his earlier observations. A seconds pendulum really did run slow nearer the equator. Richer found that, on both Gorée and Guadeloupe, he needed to shorten the pendulum by about four millimetres to keep it running to time.[11]

What could explain this variation? There was no obvious reason that a pendulum should behave differently in France as opposed to South America or West Africa. After all, the laws of physics were supposed to be constant, and Richer had carefully controlled for the influence of climate, ensuring that the pendulum itself wasn't expanding in the tropical heat. Newton, however, quickly realized the implications of what Richer had observed. Writing in the *Principia*, Newton argued that the force of gravity actually varied across the surface of the planet. There was an 'excess of gravity in these northern places over gravity at the equator', as Newton put it. This was a radical suggestion, one which seemed to go against common sense. But Newton did the calculations and showed how his equations for the gravitational force matched exactly Richer's results from Cayenne and Gorée. Gravity really was weaker nearer the equator.[12]

All this implied a second, even more controversial, conclusion. If gravity was variable, then the Earth could not be a perfect sphere. Instead, Newton argued, the Earth must be a 'spheroid', flattened at the poles rather like a pumpkin. This explained why gravity was weaker nearer the equator, where the Earth bulged outwards. 'The earth will be higher at the equator than at the poles by an excess of about seventeen miles,' explained Newton. Therefore, when Richer tested his pendulum at Gorée, it was as if he was standing atop an incredibly tall mountain (one in fact much higher than any actually existing mountain on Earth). According to Newton's inverse-square law, the force of gravity would then be diminished, as the pendulum was considerably further away from the centre of the Earth at Gorée as opposed to Paris.[13]

Newton famously claimed that 'all the world knows I make no observations myself'. Historians traditionally understood this to mean that Newton was an isolated theorist. In fact, what Newton meant was that he relied on observations made by others, stretching right across the globe. Richer's experiments near the equator are just one example of the hundreds of data points Newton relied upon in his *Principia*. Newton also collected tidal data from East India Company officers returning from China alongside observations of comets made by slaveowners in Maryland. Perhaps most tellingly, Newton owned twice as many books on travel, mostly detailing foreign voyages, as he did on astronomy. Connected to the wider world of science and empire through the Royal Society and the Royal Mint in London, Newton was able to amass a vast collection of information. It was this that allowed him to fundamentally alter how we think about the basic physical forces that govern the universe.[14]

Today, it is easy to see the *Principia* as a scientific masterpiece, the validity of which no one could deny. But at the time, Newton's ideas were incredibly controversial. Whilst most English thinkers did accept the results of the *Principia* relatively quickly, many in Continental Europe remained sceptical. Nicolaus Bernoulli, an influential Swiss mathematician, attacked Newton's theories as 'incomprehensible', whilst Gottfried Leibniz, Newton's great German rival, complained about the 'occult quality' of gravity. Many preferred the 'mechanical philosophy' of the French mathematician René Descartes. Writing in his *Principles of Philosophy* (1644), Descartes denied the possibility of any kind of invisible force like gravity, instead arguing that force could only be transferred through direct contact. Descartes also suggested that, according to his own theory of matter, the Earth should be stretched the other way, elongated like an egg rather than squashed like a pumpkin.[15]

These differences were not simply a case of national rivalry or scientific ignorance. When Newton published the *Principia* in 1687, his theories were in fact incomplete. Two major problems remained to be solved. First, there were the aforementioned conflicting reports of the shape of the Earth. And if Newton was wrong about the shape of the Earth, then he was wrong about gravity. Second, Newton's theory implied a new account of planetary motion, one in which all the planets, as well as the Sun, exerted a gravitational force on one another. (This would help explain the characteristic wobble in the orbit of the planets that astronomers had been trying to explain since the time of Ptolemy.) To confirm this decisively, astronomers needed to make new observations. In particular, they needed to know the precise distance between each of the planets. Again, this would be a key test for Newton.[16]

The history of eighteenth-century physics can therefore be read as a battle over Newton's ideas, one that continued well after his death in 1727. This was a battle which stretched from South America to the Pacific. Over the course of the eighteenth century, European states sponsored hundreds of voyages of exploration, claiming new territories and undertaking scientific observations along the way. Much as we saw in chapter 1, eighteenth-century European explorers relied on the scientific knowledge of Indigenous peoples – including Inca astronomers and Tahitian navigators – in order to make these observations as well as to find their way across new regions of the globe. Without this Indigenous knowledge, Newton's theories would have remained incomplete. Without empire – as well as the dispossession and violence associated with it – there would be no Enlightenment.[17]

II. Inca Astronomers

Charles-Marie de La Condamine could feel his gums bleeding. Climbing to the top of Pichincha, an active volcano in the Andes, the French explorer was suffering from altitude sickness. With the help of his Peruvian guides, La Condamine continued up the volcano, chewing coca leaves to stay alert. Reaching over 15,000 feet, La Condamine was now higher than any European had gone before. Once at the peak, he ordered his Peruvian guides to unpack a large crate of scientific instruments. La Condamine then set up his quadrant – a metal instrument in the shape of a quarter of a circle, divided by degrees, used to measure the angle between two objects. Down in the valley below, he spotted a small wooden pyramid, painted white. La Condamine then raised his bloodshot eyes up to the quadrant, measuring the angle between the pyramid in the valley and Mount Pambamarca, another great peak on the horizon. This was all he needed: a single data point, one that would feed into a vast survey stretching 150 miles across the Andes.[18]

Over two years earlier, in May 1735, La Condamine had left France aboard a ship bound for South America. He was part of an international team sent to undertake one of the most ambitious scientific surveys ever attempted. Their job: to measure the shape of the Earth. Backed by Louis XV, the Royal Academy of Sciences in Paris organized two major expeditions in the 1730s. The first was sent to Lapland, in the Arctic Circle. The second was sent to Quito, close to the equator in modern-day Ecuador, but at the time part of the Viceroyalty of Peru. The idea was simple in theory, although exceptionally difficult in practice. Each party would measure the exact distance, north to south, covered by one degree of latitude. The results would then be compared. If Newton was right, and the Earth was flattened at the poles, then the length of a degree of latitude near the equator would be less than that at the Arctic.[19]

The expedition to the Andes was made possible thanks to a recent alliance between France and Spain. In 1700, Philip V had been crowned King of Spain. He had been born in France at the Palace of Versailles, a grandson of Louis XIV. Philip V of Spain was therefore a member of the House of Bourbon, part of an old French dynasty stretching back to the thirteenth century. The alliance between France and Spain was quickly formalized in 1733 with the Treaty of El Escorial. It was this that allowed members of the Royal Academy of Sciences in France to work closely with their Spanish counterparts. Philip V himself granted the French permission to travel in Spanish American territories. The Viceroyalty of Peru was an ideal location to undertake the survey: close to the equator, with a chain of mountains and volcanoes that could be used as vantage points for observations.[20]

Getting to the Andes took over a year. First, La Condamine and the French expeditionary party crossed the Atlantic, stopping off for a few weeks in the West Indies in the summer of 1735. Whilst there, the team calibrated their instruments and climbed Mount Pelée, a volcano on Martinique, in order to practise the technique they would later use in South America. The French explorers also purchased a number of enslaved Africans on Saint-Domingue, in modern-day Haiti. La Condamine himself bought three slaves. Although we don't know their names, we do know that these African men accompanied La Condamine on the rest of the expedition, spending almost a decade in the forced service of the French astronomer, before being sold back into slavery at the end of the trip. These enslaved Africans, along with the Peruvian Indians the French co-opted in the Andes, provided the raw labour that the scientific expedition needed. They carried heavy instruments, led mules up steep ravines, paddled canoes, and negotiated with local people. Without this forced labour, the expedition would not have made it to Quito. After leaving the West Indies, the French team met up with two Spanish naval officers in Cartagena de Indias, in modern-day Colombia. They then crossed over to Panama on the Pacific coast before sailing down to reach the Viceroyalty of Peru. Finally, La Condamine and his guides travelled a further 150 miles upstream along the Esmeraldas River, arriving in Quito in June 1736. It was now time to begin the survey.[21]

The basic technique for conducting a survey of this kind had been pioneered in France during the seventeenth century. To begin, the team needed to construct what was known as a 'baseline'. This was a perfectly straight trench, only a few inches deep, but at least a couple of miles long. The length of this would then be measured manually, laying wooden poles end to end. Next, the surveyors would choose a point in the distance, something like a mountain top. They would then measure the angle between both ends of the baseline and this point using a quadrant. These measurements would give an imaginary triangle. With a bit of basic trigonometry, the surveyor could then calculate the lengths of the two remaining sides of the triangle, given the known length of the baseline.

Next, they needed to repeat the process. However, the surveyors didn't have to physically build another baseline. Instead, they could just use the existing imaginary triangle, beginning from the northernmost point. Again, they would measure the angle between this point and a different object in the distance, like another mountain or volcano. And again, they could work out the distance between those two points based on the length of the sides of the first triangle. Each observation, which often required scaling a mountain, therefore advanced the survey a little further along, essentially establishing a series of imaginary tessellating triangles. Once the survey had proceeded as far as it needed, perhaps over a hundred miles, the surveyors could just add up the lengths of the various imaginary triangles, and this would give a precise measure of the distance traversed. The last step was to work out how many degrees of latitude this distance represented. To do this, the surveyors would determine the latitude of the start and end points by a simple observation of the stars. Finally, by dividing the two results – the distance covered by the difference in latitude – the surveyors would get the number they were after: an exact measure of the length of one degree of latitude.[22]

This was all easier said than done. The most important thing to get right was the baseline at the start. Any error in its construction or measurement would be compounded many times over, as every other calculation depended upon it. Most of all, the baseline needed to be perfectly straight. This was made more difficult by the varied landscape found in the Andes, with mountains rising and falling one after another right along the Pacific coast. In the end, La Condamine selected a strip of land, seven miles long, on the Yaruquí plateau, just outside Quito. Relatively flat, for the Andes at least, this provided an ideal site for the baseline. Perhaps unsurprisingly, La Condamine did not build the baseline himself. The backbreaking work of digging a seven-mile trench was left to the local Peruvian Indians. They were forced to work for the European surveyors under the *mita* system. This was a regime of public service, originally developed under the Inca Empire, but adapted by the Spanish into a system of forced labour. La Condamine did not think much of the Peruvian Indians working under him, describing them as 'barely distinguishable from beasts'. Another of the French surveyors believed they were 'only capable of slavish imitation, and incapable of creating anything new'. Apparently 'all was confusion' as to the European explorers' motives.[23]

If anything, the confusion was on the part of La Condamine. The Indigenous people of the Andes were far from the ignorant 'beasts' that the Europeans described them as. In fact, Peruvian Indians possessed a sophisticated understanding of astronomy and surveying. La Condamine did not realize it, but he was relying not just on Indigenous labour, but also on Indigenous knowledge. Most significantly, the idea of building a long straight trench in order to undertake an astronomical survey was part of a well-established Andean tradition, one that dated back thousands of years. If La Condamine had travelled further south, to the desert of Nazca Pampa on the coast of Peru, he would have seen a series of incredible carved lines in the ground. These 'geoglyphs', some of which are over 2,000 years old, depict a mixture of geometric designs and more recognizable animal forms when observed from above. There is a monkey and a spider, as well as a hummingbird. These designs are all

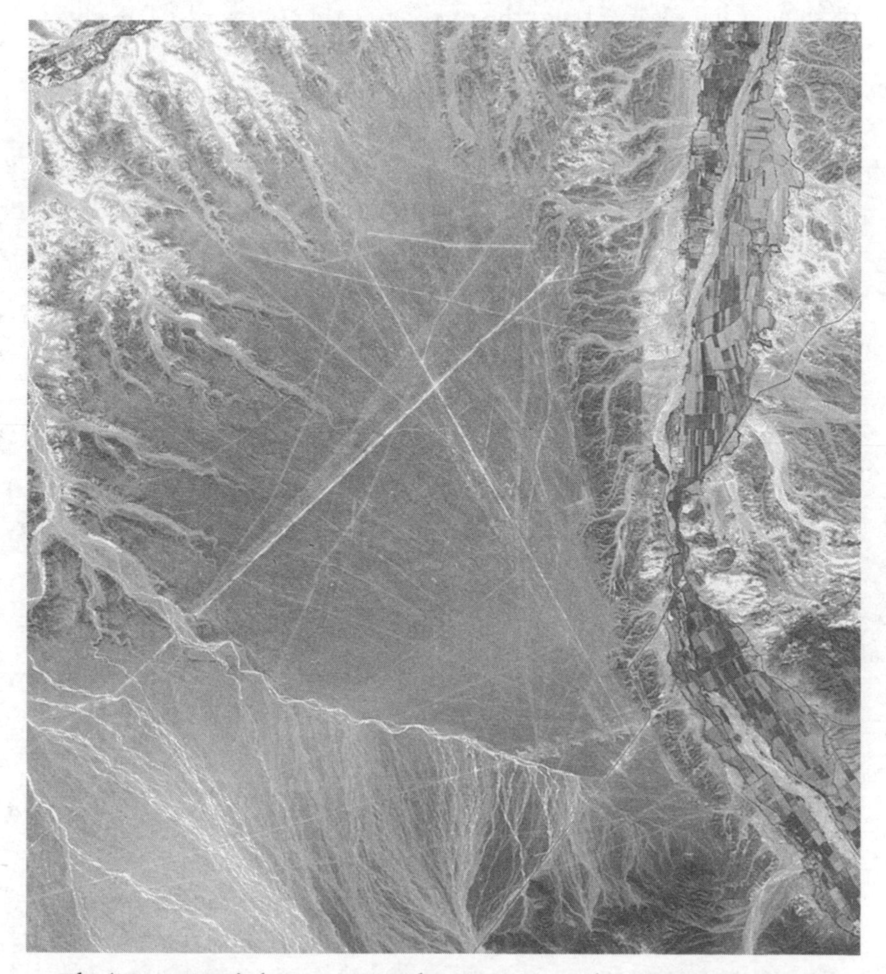

15. The 'Nazca Lines', dating to around 500 BCE, in southern Peru. Historians now think these were used to align astronomical observations.

carved into the ground, in shallow trenches around six inches deep. But not all the geoglyphs are like this. Intriguingly, some are simply long straight lines. They carry on for miles, dead straight, crossing hills and valleys. Whilst their exact function is still unclear, many historians now believe they were used to align astronomical observations, exactly as La Condamine intended with his baseline.[24]

By the fifteenth century, when the Inca Empire rose to power, this practice had developed into a complex scientific system, combining astronomy and surveying. At the heart of the Inca Empire was the Temple of the Sun, in the capital Cusco. From here, the Incas built a series of straight lines, carved into the earth, radiating outwards. These were known as *ceques*, or 'ritual lines'. In total, there were forty-one ritual lines, many of which can still be found around Cusco today, stretching out in all directions. Like the earlier geoglyphs, these shallow trenches carried on for miles, perfectly straight. The ritual lines served a variety of functions. But most significantly, they acted as an aid for astronomy and surveying. In the first instance, the ritual lines divided the Inca Empire into different regions. These regions corresponded to different social groupings, and each was also assigned to a different family or priest to oversee them. In addition to this, each of the ritual lines pointed towards a group of sacred places, known as *huacas*. In total, there were 328 sacred places, each representing a day in the Inca calendar. In some cases, these were natural sites, like a mountain top or volcano. In other cases, they were sites the Incas had selected for ritual purposes, building shrines to mark them out on the horizon.[25]

Most importantly, many of these sacred places were aligned with specific astronomical events. The ritual lines, radiating outwards from the Temple of the Sun, could then be used to align observations back in Cusco. For example, one of the most significant events in the Inca calendar was the 'Sun Festival'. This was held in June to coincide with the winter solstice in the southern hemisphere. The Incas, who called themselves the 'Children of the Sun', celebrated the end of winter, and the promise of longer days. To help plan the festival, the Incas built a series of stone pillars and pyramids, visible on the horizon from the Temple of the Sun. One of the ritual lines pointed towards these structures, allowing astronomers in Cusco to precisely record the rising of the Sun in the build-up to the solstice.[26]

The Peruvian Indians who built the baseline must have believed that La Condamine wanted to construct his own ritual line, much like the earlier Inca rulers. Working by day, and sleeping on the ground at night, the Peruvian Indians completed the seven-mile trench in just under a month. It was worth the wait. When La Condamine measured the baseline he found it to be perfectly straight. Over the following months, the same men helped La Condamine and the rest of the French explorers complete the survey, stretching 150 miles from Quito in the north to Cuenca in the south. Much of what La Condamine asked the Peruvian Indians to do made perfect sense within the older tradition of Inca astronomy. At various points, La Condamine ordered the Peruvian Indians to construct wooden pyramids at key sites, such as on the top of a mountain or at the ends of the baseline. These structures, painted white, were then used as points of measurement. The idea was that they would be easy to spot in the distance, ensuring the French surveyors picked out the right mountain peak on the horizon.[27]

The choice of a pyramid is striking. La Condamine might have been drawing on his own experience travelling in Egypt as a young man, but the Incas also built pyramids. And they did so precisely to aid in the alignment of astronomical observations. Again, the Peruvian Indians accompanying La Condamine knew exactly what to do. It was their knowledge and expertise which ultimately allowed the French explorers to undertake such an accurate survey. Indigenous knowledge, then, was not swept away by the growth of European empires in the eighteenth century. In fact, as we'll see in other examples in this chapter, Enlightenment explorers often depended upon Indigenous people in a variety of ways, even if it was rarely acknowledged. Particularly when it came to astronomy, Indigenous people – not just in South America, but also in the Pacific and Arctic – were key collaborators in the development of Newtonian science.[28]

By January 1742, the results were in. La Condamine calculated that the distance between Quito and Cuenca was exactly 344,856 metres. From observations made of the stars at both ends of the survey, La Condamine also found that the difference in latitude between Quito and Cuenca was a little over three degrees. Dividing the two, La Condamine concluded that the length of a degree of latitude at the equator was 110,613 metres. This was over 1,000 metres less than the result found by the Lapland expedition, which had recently returned to Paris. The French, unwittingly relying on Indigenous Andean science, had discovered the true shape of the Earth. It was an 'oblate spheroid', squashed at the poles and bulging at the equator. Newton was right. In the following section, we explore a parallel story, linking European empires, Indigenous knowledge, and Newtonian science in the Pacific. We begin in eighteenth-century Polynesia, where two astronomers are watching the Sun.[29]

III. Pacific Navigators

Ta'aroa peered through the telescope, watching as a small black disc crossed the face of the Sun. It was a beautiful if disconcerting sight. A strange, pale-skinned man then explained to Ta'aroa that this was a 'planet upon the sun'. The foreign man seemed obsessed, observing the planet through the telescope for over six hours. He called it 'Venus'. But Ta'aroa knew it better as 'Great Festivity'. On 3 June 1769, Ta'aroa – high chief of the island of Mo'orea – observed the transit of Venus. This is a rare astronomical phenomenon in which the planet Venus moves between the Earth and the Sun.[30]

Mo'orea is located in the Pacific Ocean, part of the great chain of islands which makes up Polynesia. Ta'aroa himself was familiar with the stars and planets. According to Polynesian legend, the deity Atea ('bringer of light') had created Venus. The planet, which appeared as one of the brightest stars in the sky, then acted as a compass. Ancient Pacific seafarers followed Venus across the ocean, settling the tranquil islands of Polynesia many moons ago. But Ta'aroa had never seen anything like this before. Transits of Venus occur in pairs, separated by eight years, but with a wait of over a hundred years after that. There had been two transits in the seventeenth century, in 1631 and 1639, and there had been an earlier one in 1761, although only partially visible from the Pacific. The next pair would not occur for at least another hundred years.[31]

The strange man who invited Ta'aroa to look through the telescope was named Joseph Banks. He would go on to become one of the most influential scientific figures of the eighteenth century, occupying the position of President of the Royal Society for over forty years. Banks had arrived in Polynesia aboard HMS *Endeavour* in April 1769. Led by Captain James Cook, the voyage of HMS *Endeavour* opened up a new era of contact between Europe and the Pacific. It was also a voyage fundamentally tied to the development of eighteenth-century science. Organized by the Royal Society in London, and sponsored by King George III, Cook's mission to the Pacific had two goals. The first was to observe the transit of Venus. The second was to locate the fabled 'Southern Continent', or *Terra Australis*, thought by Europeans to be rich in gold and silver. This was an idea that stretched back to the medieval period, and had been popularized during the fifteenth and sixteenth centuries by early European explorers travelling to Asia and the Pacific. Like the French expedition to the Andes, the voyage of HMS *Endeavour* therefore mixed imperial ambition with scientific enquiry.[32]

Venus held the key to Newton's second major problem. Astronomers had known the relative distances between the planets since the early seventeenth century. However, they had no measure of the absolute distance. This was an issue for Newton. In the *Principia*, he had demonstrated how his theory of universal gravitation could be used to explain the elliptical orbits of the planets around the Sun. Newton also suggested that the planets exerted a gravitational pull on one another, particularly when they were in close proximity, hence why the orbits sometimes appeared irregular. The same was true of the Moon and the various satellites of Jupiter. However, Newton could only talk about all this in the abstract, giving geometric proofs alongside complex mathematical formulae, but with very little concrete data. At one point in the *Principia*, Newton did describe the force of gravity exerted by the Sun on Jupiter and Saturn. But again, he was only able to do this in terms of the ratio between the two, rather than absolute values.[33]

The transit of Venus would solve this problem. In 1716, Newton's friend Edmond Halley suggested a method for measuring the exact distance between the Earth and the Sun. Halley realized that Venus would take less time to move across the face of the Sun for observers in the southern hemisphere as opposed to those in the northern hemisphere. This is an effect known as 'parallax', where the same object appears in a different position from different points of observation. (You experience the same thing when you open and close your left or right eye, and an object appears to move.) Comparing results from the northern and southern hemispheres would allow astronomers to calculate the angle between Venus and these different points. Taking this angle, and the known distance between the observers, it was then possible to use trigonometry to calculate the missing value: the distance between the Earth and the Sun. This method, essentially constructing a massive imaginary triangle between the Earth and Venus, was based on the same principles the French had used in the Andes. This time, however, it was scaled up to the size of the solar system.[34]

The distance between the Earth and the Sun, known as the 'astronomical unit', acted as a kind of cosmological yardstick. Astronomers already knew the relative distances between all the planets. They could therefore simply take this value and calculate the remaining absolute distances, giving an accurate measure of the size of the solar system for the first time. Doing so would furnish concrete proof of Newton's theories in action. Knowing the exact size of the solar system would also provide a number of practical benefits for navigation at sea. Indeed, this is why European states were willing to invest enormous sums of money into what may seem like a relatively academic question. The British were not the only ones intent on measuring the transit of Venus. The Royal Academy of Sciences in France sent observers to Saint-Domingue, whilst the Saint Petersburg Academy of Sciences in Russia sent observers to Siberia. All in all, European scientific academies sent over 250 observers to locations right across the globe, from California in the west to Beijing in the east.[35]

Since the early eighteenth century, European navigators had increasingly been encouraged to use astronomical observations to calculate their position at sea. In 1714, the British Parliament established the Board of Longitude for exactly this purpose. The Board offered a reward of up to £20,000 for an accurate method to determine longitude at sea. Some of the proposed methods relied on keeping a precise measure of time over the course of a long voyage. The clockmaker John Harrison famously adopted this approach, developing a special marine timekeeper which was tested on a voyage to Jamaica in 1761. The hope was that Harrison's timekeeper might aid in navigation between West Africa and the Caribbean, another reminder of the role of the transatlantic slave trade in shaping eighteenth-century science. However, the majority of methods favoured by the Board of Longitude were based on astronomical observations. These included measurements of the satellites of Jupiter as well as measurements of the angle between the Moon and particular stars. Longitude at sea could then be computed by comparing these results with tables produced back at the Royal Observatory in Greenwich. An accurate measure of the size of the solar system was therefore essential, not just for confirming Newton's predictions, but also for the advancement of navigation.[36]

Captain James Cook treated the transit of Venus like a military operation. In many ways it was. The Royal Society had selected the island of Tahiti as the main site for the observations. Situated in the middle of the Pacific Ocean, Tahiti was about as far away from Britain as it was possible to get. However, it was one of the few places in the southern hemisphere where it would be possible to observe the entire transit of Venus, from beginning to end. Tahiti was also of strategic interest to the Royal Navy. The Spanish explorer Ferdinand Magellan had first crossed the Pacific Ocean in the sixteenth century. However, it was only in the eighteenth century that European empires really sought to expand territorially into the region. The hope was that islands like Tahiti might act as bases for further exploration, particularly in the quest to find the great Southern Continent. The French, the Dutch, and the British all vied for control. Indeed, according to the French, Tahiti was theirs. A few years earlier, in 1767, the French explorer Louis Antoine de Bougainville had landed on the island, claiming it on behalf of Louis XV.[37]

Cook wasn't going to let any French claims deter him. On arriving on Tahiti in April 1769, he ordered the construction of a small military base, appropriately named 'Fort Venus'. 'I took care to secure ourselves in such a manner as to put it out of the power of the whole Island to drive us off,' noted Cook in his diary. Tensions between the British and the Tahitians were running high. There had been a series of violent confrontations since the arrival of HMS *Endeavour*. Cook didn't want anything disturbing his observations. Fort Venus consisted of a high wooden fence, topped with spikes and surrounded by a deep trench. A tent containing the astronomical instruments and a pendulum clock was then set up right in the centre. Cook ordered his crew to hoist a

Union Jack on top of the tent, reminding both the local Tahitian population, and the French, that the island was now considered British territory. Cook then ordered a group of marines, armed with muskets, to stand guard.[38]

The day of the transit itself was incredibly hot, reaching over 48°C. Sweating in his captain's uniform, Cook complained that he found the heat 'intolerable'. But overall he was pleased with the weather. The day before had been cloudy, and if the Sun was obscured during the transit, then the entire voyage would have been for nothing. As a precaution, Cook dispatched Joseph Banks to the nearby island of Mo'orea, to undertake additional observations, hoping he might have a clearer view there. As it happened, on the morning of 3 June 1769, the clouds cleared over Tahiti. 'The day prov'd as favourable to our purposes as we could wish, not a Cloud was to be seen the whole day and the Air was perfectly clear,' Cook recorded. As the predicted time of the transit approached, Cook watched carefully through his telescope. Sure enough, at 9.21 a.m. local time, a small black figure appeared at the edge of the Sun. Venus had arrived.[39]

But something wasn't right. Venus didn't appear as a perfect circle, but rather seemed to bleed into the edge of the Sun as it approached. Cook had been warned about this. Attempts to observe the 1761 transit of Venus had been hampered by what became known as the 'black drop effect'. One of the 1761 observers, a Russian astronomer named Mikhail Lomonosov, had realized that this was caused by the atmosphere of Venus. Before the planet itself moved in front of the Sun, the atmosphere began to refract and absorb the light, creating the visual impression that Cook described as a 'dusky shade'. Even though he was forewarned, Cook admitted it was still 'very difficult to judge precisely' when the transit of Venus actually began. To be safe, he made a drawing of what he had seen, and noted down the different times of the different points in the transit. This could then be compared with other astronomers' accounts, to ensure – as well as possible – that they were observing the same thing. After six hours, the transit was complete. Cook and Banks packed up their equipment, satisfied with what they had achieved.[40]

When HMS *Endeavour* finally returned to Britain in 1771, Cook submitted his findings to the Royal Society. By comparing the observations made in Tahiti with those made in the northern hemisphere, the

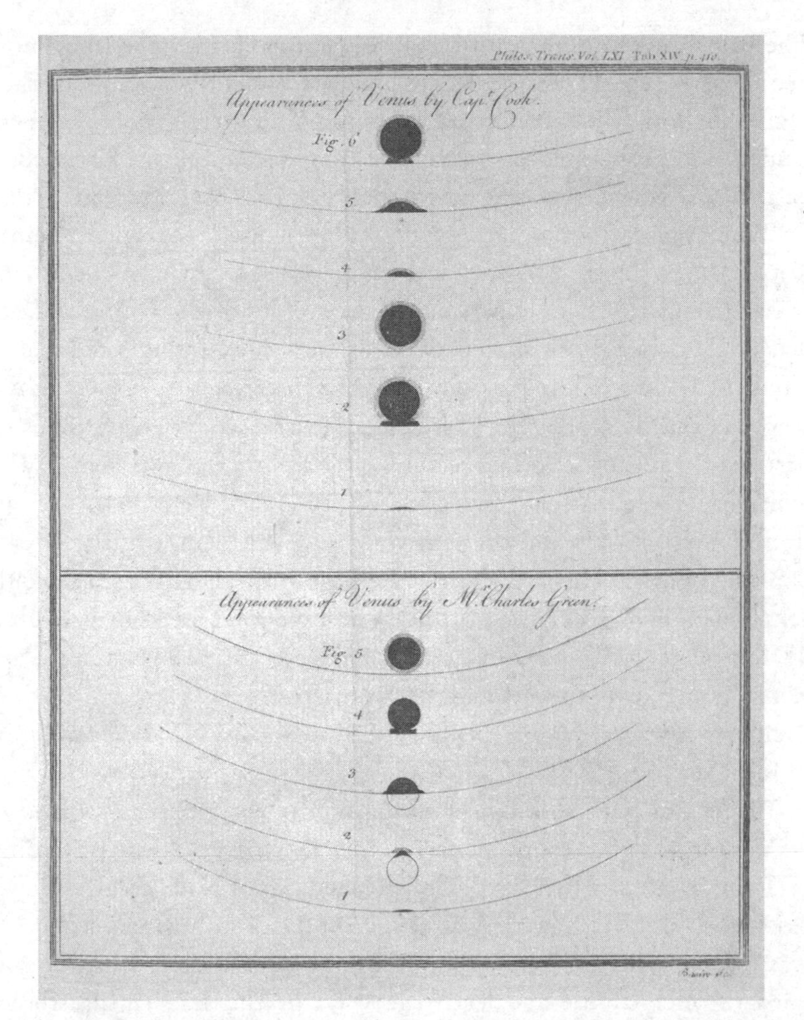

16. The transit of Venus as represented by James Cook in 1769. Note the 'black drop effect' caused by the atmosphere of the planet.

mathematicians at the Royal Society were able to calculate the distance between the Earth and the Sun. The final result: 93,726,900 miles. This was a remarkably accurate figure, within 1 per cent of the modern value of 92,955,807 miles. Finally, nearly a hundred years after the publication of the *Principia*, Newton's followers had the number they needed. By sending Cook to the Pacific, the Royal Society had been able to determine the size of the solar system.[41]

★

The British were not the only stargazers in the Pacific. The Indigenous people of Polynesia possessed a sophisticated scientific culture of their own, much of which concerned astronomy and navigation. As in the Andes, European explorers in the Pacific relied upon this knowledge, particularly when navigating across such vast expanses of ocean. Whilst on Tahiti, James Cook and Joseph Banks befriended a local priest named Tupaia. In Polynesia, religious and navigational knowledge went hand in hand. Tupaia was an expert in the geography of the region and had decades of experience sailing between islands. He also indicated that he would be willing to join the crew of HMS *Endeavour*. At first Cook was sceptical, but Banks convinced him that Tupaia would be a great asset, particularly as they set out in search of the Southern Continent. 'What makes him more than anything desirable is his experience in the navigation of these people and the knowledge of the Islands in these seas,' explained Banks. This was a rare case of a European explorer explicitly acknowledging the expertise of Indigenous people. Banks realized that, if they were going to successfully navigate uncharted waters, and return home alive, they would need someone with knowledge of the Pacific. Cook agreed, and Tupaia left Tahiti aboard HMS *Endeavour* on 13 July 1769.[42]

Tupaia was born to a high-ranking family on the nearby island of Ra'iatea in 1725. He spent his early years at the great *marae*, or temple, of Taputapuatea on the island. Constructed out of black coral, and over 1,000 years old, this temple was at the centre of Polynesian culture. It was a place where priests, diplomats, and traders came from far and wide to offer tribute and learn the ways of the sea. Tupaia studied astronomy, navigation, and history. Indeed, these three subjects all went together. Polynesian navigators needed to be able to sail, sometimes for weeks, out of sight of land. They did so without the aid of charts or navigational instruments. Instead, Tupaia was trained to memorize sailing directions, based on lists of stars, through a series of chants. These sailing directions, passed down through the generations, often recalled the voyages of ancestors. Ancient Polynesians had first left Southeast Asia around 4,000 years ago, gradually spreading across the Pacific, reaching Tahiti in around the year 1000 CE.[43]

The idea behind Polynesian navigation was simple, but extremely effective. Instead of calculating his exact position at sea, Tupaia would recall a specific route: for example, the stars to follow to travel between Tahiti and Hawai'i. (This was made slightly more complex by the fact that the position of the stars changes depending on the season, meaning that multiple lists needed to be memorized.) These 'star paths', or *aveia*, then formed the basis of navigation. The best way to understand this is the difference between knowing your exact coordinates on a GPS versus remembering directions such as 'travel down that road, straight across at the traffic lights, and then left at the next junction'. Polynesians preferred directions over coordinates. To travel between two islands, a navigator like Tupaia would first identify a specific star associated with that route. The star would need to be located relatively low in the sky, near the horizon. The navigator would then start sailing towards it. After a certain amount of time, the navigator might then need to switch to a different star, particularly if it was a long voyage. After days or even weeks at sea, they would eventually reach their destination.[44]

Polynesian navigators preferred to sail at night. But if they had to, a navigator like Tupaia could also sail by day. In the southern hemisphere, a shadow cast by the Sun at midday points due north. It was therefore relatively easy to get a good sense of the direction you were travelling in based on the position of the Sun. But Polynesian navigators didn't just watch the heavens. They also paid close attention to the rise and fall of the ocean. These ocean swells are modified by land, bouncing back off a large island or sometimes bending around it. Polynesian navigators were trained to recognize these subtle differences in the rise and fall of the ocean, and the interaction between different swell patterns. In the Marshall Islands, navigators even made charts, constructed out of palm ribs and bound with coconut fibres, representing ocean swells. Islands were then represented by tiny shells, tied to the chart. These 'stick charts', or *mattang*, weren't actually used at sea. Instead, they were used as training devices in temples, helping young navigators memorize the different tidal and swell patterns. It is worth noting that, in the eighteenth century, European navigators did not have a workable theory of oceanic swells, whilst Newton himself had only just started to develop a more complex theory of the tides in his *Principia*.[45]

Ultimately, whilst Europeans tended to think of the Pacific as a vast, empty space, dotted with tiny islands, Polynesians understood the ocean itself as a kind of terrain. The Pacific was full of texture, swells and

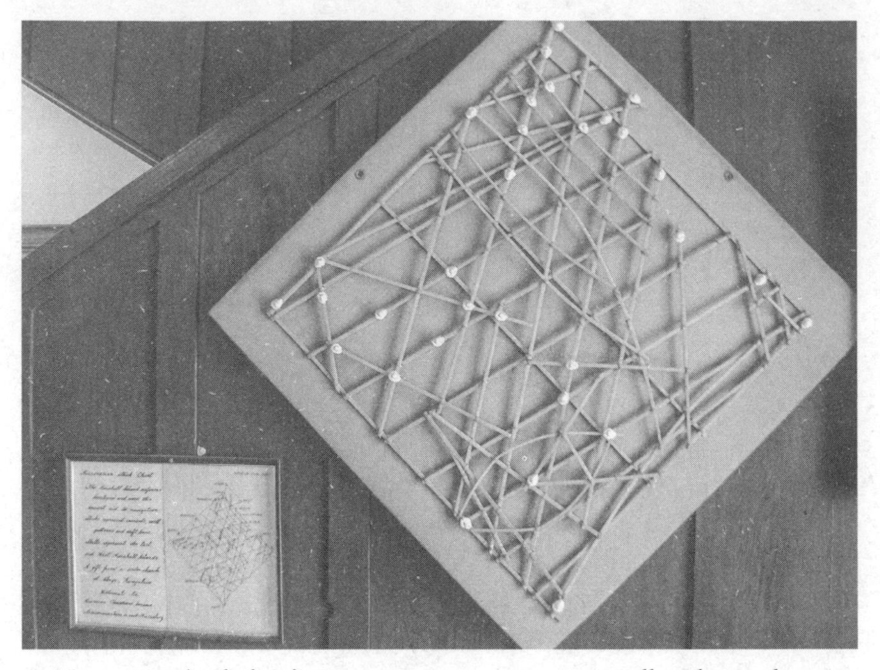

17. Micronesian 'stick chart', or *mattang*, representing ocean swells and routes between islands.

currents forming the equivalent of hills and valleys. Recognizing this terrain, and memorizing the stars which passed overhead, provided the key to navigating this unforgiving expanse.

At the age of twelve, Tupaia received his first tattoos. Covering his legs and lower back, these markings indicated Tupaia's new status as a person of great learning. He had mastered the basic techniques of navigation and was ready to join a society of travellers, known as the *ariori*, dedicated to the war god 'Oro, patron of the temple at Taputapuatea. Tupaia then began to travel between islands, putting what he had learned into practice, crossing large expanses of ocean in order to spread the teachings of 'Oro. On arriving at an island, members of the *ariori* would perform a dance on the beach, demanding tribute and sometimes even human sacrifice. The *ariori* would then stay for a few months, cementing religious and diplomatic ties between different islands, before heading out to sea once more.[46]

Tupaia spent nearly twenty years in the society of travellers, developing an exceptional understanding of the geography of the Pacific Ocean. However, he did experience one major setback. In 1757, warriors from the island of Borabora invaded Ra'iatea. Hundreds of people were killed, including the high chief. Tupaia, a wealthy man at this point, lost his land and was forced to flee. In the dead of night, cradling the sacred relics of 'Oro, Tupaia set out in a canoe. Alone, he covered well over a hundred miles of open sea, before finding refuge on Tahiti. Once there, he became a favourite of the local queen, a convert to the cult of 'Oro. Tupaia quickly established himself as high priest once more, providing religious and political guidance to the queen and her husband. When the British arrived, Tupaia therefore acted as a kind of diplomat. He accompanied the Tahitian queen aboard HMS *Endeavour*, negotiating with James Cook and Joseph Banks before allowing the British ashore. Tupaia was clearly fascinated by the British ship, described by Tahitians as a 'canoe without an outrigger'. He also shared Cook's interest in the heavens, pointing out the different stars as they travelled around the island together. But most of all, Tupaia hoped the British would help him return to Ra'iatea.[47]

After leaving Tahiti, Cook put his faith in Tupaia. He asked the Polynesian priest to act as chief navigator of HMS *Endeavour*. A decade earlier, Tupaia had been forced to flee his homeland in a tiny canoe. Now, he was back at sea, putting his knowledge of the stars and swells to good use. As the waves crashed against the side of the ship, Tupaia could be found calling out a prayer from the stern, 'O Tane, ara mai matai, ara mai matai!' ('O Tane, bring me a fair wind!') He followed the stars by night, paying careful attention to the rise and fall of the ocean along the way. Cook himself started to understand some of the subtleties of Tupaia's approach. He was impressed, writing, 'these people sail those seas from Island to Island for several hundred Leagues, the Sun serving for them as a compass by day, and the Moon by night'. Another member of the *Endeavour* crew described how 'their whole art of navigation depends upon their minutely observing the motions of the heavenly bodies'. He continued, noting, 'it is astonishing with what exactness their navigators can describe the motions and changes of those luminaries'. Tupaia was 'a man of real genius' according to the Englishman.[48]

Over the following weeks, Tupaia guided HMS *Endeavour* to

Ra'iatea, 150 miles northwest of Tahiti. Things had calmed down on the island, and Tupaia was able to visit the sacred temple at Taputapuatea. This was the place where he had learned to navigate as a boy. Entering the coral temple, Tupaia offered a prayer to the gods, blessing Cook and the crew in preparation for the next phase of the journey. Tupaia then agreed to help Cook chart the islands of the Pacific, before heading south in search of the great unknown continent. On 9 August 1769, Tupaia left his homeland for the last time. 'We again Launched out into the Ocean in search of what chance and Tupia [sic] might direct us to,' Cook recorded in his diary. With the help of Tupaia, HMS *Endeavour* sailed a further 500 miles south, reaching the Austral Islands. It was here that Tupaia produced one of the most incredible artefacts in the history of science, a true example of cultural exchange in action.[49]

Sitting at the chart table of HMS *Endeavour*, Tupaia began to draw a map. As we've seen, Polynesian navigators tended not to use maps or charts; they simply memorized star paths. So the idea of drawing a map, divided by a grid representing lines of longitude and latitude was largely alien to Tupaia. Nevertheless, at Cook's behest, he gave it a go. Tupaia sketched a total of seventy-four islands, representing an area equivalent

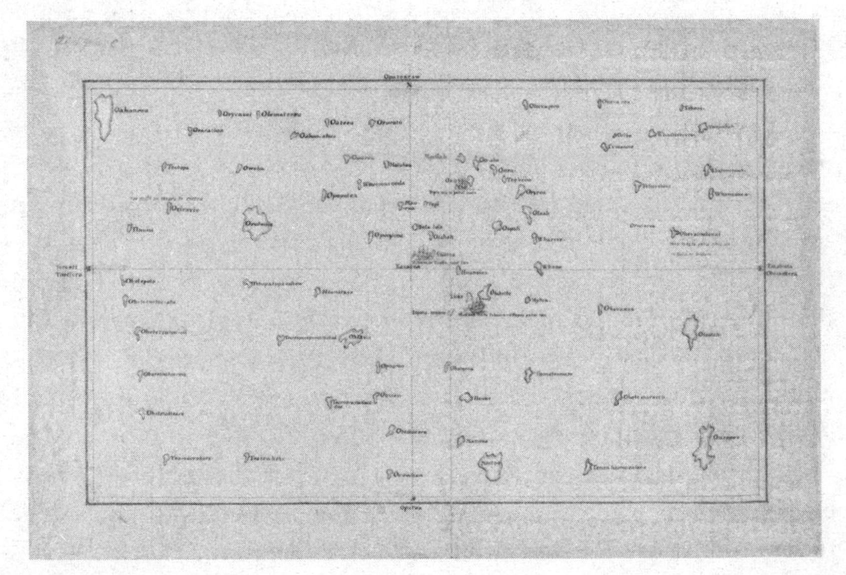

18. Tupaia's chart of the Society Islands, copied on paper in ink by Captain James Cook in 1769.

to the size of the entire continental United States today. This was an enormous expanse of ocean, a testament to the depth of Polynesian navigational knowledge.[50]

At first glance, the map doesn't look too different from a typical European sea chart. However, look a little closer, and we can see how Tupaia subtly adapted the format to suit his own needs. There are a few Polynesian words written on the map. Right in the centre, where the grid-lines cross, there is the word *eavatea*, which translates as 'noon' in Tahitian. Tupaia also replaced the directions of the compass – north, south, east, and west – with the Tahitian words for sunset and sunrise, as well as the Tahitian words for northerly and southerly winds. The chart now started to resemble a Polynesian understanding of navigation. This is how it worked. First, you would identify which island you were on, say, Tahiti. You would then draw a straight line to *eavatea* in the centre, representing 'noon'. Then, you would draw another straight line from your starting point to the island you wanted to travel to, say, Ra'iatea. The angle between these two lines would give you your bearing. Ingeniously, this bearing was already corrected for winds and currents. All you had to do, was follow that same angle relative to the shadow cast by the Sun on the ship's mast at noon. What Tupaia had created was utterly unique. He had combined European and Polynesian navigational techniques to create, not only a chart of the Pacific, but one that effectively acted as a calculating device.[51]

There was another subtle difference between Tupaia's map and a typical European chart. When historians first started to study the map, they were somewhat disappointed. Although the islands Tupaia identified certainly all existed, the relative distances between them seemed way off. But reading the map like this completely misses the point. Tupaia wasn't trying to represent the islands in absolute space, with fixed distances between them. Instead, the distances between the islands actually represent time, not space. This makes a lot of sense. For practical purposes, it makes no difference whether two islands are 100 or 300 miles apart. What matters is how many days it might take to sail between them. As anyone who has ever taken a long-distance flight knows, it takes longer to go in one direction than in the other. The same is true at sea. The winds and currents mean that sailing times are always relative to the direction you're travelling in. Once again, a Polynesian approach to navigation was perfectly adapted to the Pacific, with distances indicating sailing times rather than the number of miles. Understood like this, historians have recently shown that Tupaia's map provides an incredibly accurate guide to many of the major island groups of the Pacific Ocean.[52]

Map in hand, Captain James Cook headed further south. In October 1769, HMS *Endeavour* reached New Zealand. After a few months charting the coast, Cook finally found what he had been searching for. HMS *Endeavour* crossed the Tasman Sea, landing at Botany Bay in Australia on 29 April 1770. This was the great Southern Continent. Having completed his mission, Cook spent a few months exploring the Australian coast, before beginning the return voyage to Britain. Sadly, Tupaia – the great Polynesian navigator – passed away en route. He died, most likely from malarial fever, whilst HMS *Endeavour* docked at Batavia.[53]

Nonetheless, Tupaia's impressive knowledge of the Pacific Ocean lived on. His map was used on a number of subsequent voyages. Cook took a copy of the chart with him on his second voyage to the Pacific. Between 1772 and 1775, Cook followed Tupaia's map, visiting many of the islands he had missed during the first voyage, and claiming them on behalf of the British. An engraving of the map was then printed on Cook's return to London. Readers were invited to look upon 'A Chart representing the Isles of the South-Sea according to the Notions of the Inhabitants of O-Taheitee . . . chiefly collected from the accounts of Tupaya [sic]'. The printing of Tupaia's map in London signalled a significant transformation in the history of science. At the beginning of the eighteenth century, the Pacific Ocean was largely unknown to Europeans. However, by the end of the century, anyone in London with a bit of money to spare could purchase a copy of a map drawn by a Polynesian master navigator.[54]

In Tupaia's map, we can ultimately see both sides of eighteenth-century science. On the one hand, European explorers increasingly depended upon Indigenous knowledge, particularly when it came to astronomy and navigation. At the same time, it was that very knowledge which allowed European empires to expand, and ultimately conquer, these previously unknown regions. Empire and Enlightenment always seemed to go together. And now, moving from the tropical heat of the South Pacific to the frozen landscape of the Russian Arctic, we explore another side to this story.

IV. Newton in Russia

For much of the seventeenth century, Russia seemed stuck in the past. Even the most well-educated Russians still believed that the Earth was at the centre of the universe. There were no scientific academies or universities, and academic learning consisted of little more than ancient Greek philosophy mixed with the theology of the Russian Orthodox Church. When Peter the Great came to power in 1682, he was determined to change all this. Within the space of a few decades, Peter transformed Russia into a hub of Enlightenment science.[55]

For Peter, nothing represented progress quite like Isaac Newton and his *Principia*. It is almost certain that the Russian Tsar actually met Newton himself. In January 1698, Peter the Great arrived in London. He was travelling as part of a diplomatic mission, hoping to gather support amongst European powers for war against the Ottoman Empire. At the same time, Peter took the opportunity to learn about all the new science that was being done in countries across Europe. In London, Peter visited the Royal Observatory as well as the Royal Society, where he witnessed 'all sorts of amazing things' – air pumps, microscopes, and glass prisms bending light. Most significantly, Peter visited the Royal Mint at exactly the same time as Newton was working there. In February 1698, Newton received a letter informing him that 'the Czar intends to be here tomorrow . . . he likewise expects to see you'. Whilst neither Newton nor Peter personally made a record of this meeting, we do know that Newton held the Tsar in high regard, sending copies of his later publications to the Russian court. Peter the Great even acquired a copy of Newton's *Principia* for his personal library.[56]

Peter returned to Russia in 1698 with a renewed enthusiasm for Newtonian science. He quickly established a string of institutions designed to modernize scientific research and education in Russia. The first of these was the School of Mathematics and Navigation, founded in Moscow in 1701. Russian engineers and naval officers were now taught mathematical sciences according to Newtonian principles. Peter the

Great also ordered a move away from traditional Russian numerals written in the Cyrillic script. Instead, Russian students would be required to use the Arabic numerals employed by European mathematicians. Finally, and most significantly, Peter established the Saint Petersburg Academy of Sciences in 1724. This would function as a Russian equivalent of the Royal Society, a national scientific academy with weekly meetings and regular publications. In Peter's own words, the Academy 'would earn us respect and honour in Europe' and challenge the idea that 'we are barbarians who disregard science'.[57]

At the time, there were still very few Russians with any kind of advanced scientific education. The Saint Petersburg Academy of Sciences was therefore initially staffed almost entirely by foreigners. Peter the Great had been able to convince some of the leading minds in Europe to relocate to Russia. They were attracted by promises of high pay and access to specialist scientific equipment. The Saint Petersburg Academy of Sciences even featured a dedicated astronomical observatory, built atop a three-storey tower on Vasilyevsky Island. Early members of the Academy included the leading Swiss mathematicians Leonhard Euler and Daniel Bernoulli. However, by the 1730s, Russians started to join the Academy too. Amongst these was Mikhail Lomonosov, who discovered the atmosphere of Venus, as well as Stepan Rumovsky, who made observations of the 1769 transit of Venus in the Arctic Circle. In many ways, the Saint Petersburg Academy of Sciences acted as a microcosm of the Enlightenment. British, French, German, Swiss, and Russian thinkers all came together in one place to discuss and debate the latest scientific theories. As with the Enlightenment more generally, opinion was initially split over Newton's theory of universal gravitation. Bernoulli supported Newton's ideas, whilst Euler and Lomonosov were much more sceptical.[58]

Fittingly, the first official correspondence issued by the Saint Petersburg Academy of Sciences was sent to Newton himself. The secretary of the Academy informed Newton that 'we hope that our observations will be useful above all for the development of astronomy'. Newton had already shown an interest in Russian science. As President of the Royal Society, he had helped establish the 'Committee for Russia' in 1713, exchanging information and letters with Russian academics and explorers. In particular, European astronomers like Newton needed more data based on scientific observations made in the far north, around the Arctic Circle. Indeed, whilst Newton's *Principia* did make use of information arriving from across the globe, the majority of his data was from regions around the equator, such as the West Indies, West Africa, and Southeast Asia. What Newton and his followers really needed was an equally accurate set of results from the far north. This would allow them, as we saw earlier, to compare results from the northern and southern hemispheres, establishing the size of the solar system and the true shape of the Earth.[59]

Over the course of the eighteenth century, Russian astronomers and explorers contributed to a number of international scientific endeavours. At the same time, Russia began to transform into a significant imperial power. For much of the sixteenth and seventeenth centuries, the lands east of the Ural Mountains had only loosely been under Russian control. Small groups of Cossacks occupied forts across Siberia, whilst traders headed farther east in search of furs to sell back in Europe. In the early seventeenth century, Russian explorers did reach the Pacific coast, establishing a small fort at Okhotsk. However, this was later attacked and burned down by the local Indigenous population. Even by the early eighteenth century, there were still no accurate maps of the Russian Far East. From the perspective of earlier Tsars, this region was simply an unknown and untamed wilderness. Peter the Great, however, had other ideas. He intended to transform Russia, not only into a modern scientific nation, but also a powerful and confident empire, one that stretched from Europe in the west to America in the east.[60]

The Saint Petersburg Academy of Sciences played an important role in supporting the territorial expansion of the Russian Empire. During the eighteenth century, the Academy helped to organize a number of scientific expeditions into Siberia and the Pacific Northwest. The most famous of these expeditions was led by Vitus Bering. Peter the Great personally appointed Bering, a Danish navigator, to take charge of what became known as the First Kamchatka Expedition of 1724–32. Bering's mission was to explore the land and sea to the north of the Kamchatka Peninsula in the Russian Far East. He was then to 'search for a place where that land might be joined to America'. Finally, Bering was asked to produce an accurate map of everything he discovered.[61]

As we saw in chapter 1, the question of whether Asia and America might be connected had puzzled European geographers since the 'discovery' of the New World in the fifteenth century. There had been unconfirmed reports of a Cossack navigator, named Semen Dezhnev, who had managed to sail from northern Siberia down into the Pacific Ocean. But most people were still unsure about the existence of any such strait. Settling this question once and for all would certainly enhance the scientific standing of Russia in the eyes of Europe. Peter the Great also understood the strategic importance of such an expedition. Accurate knowledge of the geography of Siberia and the Pacific Northwest would allow Russia to control the lucrative fur trade, as well as forge wider connections across the Pacific Ocean, particularly with Spanish America and Japan. Most importantly, Peter the Great hoped that Bering might secure territory for Russia on the American continent itself.[62]

Bering left Saint Petersburg in February 1725, travelling 6,000 miles across land and snow. It took over three years just to reach the Kamchatka Peninsula. Once there, Bering launched into the Pacific Ocean aboard the *Archangel Gabriel*. Sailing north, he was finally able to confirm that Asia and America are not connected. A narrow strip of sea, just over fifty miles wide, separates the two continents. Today, this sea is known as the Bering Strait. However, Bering was unable to sight the American continent itself. He returned to Saint Petersburg in 1732, determined to secure support for a second, more ambitious, venture.[63]

By this time, Peter the Great had passed away. Nonetheless, the rulers who followed were equally intent on expanding the Russian Empire to the east. Peter the Great's successor, Anna of Russia, certainly felt this way. The new empress ordered Bering to return to Kamchatka with an even larger party composed of over 3,000 personnel. The Saint Petersburg Academy of Sciences also supplied precise instructions on how to undertake a survey of the area. Bering was told to make astronomical observations every twenty-four hours, calculating his latitude and longitude at sea and plotting this on a chart. Additionally, he was advised on how to undertake the equivalent of a land survey at sea, hopping between different islands and measuring the angles between them with a quadrant. Finally, the Academy sent a number of its leading members to join the expedition and assist with the survey. Amongst these was the French astronomer Louis de l'Isle de la Croyère, an expert in Newtonian physics who had previously undertaken experiments on the force of gravity in northern Russia.[64]

The Second Kamchatka Expedition left Saint Petersburg in April 1733. Bering himself never returned, dying – probably from scurvy – on a small island off the coast of Kamchatka in December 1741. Despite his death, Bering was nonetheless successful in his mission. On 16 July 1741, just a few months before he died, Bering sighted the American coast. On the horizon, he spotted an immense mountain range, known today as the Saint Elias Mountains. A few days later, Bering and his party landed on a nearby island, becoming the first Europeans to reach Alaska. After taking a series of astronomical observations, Bering's Russian navigator, with the aid of Louis de l'Isle, was able to pinpoint their exact position on a map.[65]

The success of Vitus Bering's expedition sparked a wave of new voyages sponsored by the Russian state. Over the course of the eighteenth century, there were a total of five major expeditions, reaching as far north as the Arctic Circle and as far south as the islands around Japan. The most significant of these expeditions was commissioned by Catherine the Great in 1785. At the time, Catherine was concerned about the growing British presence in the Pacific Northwest. During his third voyage of discovery, Captain James Cook himself had reached the Bering Strait, landing on an island off the coast of Alaska in 1778. The French were also voyaging further north, whilst the Spanish continued to push up the coast from California. Given the competition between different European empires, Catherine realized that she needed to secure the Russian presence around the Bering Strait. She also understood that the best way to achieve this was through a major scientific survey, sending in military personnel and mapping the area as they went. This survey became known as the Great North-Eastern Geographical and Astronomical Expedition. It was led by Joseph Billings, an English navigator who had actually already travelled to Alaska as an assistant astronomer on Cook's third voyage. He was joined by a Russian naval officer named Gavril Sarychev, who undertook much of the survey work.[66]

As in Polynesia, European explorers in the Arctic combined Newtonian science with Indigenous knowledge. The Saint Petersburg Academy of Sciences made this explicit in the official instructions for the Great

North-Eastern Geographical and Astronomical Expedition. By the 1780s, Newton's theories were widely accepted in Russia. Billings and Sarychev were therefore asked to 'determine the degrees of longitude and latitude' by astronomical observation. The hope was that this would allow for a more precise measurement of the width of the Bering Strait. At the same time, however, the explorers were also instructed to ask Indigenous people about the local geography. The Academy even produced a list of questions, including, 'what are the names of the places they are in the habit of visiting, and at what bearings and distances these lands or islands lie in respect?' The instructions went on, explaining that 'when they use their hands to indicate, you should measure them in a secretive and precise way by means of a compass'.[67]

Billings went a step further in this respect. He wasn't content just to ask Indigenous people about the geography of the region. Instead, he actually recruited an Indigenous man to join the expedition. Born around 1730, Nikolai Daurkin was a member of an Indigenous group called the Chukchi. These people had been living around the coast in the far northeast of Siberia for thousands of years. They possessed an exceptional knowledge of the geography of the region, and were of course aware of the Bering Strait well before Bering himself.[68]

Chukchi navigational techniques shared much in common with those of Polynesia. Like most Indigenous people in the Arctic, the Chukchi watched the stars, memorizing sequences in order to use them as bearings between particular islands. However, there were also a number of subtle differences between Polynesian and Arctic navigation. For a start, the seasons are much more extreme in the far north. In the summer months, the Sun does not set, whilst for several weeks in the winter, it does not rise. Even more confusingly, at least for European navigators, the position of sunrise and sunset in the Arctic changes radically over the course of the year. In March, the Sun rises in the east and sets in the west, as you might expect. But in May, the Sun actually rises in the north, and sets in the south. This makes navigation based on the position of the Sun exceptionally difficult.[69]

The Indigenous people of the Arctic developed a range of techniques to solve these problems. To begin with, Chukchi navigators like Daurkin would spend a lot of time using the Sun and stars to ascertain the precise time of the year. Particular stars move across the sky during the run-up to particular seasons. For example, Aquila, known as *Peggitlyn* by the Chukchi, appears in the sky just before dawn during the winter months. Similarly, Orion moves to the south as the days get longer. Knowing the precise time of year then allowed Chukchi navigators to make proper use of the Sun, despite its variable position. For example, if you know it is the middle of May, then the sunrise actually provides a good northward bearing. But if you don't know the time of year, you might be following the Sun north when you think you're travelling east.[70]

Alongside the stars, Indigenous people in the Arctic paid close attention to the water, snow, and ice. Much like Tupaia in Polynesia, Daurkin would have read the ocean swells for signs of nearby land. The Chukchi also watched the flow of seaweed and ice in order to get a good sense of the current. Finally, and most inventively, Indigenous people in the Arctic studied patterns in the snow. Men like Daurkin needed to be able to navigate even in a blizzard. Visibility could be reduced to a couple of feet, if that. In such cases, knowledge of the stars was useless. Instead, whilst on land, the Chukchi identified bearings based on the snow at their feet. In the Arctic, wind erosion forms long ridges of snow known as *sastrugi*. These run north to south, in the same direction as the 'Northern Master' wind that blows throughout Siberia. By feeling these

19. Snow ridges known as *sastrugi* are formed by wind erosion in the Arctic. Indigenous people, such as the Chukchi, use the *sastrugi* to find their bearings in low visibility.

snow ridges, the Chukchi could be sure of the direction of north, even in zero visibility.[71]

Daurkin was unusual in that he crossed between Chukchi and Russian culture. As a young boy, he had been captured by a Russian explorer and sent to Yakutsk, a Siberian port thousands of miles from his homeland. Once there, he was baptized and learned to read and write Russian. Daurkin then trained at the Irkutsk Navigational School in Siberia, one of the new scientific institutions established following the reforms of Peter the Great. After his studies, Daurkin spent the early 1760s paddling around the Bering Strait in a small canoe, interviewing local Chukchi and surveying the region. In doing so, Daurkin combined his training in navigational science with Indigenous knowledge drawn from the Chukchi. The result was a map of the area around the Bering Strait, the first to include any detail of the northern coast of Alaska. (It is worth noting that Daurkin's map was completed in 1765, a decade before Cook, who is usually credited with being the first to map the region, had even arrived in Alaska.)[72]

Billings learned about Daurkin's map whilst preparing for the Great North-Eastern Geographical and Astronomical Expedition in Saint Petersburg. He was impressed, and instantly recognized how useful it would be to have an Indigenous navigator as part of the crew. Daurkin, who was still working at the Irkutsk Navigational School, agreed to join the expedition. In May 1790, Billings, Sarychev, and Daurkin launched into the Pacific Ocean aboard the *Glory of Russia*. The crew exemplified the world of eighteenth-century science. The captain, Billings, was English. The surveyor, Sarychev, was Russian. And the navigator, Daurkin, was Chukchi. Together, these men spent the next three years charting the islands of the Bering Strait. In total, the Great North-Eastern Geographical and Astronomical Expedition produced over fifty new maps, stretching from Siberia in the west to Alaska in the east. The message was clear. America was now part of the Russian Empire.[73]

V. Conclusion

The publication of Isaac Newton's *Principia* in 1687 is usually understood as marking the beginning of the Enlightenment. In this narrative,

Newton is typically portrayed as an isolated genius, applying the principles of reason. This, however, is inaccurate, as becomes obvious on reading the *Principia* itself. In this chapter, I've argued that Newton represents the beginning of the Enlightenment, not because he was isolated, but because he was so well connected. Newton was able to make a major scientific breakthrough only by virtue of his connections to the wider world of empire, slavery, and war. In developing his theory of universal gravitation, Newton relied on data collected from French astronomers travelling aboard slave ships, as well as East India Company officers trading in China. This is something that people at the time were well aware of, even if it is often forgotten today. Voltaire, possibly the most famous of the French Enlightenment philosophers, wrote that 'without the voyage and experiments of those sent by Louis XIV . . . never would Newton have made his discoveries concerning attraction'.[74]

Taking Newton as a starting point, this chapter presented a new history of Enlightenment science. Over the course of the eighteenth century, scientific academies in Europe organized a series of state-sponsored voyages of exploration. These voyages provided Newton and his followers with the data they needed to answer some of the most fundamental questions in the physical sciences. The French expedition to the Andes proved that Newton was right about the shape of the Earth, whilst Captain James Cook's voyage to the Pacific finally established the absolute size of the solar system. Alongside these more theoretical questions, the eighteenth century saw the development of a number of related practical sciences, such as navigation and surveying. Deploying the latest Newtonian science, the British, French, and Russian empires expanded into new territory. Cook sailed south from Tahiti, reaching as far as Australia, whilst Vitus Bering was able to map the coast of Alaska, incorporating part of the American continent into the Russian Empire for the first time.

This, however, is not simply the story of the triumph of European science. As they crossed strange seas and climbed spectacular mountains, European explorers constantly relied upon the existing knowledge of Indigenous people, many of whom possessed advanced scientific cultures of their own. In Peru, French surveyors unwittingly depended upon Inca astronomical traditions. In the Pacific, Captain Cook relied on the navigational expertise of a Polynesian priest. And in the Arctic,

Russian explorers recruited Indigenous people to guide them across the frozen landscape. Recognizing the contributions of these individuals helps to paint a very different picture of eighteenth-century science. Ultimately, the development of Enlightenment science needs to be understood as part of a global history, one that incorporates the history of slavery and empire, but also the history of Indigenous knowledge. Newton might have been a genius. But he wasn't alone.[75]

We started this chapter with Newton's investment in the slave trade, but there is another side to this story, one that is also often forgotten today. In 1745, a man named Francis Williams posed for a portrait in his study in Jamaica. In many respects, the portrait looks like that of a typical eighteenth-century scholar. There is a copy of *Newton's Philosophy* on the table in front of Williams, alongside a compass and globe. However, in one very important respect, this portrait is remarkable, particularly in light of the traditional narrative in the history of science, in which people of African descent are often misleadingly excluded. Williams, after all, was Black. Just before Williams was born, his father, an enslaved African, had been granted his freedom. Williams was therefore a free man as well. It seems that he was quite wealthy, later inheriting both land and slaves in Jamaica. By around 1720, Williams was sufficiently well off to be able to travel to Britain, where he enrolled at the University of Cambridge, studying mathematics and classics. It was here, just around the time of Newton's death, that Williams learned about the *Principia*. He returned to Jamaica a few years later in order to set up a school, bringing with him many of the latest scientific books, including some written by Newton. To be clear, Williams was far from typical. Most of the Black population of the Caribbean at that time had no opportunity to learn about Newtonian science. But he is nonetheless an important reminder of another side to the history of science in the age of slavery. In the next chapter, we follow this theme in more detail, exploring how, even in the most desperate of circumstances, and despite their later erasure from the story, enslaved Africans and their descendants continued to contribute to the making of modern science.[76]

4. Economy of Nature

Foraging at the edge of the plantation, Graman Kwasi came across a plant he had never seen before. The bright pink flowers caught his eye. Cutting a sample of the small shrub, Kwasi took it back to his hut and stowed it away. He didn't know it at the time, but this plant would ultimately change the course of his life. Graman Kwasi was born around 1690 in West Africa, a member of an Akan-speaking tribe in what is part of modern-day Ghana. Aged just ten, he was captured during a raid by African slave traders from a rival tribe. Kwasi was then marched in chains to the coast. Once there, he was purchased by a Dutch captain, and shipped across the Atlantic Ocean – just one of the six million enslaved Africans transported to the Americas during the eighteenth century. On arriving in South America, Kwasi was put to work on a sugar plantation in the Dutch colony of Surinam. As a child, he was forced to spend all day in the baking heat, picking weeds from the ground. As he grew into a young man, Kwasi then took part in the backbreaking work of the harvest, cutting sugar cane by hand using a machete.[1]

Kwasi, however, was much more talented than his Dutch masters initially gave him credit for. Amidst the diverse flora and fauna of South America, he began to develop an intimate knowledge of the natural world. Fusing healing traditions from both Africa and the Americas, Kwasi collected plants and prepared medicines. He treated both Africans and Europeans on the plantation, earning small sums of money. However, one plant in particular brought Kwasi great fame. The small shrub with pink flowers that he collected on the plantation in Surinam turned out to have incredible healing properties. The bark, when boiled in water to make a bitter tea, acted as an effective treatment against malarial fever. It also seemed to strengthen the stomach and restore appetite. Kwasi most likely learned about the medicinal properties of the plant from an Amerindian slave on the same plantation, as the shrub was used in existing South American herbal traditions – exactly the kind of medical knowledge we learned about in chapter 1. Before long, word of Kwasi's discovery spread across Surinam, and then on to Europe. At the time, the only effective treatment for malaria was derived from the bark of the cinchona tree, known as 'Peruvian bark'. However, the Spanish had a monopoly on this valuable product, which could only be found in the Viceroyalty of Peru, hence the name by which it was known. In fact, at the beginning of the eighteenth century, cinchona bark was the most expensive commodity in the world. It was literally worth more than its weight in gold. An alternative treatment for malaria was therefore an extremely lucrative prospect.[2]

In 1761, a sample of the shrub discovered by Kwasi reached Carl Linnaeus, one of the most influential scientific thinkers in Europe at the time. Linnaeus, who was Professor of Medicine and Botany at the University of Uppsala in Sweden, had transformed the study of the natural world through his new taxonomic system. This was first set out in his *System of Nature* (1735). In this book, Linnaeus divided the natural world into three major kingdoms: the animal, mineral, and vegetable. Below this, there were four more levels of classification, each one more precisely identifying a particular animal or plant. These went from the class, through the order, genus, and finally, species. In this system, everything in the natural world had its place. Following this, Linnaeus proposed that animals and plants each be given an official 'binomial' or 'two-part' name consisting of the genus and species. For example, the scientific name for the lion is *Panthera leo*, indicating the lion is a member of the genus *Panthera* (which includes tigers, leopards, and jaguars) and the species *leo* (which includes the different sub-species of lion in Africa and Asia). The advantage of this system was that it provided a straightforward and uniform way to classify the natural world. It also allowed naturalists to express the similarities between different species of animals or plants, as in the example of the lion being part of the same genus as the tiger. Linnaeus's binomial system still forms the basis of all modern biological classification systems today.[3]

Linnaeus was sent a sample of the plant by a Swedish plantation owner in Surinam. On confirming the medicinal properties of the plant, Linnaeus was impressed. He duly recorded the discovery in the new edition of his *System of Nature*, not only as a previously unknown species, but an entirely new genus. In honour of Kwasi, Linnaeus named the plant *Quassia amara*. ('Quassi' was the Latinized version of Kwasi's Akan name.

'Amara' means bitter in Latin, referring to the taste of the medicine.) With the backing of Linnaeus, and having discovered this revolutionary treatment, Kwasi found his life transformed. As knowledge of the plant spread, *Quassia amara* became a major export crop for planters in Surinam, cultivated for sale as an alternative to the more expensive cinchona bark. Kwasi was soon granted his freedom. He was then invited to Holland to meet William V, Prince of Orange, who, in recognition of Kwasi's achievements, presented him with an ornate coat and a gold medal. On returning to Surinam, Kwasi was given his own small plantation, complete with enslaved people to work the land. He also began receiving letters from European naturalists, eager to learn more about the plants of South America. Some of these letters were even addressed to Kwasi as 'Professor of Herbology in Surinam'. Somehow, against all the odds, Graman Kwasi had escaped slavery and found himself as a respected authority on the medicinal properties of South American plants.[4]

Graman Kwasi's story is exceptional in many ways. During the eighteenth century, it was extremely rare for an enslaved African to be publicly recognized in Europe as a source of scientific knowledge. When it came to natural history, plants were typically named after the European men credited with discovering them. For the most part, Europeans saw Africans as little more than commodities, to be bought and sold in order to work on plantations. Kwasi was unusual in that, through his knowledge of the healing properties of plants, he managed to escape this world, or at least end up on the other side of it. However, in another sense, Graman Kwasi is an example of something much more widespread.

The traditional story of natural history in the Enlightenment focuses almost exclusively on the achievements of European men like Carl Linnaeus, who are celebrated for 'discovering' new plants and inventing new classificatory systems. But this is misleading. Often ignored in the history of science, a diverse range of people from across Africa, Asia, and the Americas contributed to the development of eighteenth-century natural history. They brought with them their own scientific traditions, upon which Europeans often relied in order to understand and classify foreign environments. In some cases, this was straightforward appropriation, as with much of the botanical information extracted under the threat of violence from enslaved Africans. But in other cases, this scientific relationship was more collaborative, as we'll see in the case of Tokugawa Japan. From African healers to Indian priests, this chapter uncovers the forgotten contributions of people like Graman Kwasi to the development of natural history during the Enlightenment.

Whilst the previous chapter focused on state-sponsored voyages of exploration, this one uncovers the role of global trade in the development of Enlightenment science. Over the course of the seventeenth and eighteenth centuries, the world was transformed by the expansion of European trading companies: the Dutch East India Company in Southeast Asia and Japan, the Royal African Company in the Atlantic, and, most famously, the British East India Company in India and China. These lucrative enterprises made immense profits by controlling the supply of goods: sugar, spices, tea, and indigo all reached Europe aboard trading company ships. Crucially, much of this trade was in products derived from the natural world. This provided the impetus for a more detailed study of natural history, as trading companies needed to be able to classify and assess the goods they were dealing with.

To give a sense of the scale of change: at the beginning of the seventeenth century, European naturalists had identified around 6,000 different species of plant. By the end of the eighteenth century, they had identified over 50,000 species, the majority of which originated outside of Europe. As we saw in the previous chapter, trading companies like the Royal African Company and the British East India Company maintained close links to the major scientific institutions of the day, such as the Royal Society in London. Knowing the difference between gold and platinum, or cinnamon and nutmeg, was a key commercial concern, not just a scientific one. In some cases, trading companies even commissioned chemical tests, using the latest laboratory techniques in order to ascertain the purity of a metal or a dye.[5]

Natural history during the Enlightenment was therefore an economic science as much as a biological one. Linnaeus himself certainly saw his work this way. Like many others, he worried that global trade was weakening European economies, making them dependent on others for goods. In particular, Linnaeus feared that the 'balance of trade' was not in Europe's favour – countries like his native Sweden imported far more than they exported. In response to this, Linnaeus suggested that Sweden start to cultivate alternative crops, or even try and grow locally the products it was importing. 'Nature has arranged itself in such a way, that each country produces something especially useful; the task of economies is to collect from other places and cultivate such things that don't want to grow,' argued Linnaeus. For him, this was the point of natural history: not simply to catalogue the world, but to find a way to tip the balance of trade in favour of Europe. Linnaeus even suggested it might be possible to grow mulberry trees in Sweden, reducing the reliance on Chinese imports of silk.[6]

Unsurprisingly, Linnaeus found it difficult to cultivate tropical plants in Sweden, with its bitterly cold winters. But countries with larger empires were much more successful. Over the course of the eighteenth century, European naturalists helped to establish hundreds of botanical gardens across the colonial world. These were created with the explicit goal of growing tropical plants in order to reduce reliance on imports. For example, in 1735 the French East India Company established a botanical garden on Isle de France, modern-day Mauritius. The French naturalists there were tasked with growing pepper, cinnamon, and nutmeg with the hope of breaking the Dutch monopoly on the spice trade. (At the time, the only place anyone in Europe could get these spices was from the territories in Southeast Asia controlled by the Dutch East India Company.) The French East India Company even employed a missionary – named Pierre Poivre, no less – to smuggle seeds and saplings out of Southeast Asia to be grown in the new garden. The British did a similar thing in India, establishing a botanical garden in Calcutta in 1786 with the hope of growing cinnamon in order to break the Dutch monopoly. By the end of the eighteenth century, most European colonies – including Jamaica, New South Wales, and the Cape Colony – had a botanical garden. These, connected to the major botanical gardens in Europe, such as Kew Gardens in London, acted as important sources of information on the natural history of the world.[7]

I. Slavery and Botany

Arriving in Jamaica in 1687, Hans Sloane headed for the mountains. Riding on horseback, and accompanied by an enslaved African guide,

Sloane began collecting as many plants as he could: ferns, orchids, and grasses filled his bag. Sloane needed to be careful. The mountains were a dangerous place for Europeans to travel, as runaway slaves and pirates might attack. But the risk was worth it. Over the following year, Sloane managed to collect over 800 plant specimens, each of which was carefully dried and stuck into a bound volume. Officially, Sloane had come to Jamaica to act as personal physician to the new governor of the island, the Duke of Albemarle. But Sloane wasn't particularly interested in the governor's health. (In fact, the governor died less than a year after Sloane's arrival.) What Sloane really wanted to do was study the natural history of the island. On his return to London in 1689, Sloane began to write up an account of what he had discovered. This was published in two large illustrated volumes entitled *The Natural History of Jamaica* (1707–25).[8]

Sloane went on to become one of the most influential naturalists of the early eighteenth century. Following the publication of his book, Sloane was elected as both President of the Royal Society and President of the Royal College of Physicians. Carl Linnaeus also consulted Sloane, visiting him in London, and incorporating some of the information in *The Natural History of Jamaica* into his *System of Nature*. When Sloane died in 1753, his entire collection – by that point consisting of over 70,000 specimens of plants, animals, minerals, and antiquities – was purchased by Parliament, forming the basis of the British Museum and later the Natural History Museum in London. Sloane was successful in large part because he understood the relationship between natural history and the economy. On the opening page of his book on Jamaica, Sloane reminded his readers that the island was 'The Largest and Most Considerable of Her Majesty's Plantations in the Americas'. *The Natural History of Jamaica* described all kinds of valuable crops, just at the time when the British, through the expansion of slavery, were transforming the West Indies into a full-blown plantation economy. Sloane himself profited from this world. Through marriage he had access to one third of the profits of a large sugar plantation on Jamaica. He also invested in a number of financial schemes in the Americas, including one to sell 'Jamaican bark', another possible alternative to cinchona.[9]

Sloane's success also depended on the enslaved Africans he met in the West Indies. However, this was often only partially recognized at the time. Like many of the European naturalists we will encounter in this chapter, Sloane's approach towards African knowledge, as well as his use of language, reflected the typical racist attitudes of the period. In *The Natural History of Jamaica*, Sloane described how he had asked for botanical information from 'the Inhabitants, either Europeans, Indians, or Blacks'. One plant in particular caught Sloane's attention. 'It is called Bichy by the *Coromantin* Negroe's [sic] and is both eaten and used for Physick in Pains of the Belly,' explained Sloane. The kola nut, or 'Bichy' as it was known in Jamaica, acted as a stimulant. It also seemed to make stale water taste fresh, and calmed the stomach. Later, in the nineteenth century, the kola nut formed one of the original ingredients of the soft drink Coca-Cola. Despite appearing in Sloane's *Natural History*, this nut was not in fact native to Jamaica. Instead, it was originally from West Africa. Sloane himself realized this, noting that the kola nut was grown from 'seed brought in a Guinea ship'. In West Africa, these nuts had long been used medicinally, as well as exchanged between neighbours or guests at ceremonies, as a symbol of goodwill. 'Who brings kola nut, brings life,' went a typical saying amongst the Igbo people of West Africa. It is a grim irony, then, that the kola nut, a traditional token of friendship, found its way to Jamaica. Enslaved Africans would chew the nut, trying to keep going in the unbearable conditions.[10]

Sloane soon noticed that many other plants in Jamaica were in fact native to Africa. Typically, Sloane came across these on the 'provisioning grounds' assigned to enslaved people. Rather than providing proper food supplies, European plantation managers would simply allocate a small plot of unproductive land on which the enslaved were expected to grow their own food. In Jamaica, Sloane spent a good amount of time investigating these 'Negro Plantations', as he called them. He would interview the Africans working on the plots, finding out about the different foods they had brought from their native countries. In Jamaica, then, Sloane was learning as much about African botany as he was about the West Indies. In the provisioning grounds he was shown yams, millet, and black-eyed peas, all crops that had been brought across the Atlantic aboard slave ships. For the enslaved Africans in Jamaica, these vegetables provided a taste of home, even in the most desperate of circumstances.[11]

*

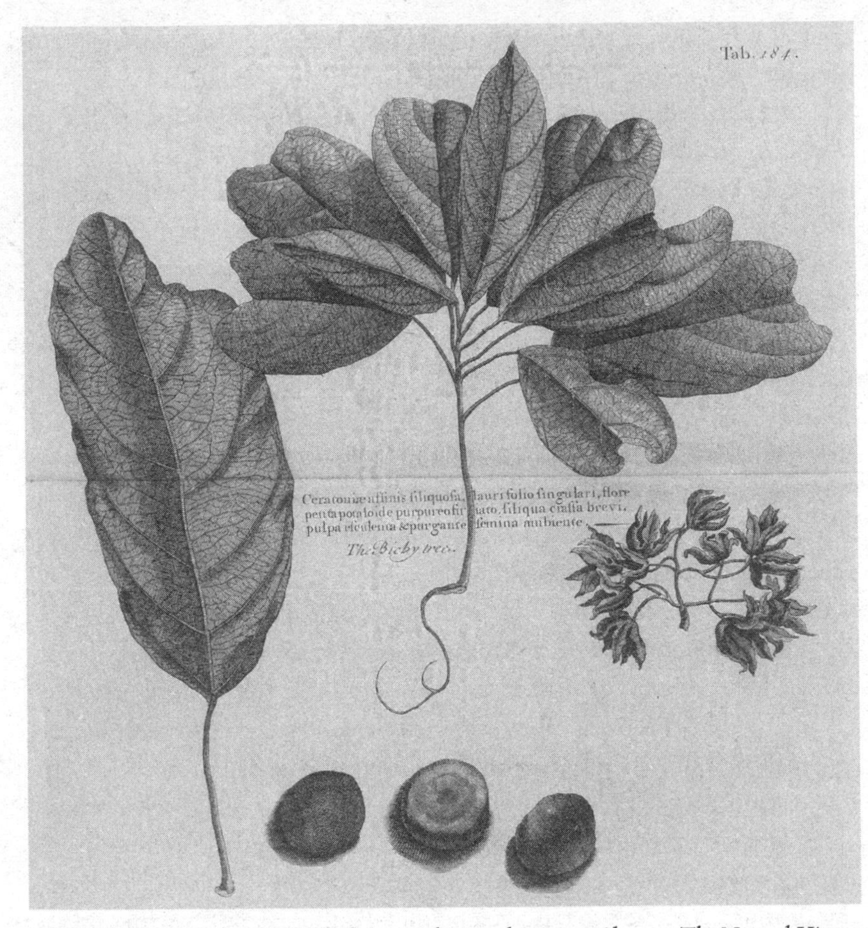

20. The kola nut from the 'Bichy' tree as depicted in Hans Sloane, *The Natural History of Jamaica* (1707–25).

Across the Americas, European naturalists quizzed enslaved people, hoping to discover new plants that might turn a profit, particularly if they had medicinal qualities. It is important to remember the violent power dynamics at play here. European slaveowners treated both African people and African knowledge as property to be exploited. In 1773, the Scottish planter Alexander J. Alexander described conducting a number of experiments on what he called a 'Negro Dr's Materia Medica', referring to the medicinal plants used by enslaved Africans. Alexander, who had studied chemistry at the University of Edinburgh, had learned of a bark used by the enslaved people on his plantation in

Grenada. It was said to act as an effective treatment against the yaws, a painful and widespread infection of the skin. 'The Negroes Method [sic] is making them stand in a Cask where there is a little fire in a pot & sweating them powerfully in it twice a day giving them decoctions of 2 woods in this country called Bois Royale & Bois fer,' explained Alexander. In a letter to Joseph Black, Professor of Chemistry at the University of Edinburgh, Alexander reported the 'astonishing' results. All those treated with the medicine were cured within a fortnight. Alexander sent a sample to Black, suggesting he conduct some chemical tests on the composition of the bark.[12]

The physician Henry Barham, who exchanged letters with Sloane, reported a similar experience in Jamaica. After suffering a severe fever and inflammation of the legs, Barham had almost given up hope. However, one of the enslaved Africans on the plantation suggested using the bark of a tree known as 'hog plum'. Barham recalled that 'a negro going through the house when I was bathing . . . said, "Master, I can cure you". Immediately he brought me bark of this tree, with some of the leaves, and bid me bathe with that.' According to Barham, after bathing in the solution, 'I was perfectly recovered, and had the full strength and use of my legs as well as ever.' Similarly, Patrick Browne, another physician in Jamaica, described the curative properties of 'worm grass'. 'This vegetable has long been in use among the Negroes and Indians, who were first acquainted with its virtues, and it takes its present denomination from its peculiar efficacy in destroying of worms,' reported Browne. Naturalists back in Europe also paid attention to what Africans knew about plants. James Petiver, an influential naturalist in London at the beginning of the eighteenth century, published an account of 'Some Guinea-Plants' collected by a Royal African Company employee in West Africa. Petiver listed the African names and medicinal uses of each of the plants, including 'concon' for killing worms and 'acroe', a tonic for restoring strength.[13]

By the end of the eighteenth century, some European physicians began to admit, tentatively, that Africans might know more than them about certain plants. In Surinam, one Dutch doctor wrote that 'the Negroes and Negresses . . . know the virtues of plants and offer cures that put to shame physicians coming from Europe'. Others were less convinced. Some argued that, whilst Africans clearly did know a great deal about plants, they nonetheless lacked a systematic approach based on classification. Edward Long, a notorious planter in Jamaica, made exactly this argument, claiming that 'brutes are botanists by instinct'. Long, however, was wrong. African botanical knowledge was rarely written down, but it was nonetheless systematic. Igbo healers in West Africa categorized plants by habitat, distinguishing between those growing in the 'forest' and the 'savannah'. This taxonomy of plants then mapped on to the classification of disease, with different illnesses requiring plants from particular environments. Despite what Long suggested, and many subsequent historians have repeated, Africans not only knew about the healing properties of plants, but also integrated this knowledge into a complex classificatory system.[14]

Not all plants were used for healing. In 1705, the German naturalist Maria Sibylla Merian published an account of a plant used to induce abortion in Surinam. As a European woman, Merian was unusual. Very few women in the eighteenth century were able to travel such long distances, as employment in trading companies was reserved for men. Merian, who had earlier divorced her husband, travelled to Surinam in 1699, accompanied by her youngest daughter. She supported herself by selling subscriptions to a book she intended to write on her return, entitled *The Metamorphosis of the Insects of Suriname* (1705). (Many of the most famous naturalists of the age, including Carl Linnaeus and Hans Sloane, later consulted Merian's book.) Over the next two years, Merian and her daughter travelled across Surinam, staying on plantations and collecting plants and insects. In her book, Merian described learning about a plant called the 'peacock flower' from some of the 'slave women' on a plantation. According to Merian, enslaved women in Surinam used the seeds of the peacock flower in order to 'abort their children, so that their children will not become slaves like they are'. She also described how enslaved Africans, both men and women, used the roots of the peacock flower to commit suicide – an act of resistance against the institution of slavery, as well as another reminder of the hopeless condition that had been forced upon them. According to Merian, 'they believe they will be born again, free and living in their own land'.[15]

Reports of dangerous plants frightened European doctors in the Americas. After all, if a flower could be used to induce abortion, or

1. A map of Oaxtepec, Mexico, produced by an Aztec artist and sent to the Council of the Indies as part of the Geographical Reports in 1580.

2. An illustration of the people, plants, and animals of Mexico, produced by an Aztec artist for the *General History of the Things of New Spain* (1578).

3. The Istanbul observatory, originally built in 1577. The chief astronomer, Taqi al-Din, is holding an astrolabe. Note the collection of scientific instruments on the table in front of him, including a mechanical clock.

4. An Arabic astronomical manuscript written in early eighteenth-century Timbuktu.

5. The Astronomical Bureau in Beijing, including a number of seventeenth-century scientific instruments.

6. The Jantar Mantar astronomical observatory, overlooking the River Ganges, built in 1737 in Varanasi, India.

7. An oil painting of Tahitian boats in Matavai Bay, Tahiti. The Polynesian navigator Tupaia joined the crew of HMS *Endeavour* in Tahiti in 1769.

8. Scientific exchange between Chinese, Japanese, and Dutch scholars in the eighteenth century. Note the anatomy textbook and natural history specimens on the table.

9. An oil painting of Francis Williams in his study in Spanish Town, Jamaica, in 1745. On the table in front of him is a copy of *Newton's Philosophy*.

10. Gorée, a former slave-trading station off the coast of Senegal. It was in this fort that the French astronomer Jean Richer conducted experiments which Isaac Newton later cited in his *Principia Mathematica* (1687).

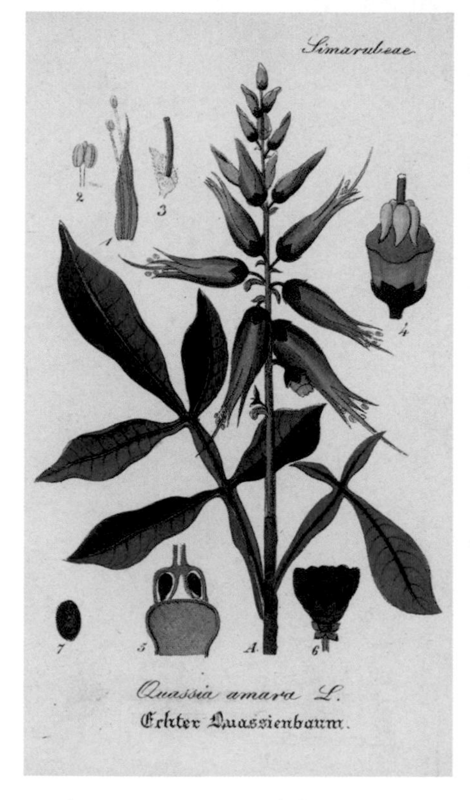

11. The *Quassia amara*, named after Graman Kwasi, an enslaved African who discovered the plant in Dutch Surinam at the beginning of the eighteenth century.

12. A sixteenth-century Mughal natural history manuscript depicting a screw pine (*bottom*) and oleander (*top*).

13. A Japanese manuscript illustrating the delivery of a Vietnamese elephant to the shogun in Edo in 1729.

commit suicide, then it might also be used as a poison. In 1701, Henry Barham described a fellow physician in Jamaica who was 'poisoned . . . by his negro woman'. After ingesting a tea laced with the juice of a savanna flower, the man was 'seized with violent griping, inclining to vomit . . . he had small convulsions in several parts of him'. African botanical knowledge therefore formed part of the resistance against slavery. However, European fear of poisoning created a somewhat paradoxical situation. As we've seen, European naturalists relied on Africans for knowledge about many of the plants they found in the Americas. Yet at the same time, colonial laws were being passed that effectively barred Africans from working with medicinal plants. In 1764, the French colonial government in Saint-Domingue, modern-day Haiti, outlawed all persons of African descent from 'exercising medicine or surgery and from treating any illness under any circumstance'. A similar law was passed in South Carolina, recommending the death penalty 'in case any slave shall teach or instruct another slave in the knowledge of any poisonous root, plant, herb, or poison whatever'. Laws such as these are one of the reasons that Africans have been excluded from the history of science as traditionally written, although of course there is also the deeper problem of structural racism at work here too. Understandably, many enslaved people chose to keep their botanical knowledge hidden, for fear of punishment. It is only recently that we've begun to uncover what one historian has called 'the secret cures of slaves'.[16]

The growth of Atlantic slavery in the seventeenth and eighteenth centuries had a profound effect on the development of European society. The wealth generated from the forced labour of enslaved Africans funded everything from art and architecture to ports and factories. Slavery also transformed the world of science. As we saw in the previous chapter, Isaac Newton and his followers relied on astronomical observations made by those travelling aboard slave ships. And in this chapter, we've seen how famous European naturalists, such as Carl Linnaeus and Hans Sloane, depended upon enslaved Africans to tell them about the plants of the West Indies and South America. Slavery was a deeply exploitative system, one that relied on the constant threat of violence. The same is true of empire more generally, a theme we explore in more detail in the rest of this chapter. As European trading empires expanded, so too did interest in Asian natural history. In some cases, scientific exchange was on a slightly more even footing. At other times, European naturalists still relied on coercion. Nonetheless, wherever we look, the development of natural history in this period cannot be separated from the commercial world of trade and empire. In the following section, we explore how this relationship between empire and natural history played out in the East Indies. We begin with a Dutch military commander and his Indian servant.

II. Natural History in the East Indies

Hendrik van Rheede watched as his Indian servant climbed a nearby palm tree. On reaching the top of the palm, nearly thirty metres high, the servant pulled out a knife. Cutting into the shoots, he began to collect some of the sap. On climbing back down, the Indian servant told

21. The 'Carim-pana', or palmyra palm, in Hendrik van Rheede, *The Garden of Malabar* (1678–93). The name of the palm is given in three different languages (written in four different scripts) at the top.

Van Rheede that this particular tree was called the 'Carim-pana'. The sap was used to make a kind of alcoholic beverage, known as palm wine or 'toddy'. Van Rheede wrote down the name of the tree, along with its use, adding a cutting to his growing collection of Indian plants. The 'Carim-pana', or palmyra palm as it is known today, was just one of the 780 species of plant listed in Van Rheede's monumental *Garden of Malabar* (1678–93). Consisting of twelve volumes, and containing over 700 illustrations, this book was the first European work to provide a comprehensive account of Indian botany. *The Garden of Malabar* was consulted by many of the most important naturalists of the Enlightenment, including Carl Linnaeus. It was also a work that drew heavily on Indian scientific and medical traditions.[17]

Van Rheede had arrived in India, not as a naturalist, but as a military commander. Born in Utrecht to a wealthy merchant family, he had joined the Dutch East India Company at the age of just fourteen. In 1670, after rising through the ranks, Van Rheede was appointed Commander of Malabar, a Dutch colony on the southwest tip of India. He was astonished by the lush landscape he found himself in, filled with palm trees and spices. 'There was no place, not even the smallest, which did not display some plants,' recalled Van Rheede. The 'large, lofty and dense forests' of Malabar 'radiated such fertility'. He concluded that 'this part of India was truly and rightly the most fertile part of the whole world'. From coconuts and bananas to cardamom and pepper, Malabar was a luxurious environment, one which the Dutch East India Company was keen to exploit commercially.[18]

With this in mind, Van Rheede began an ambitious project to collect, sketch, and describe all the different species of plant in Malabar. This was not something that Van Rheede could accomplish on his own. As in the Americas and Africa, European naturalists in the East Indies relied on the existing knowledge of local people in order to understand the flora and fauna of the region. These individuals, after all, had far greater expertise in the natural history of South Asia than any European could possibly hope to possess. In the first instance, Van Rheede enlisted an army of over 200 Indian collectors, sent out far and wide searching for different plants. As a military governor, he had the power to get what he wanted, by force if necessary. Van Rheede also made use of his diplomatic contacts, writing to local Indian princes, asking them to forward specimens. The Raja of Cochin and the Raja of Tekkumkur both obliged, sending a considerable number of rare plants. Van Rheede then employed three Indian artists to make sketches of the different specimens. It was these Indian drawings which later featured in *The Garden of Malabar* when it was published in Amsterdam. Most significantly, Van Rheede assembled a group of Indian scholars in order to name and identify the uses of the different plants. The group consisted of three Brahmin priests – Ranga Bhatt, Vinayaka Bhatt, and Apu Bhatt, high-caste Hindus with expertise in ancient religious and scientific texts. Alongside the Brahmin priests, Van Rheede also employed a local doctor named Itti Achuden. Trained in the traditional Indian medical system known as Ayurveda, Achuden was an expert in identifying the different healing properties of Malabar plants.[19]

Unlike in Africa, much of this knowledge was written down. Achuden kept what Van Rheede described as a 'famed medical book', another reminder of the existing scientific knowledge of South Asian peoples. This, however, wasn't your typical printed book. In seventeenth-century southern India, people didn't write on paper. Instead, they wrote on dried palm leaves, bound together with string. This had the advantage that you could always add to an existing text, simply by tying another palm leaf to the collection. Itti Achuden's medical text, written in the local language of Malayalam, had been passed down through the generations. It contained hundreds of palm leaves detailing the different medicinal uses of local plants. Similarly, the Brahmin priests drew on their knowledge of the *Vedas*, a series of ancient Hindu texts. A number of these texts, which were written in Sanskrit verse, described the medicinal uses of plants. For example, the *Atharva Veda*, originally composed sometime in the second millennium BCE, contains a description of 288 plants. These include the flannel weed, said to help heal wounds, as well as 'goat's horn', a shrub said to drive away mosquitos when burned.[20]

Van Rheede valued the knowledge contained in the *Vedas*, noting that 'as regards medicine and botany, the knowledge of these sciences is preserved in verses'. There was clearly a wealth of information contained within these ancient texts. 'The first line . . . begins with the proper name of the plant, whose species, properties, accidents, forms, parts, location, season, curative virtues, use, and the like they described highly accurately,' Van Rheede explained. Consulting with the Brahmin priests, he began to understand how the naming of plants reflected an Indian classificatory system. Plants were typically assigned names with a suffix that indicated the species. For example, 'Atyl-alu', 'Ittyalu', and 'Are-alu' were all local names for different types of fig tree, identified by the suffix '-alu'. These names were then reproduced in *The Garden of Malabar*. In the final work, the plants are listed in three different languages: Malayalam (written in both the Arabic script and the local Aryaezuthu script), Konkani (written in the Devanagari script used for Sanskrit religious texts such as the *Vedas*), and Latin (written in the Roman script).[21]

The Garden of Malabar was a typical work of Enlightenment science. It brought together the scientific traditions of different cultures, presenting a unique view of the natural history of southern India. At the same time, *The Garden of Malabar* was a work that reflected the growing influence of European trading companies. Van Rheede's book listed all kinds of valuable commodities: sandalwood, cardamom, ginger, and black pepper. It was this economic concern which motivated a renewed interest in the study of natural history at the end of the seventeenth century.

Georg Eberhard Rumphius could feel the ground trembling. At first, it was just a little, but then the whole house began to shake violently. On 17 February 1674, 'the most terrible earthquake' struck the small island of Ambon, part of what is now Indonesia. Rumphius, a Dutch East India Company merchant, had been living on the island for over twenty years. He had never experienced anything like this before, but there was worse to come. After the initial tremors, Rumphius spotted something on the horizon. 'Three dreadful waves . . . they stood tall like walls,' he later recalled. It was a tsunami. The population of Ambon was devastated. Entire villages were washed away, with over 2,000 people reported dead, mostly local Ambonese. For Rumphius, it was a particularly tragic day as his wife, Susanna, along with two of their children, was killed. He decided to name a flower in memory of his wife. After all, the two had often collected plants together on the island. The flower Rumphius chose was a white orchid and he named it *Flos susannae*, 'to commemorate the person who, when alive, was my first Companion and Helpmate in the gathering of herbs and plants, and who also was the first to show me this flower'.[22]

At the time of the earthquake, Rumphius was in the middle of a major study of the natural history of Ambon. This was later published in two parts. The first, which covered shellfish and minerals, was titled *The Ambonese Curiosity Cabinet* (1705). The second, which covered plants, was titled *The Ambonese Herbal* (1741–50). Both works were beautifully illustrated, including hundreds of plates depicting all kinds of plants and animals, from horseshoe crabs to durian fruits. Carl Linnaeus, who consulted both works, even copied a number of the images from *The Ambonese Curiosity Cabinet* into his influential *System of Nature*.[23]

As with Hendrik van Rheede in Malabar, Rumphius believed that understanding the natural history of Ambon would aid the Dutch East India Company. European mortality in Southeast Asia was notoriously high, with medicines hard to come by. 'We experience every day, and to our detriment, that the European medicaments which the Company dispatches at considerable cost, are either obsolete or spoiled,' noted Rumphius. Instead, he suggested that Europeans study the properties of local medicinal plants. These, Rumphius claimed, were not only more accessible, but more likely to be effective against local diseases. 'All countries have their particular diseases, which should be cured with native remedies,' he argued. At the same time, many plants in Southeast Asia were known to be especially valuable. The Dutch already controlled the supply of cloves, nutmeg, and mace from the Moluccas. Rumphius was in search of other potentially valuable commodities.[24]

Rumphius relied on local people to teach him about the flora and fauna of Southeast Asia. In the first instance, Rumphius learned a lot from his wife. Although she had a European name, Susanna was in fact a native of Ambon. Most likely of mixed ancestry, Susanna converted to Christianity and married Rumphius shortly after he arrived on the island in 1653. She knew a lot about local botany, as Indonesian women often acted as healers and herbalists, another reflection of the existing scientific expertise of local people. It was Susanna who first led Rumphius around Ambon, pointing out which plants might be worth including in his *Herbal*. By this point, Rumphius had started to lose his eyesight, so he was completely reliant on Susanna, as well as other

Ambonese guides, to identify, collect, and even sketch the plants that later appeared in his book. When Susanna died, Rumphius lost, not only his soulmate, but also a major source of botanical information.[25]

Much like Van Rheede, Rumphius listed all the plants he discovered in multiple languages. *The Ambonese Herbal* gave the names of plants in Latin, Dutch, Ambonese, and Malay. In some cases, Rumphius also noted down the Chinese, Javanese, Hindustani, or Portuguese name for the same plant. This reflected the diversity of people and cultures found in Southeast Asia at the end of the seventeenth century. Alongside the Dutch, rulers from China, India, and Africa sent merchants to Southeast Asia in order to acquire spices. Knowing the different local names was therefore crucial, not just for scientific purposes, but also for trade.[26]

When he wasn't collecting plants in the countryside, Rumphius would head to the market. In the bazaars of Ambon, Rumphius – who could speak several Asian languages – learned a great deal about local wildlife, simply from chatting to merchants and travellers. Local fishermen taught him about a species of giant octopus, known as the Greater Argonaut, or the 'Ruma gorita' in Malay. The female of the species produced an intricate spiralled eggcase, which looked rather like a shell. 'Fishermen consider it a great boon if they catch one,' explained Rumphius. 'This whelk is found so rarely that it is priced very highly, even in the Indies,' he noted. Similarly, on the nearby island of Buru, a Muslim priest taught Rumphius how to distil oil from the wood of a local tree. Rumphius also reported how Chinese merchants in Manilla sold candied orchid roots, most likely as an aphrodisiac.[27]

Before long, Rumphius had compiled a catalogue of many of the most valuable natural goods in Southeast Asia. In fact, Rumphius's work was considered so economically important that the Dutch East India Company initially declared *The Ambonese Herbal* a 'secret document'. This delayed its printing until after Rumphius's death. The Dutch East India Company, keen to maintain its monopoly on the spice trade, didn't want word getting out about all these other potential commodities. When *The Ambonese Herbal* was finally published, the Dutch East India Company only agreed on condition that certain sections, including those detailing the harvest of nutmeg, were censored.[28]

*

22. The 'Ruma gorita', or Greater Argonaut, depicted alongside its eggcase, in Georg Eberhard Rumphius, *The Ambonese Curiosity Cabinet* (1705).

The Dutch were right to be worried about competition. In the seventeenth century, a variety of European trading companies operated in Asia. However, over the course of the eighteenth century, the British came to dominate, particularly in India. Through a series of military conquests, the British East India Company seized control of a significant portion of the Indian subcontinent. By the late eighteenth century, the Dutch and French were largely forced out, confined to tiny trading stations. Even the Mughals, the rulers of much of India for the previous 200 years, were eventually defeated by the British. The expansion of the

British East India Company was in part fuelled by the new scientific work being done in the field of natural history. The British saw what the Dutch had achieved, and wanted to emulate it. The idea was to transform India into a tropical plantation economy, one that could supply all the different commodities Asia had to offer, from spices and sugar to timber and tea.

With this in mind, the British East India Company established the Calcutta Botanical Garden in 1786. Calcutta, in northeastern India, was the capital of Bengal, a territory that the British East India Company had recently acquired following the defeat of the local ruler. Appropriately, the first director of the garden, Robert Kyd, was a military officer. He explained the purpose of the new botanical garden to the directors of the British East India Company back in London. The Calcutta Botanical Garden was 'not for the purpose of collecting rare plants . . . as things of mere curiosity'. Rather, it was 'for establishing a stock for disseminating such articles as may prove beneficial to the inhabitants, as well of the natives of Great Britain'. Crucially, Kyd believed that such a garden, stocked with 'useful' plants, would 'ultimately tend to the extension of national commerce and riches'.[29]

The Calcutta Botanical Garden was therefore an economic initiative as much as a scientific one. It was designed to cement the status of the British East India Company in Bengal, as well as provide a source of valuable plants that might then be cultivated in plantations across India. Kyd immediately set to work. He sent for black pepper from Malabar and cinnamon from Southeast Asia. In each case, the idea was to break existing monopolies and reduce British reliance on foreign imports. By growing these valuable plants itself, the British East India Company hoped to lower costs and thus increase profit margins. By 1790, the Calcutta Botanical Garden housed over 4,000 plants representing 350 different species, most of which were not native to Bengal.[30]

When Kyd died in 1793, the position of director of the Calcutta Botanical Garden was taken up by a Scottish surgeon named William Roxburgh. Unlike Kyd, Roxburgh was trained in natural history and medicine. As a student at the University of Edinburgh, he had learned how to dissect plants and identify different species based on Linnaean classification. In 1776, Roxburgh arrived in India, employed as an assistant surgeon. Before being transferred to Calcutta, he set up a small experimental plantation in Samalcottah, part of the Madras Presidency in southern India. On the plantation, Roxburgh grew black pepper, coffee, and cinnamon. He also experimented with growing breadfruit, imported all the way from Tahiti, a plant that many naturalists thought might provide a cheap and high-energy source of food.[31]

Alongside this, Roxburgh identified an alternative source of indigo dye, another valuable commodity. Traditionally, this deep blue dye was manufactured from the leaves of the indigo plant. At the time, the majority of indigo was grown in the Americas, and the trade was largely controlled by the Spanish. Indigo had been cultivated in India, but not on a large scale or with much success. Roxburgh was therefore keen to promote an indigenous alternative. He claimed to have discovered a totally different species of plant, classified by Carl Linnaeus as a *Nerium*, whose leaves also seemed to secrete a similar blue dye. Roxburgh quickly wrote to the British East India Company directors in London, sending a sample of his 'Nerium Indigo' for chemical testing, and suggesting it might prove 'infinitely profitable'.[32]

On the back of this, Roxburgh was the obvious candidate to take over the Calcutta Botanical Garden. He combined Kyd's emphasis on commerce with a deep understanding of the latest scientific work on biological classification. On taking up the position, Roxburgh set about expanding the garden. He began cultivating a variety of other tropical plants, many of which originated far from India. These included Jamaican all-spice, as well as sweet potato and papaya from South America. Roxburgh also sent collectors to the Moluccas, charged with smuggling out samples of nutmeg and cloves. With such a large number of different plants, Roxburgh began to expand the staff in the garden. Many of the foreign species of plant needed expert care. Like other European naturalists, Roxburgh quickly realized that the most knowledgeable people with regards to Asian plants were those from the region. To this end, Roxburgh recruited 'two Malay Gardeners' from Ambon, named Mahomed and Gorung, most likely because of their expertise as either herbalists or spice farmers. These two men were employed solely to care for the nutmeg, which proved exceptionally difficult to grow outside of Southeast Asia. Similarly, Roxburgh employed a number of Chinese gardeners in order to help grow tea trees, as well as Tamils, in order to cultivate spices from southern India.[33]

This diversity of cultures was reflected in Roxburgh's first major scientific publication. With the support of the British East India Company, Roxburgh published *Plants of the Coast of Coromandel* (1795). This book detailed many of his early botanical findings. Plant names were given in English, Latin, and the local Indian language of Telugu. The book also contained over 300 life-size, hand-coloured illustrations, depicting each of the plants Roxburgh described. These illustrations, however, weren't done by Roxburgh himself. Rather, they were produced by 'two native artists'. Since its foundation, the Calcutta Botanical Garden had employed Indian artists to sketch and catalogue the different species of plant. The British employed them because of their familiarity with the local environment, as well as their skill in depicting what, to Europeans, were previously unknown species. These artists typically combined both European and Indian traditions, developing a style that came to be known as 'Company School'. Many of the artists employed in Calcutta had previously worked for the Mughals, producing illustrated manuscripts, often with botanical or zoological themes. In some aspects, then, the illustrations in *Plants of the Coast of Coromandel* looked like a typical Mughal court painting, with clear blocks of colour and a relatively flat appearance. But at the same time, these illustrations reflected the demands of Linnaean classification. Roxburgh ensured that the Indian artists carefully separated out the sexual organs of the plants, as well as the seeds, as these were crucial for identifying different species under the Linnaean system.[34]

In the Calcutta Botanical Garden we can ultimately see Enlightenment science in microcosm. It was an institution, established by the expanding British Empire, for the purpose of economic gain. And it was built on land that the British had seized, by military force, from local Indian rulers. At the same time, the Calcutta Botanical Garden was also a place in which a diversity of cultures and scientific traditions came together, from Scottish surgeons to Indian artists. In the following section, we explore the history of natural history in seventeenth- and eighteenth-century China. This was a region that the British East India Company was also trying to expand into, although with much greater difficulty. And there was one Chinese plant that British merchants and naturalists were desperate to get their hands on.

III. The China Drink

In 1658 an exotic new drug reached the streets of London. Some doctors promoted it as a miracle cure, using it to treat diseases ranging from kidney stones to depression. Others, however, thought it might be a harmful intoxicant, potentially as dangerous as alcohol or even opium. The British certainly seemed addicted. One doctor claimed that the drug was responsible for 'the introduction of a numerous class of nervous ailments'. Another that 'sipping this decoction, one may sometimes spend entire nights working . . . without being otherwise overcome by the need for sleep'. What was this controversial new remedy? Samuel Pepys, the famous diarist, called it 'the China drink'. We know it better as tea.[35]

When tea first arrived in Britain in the middle of the seventeenth century, it was an exotic commodity. Imported all the way from China, it cost ten times as much as coffee per pound weight. However, by the end of the eighteenth century, tea had become an article of everyday consumption. The British were considered a 'tea-drinking nation', with people from all walks of life partaking in the habit. The first tea leaves arrived in Europe in 1610, aboard a Dutch East India Company ship. Initially, the British purchased their tea from the Dutch. However, as demand continued to grow, the British East India Company focused its efforts on acquiring tea directly from China, bringing back the first shipment in 1713. Alongside tea, European trading companies imported vast quantities of Chinese silk and ceramics, as well as other medicinal herbs such as gingko. In fact, the eighteenth century witnessed a craze for all things Chinese. European doctors experimented with acupuncture, whilst British gardens filled with Chinese shrubs including peonies and magnolias.[36]

The growth in trade with China also sparked interest amongst European naturalists. Tea in particular caused much scientific debate, as people puzzled over how to classify it. The tea that Europeans purchased in China came in a number of different varieties. In the eighteenth century, these were referred to as 'bohea' (black tea), 'singlo' (green tea), and 'bing' (imperial tea). The leaves of each variety were of a different colour, and produced distinct tastes when made into an infusion.

However, at this time very few Europeans had actually seen a tea tree growing in its native environment. Rather, tea was purchased at Chinese coastal ports – such as Canton and Amoy – once it had already been processed. This processing involved repeated stages of drying and rolling the leaves by hand. European naturalists were therefore unsure if these different varieties of tea came from one plant or many different species. Carl Linnaeus co-authored a book, titled *The Tea Drink* (1765), addressing this very problem. He incorrectly argued that the different varieties of tea must represent different species. (In fact, all tea comes from the same plant, something European naturalists weren't completely sure of until well into the nineteenth century.)[37]

As we've seen elsewhere, these scientific questions also had a commercial dimension. European traders arriving in China needed to be able to distinguish between different teas, as well as spot a fake. This was crucial in order to avoid getting ripped off. You didn't want to pay for an expensive imperial tea, if in fact you were just getting a regular green tea. Some British East India Company officers even reported finding sage leaves or other cheap alternatives mixed in amongst the tea crates. There was also a big financial incentive to experiment with growing tea in Europe. Linnaeus himself promoted this idea, complaining of the vast sums of money that left Europe in exchange for Chinese goods. 'Let us bring the Tea-tree here from China,' Linnaeus wrote. He hoped that 'in the future not a pence would leave us for those leaves'. This was all part of an argument we saw earlier concerning the 'balance of trade'. The Chinese only accepted payment in silver bullion, and Linnaeus – along with many others – worried that trade with China was weakening European economies. Imported goods, like tea, far exceeded those exported.[38]

With this in mind, European naturalists devoted considerable effort to studying Chinese plants. In 1699, James Ovington published the first detailed account of tea written in the English language. In *An Essay upon the Nature and Qualities of Tea*, Ovington described the different varieties as well as the cultivation of the tea tree. However, he hadn't actually seen tea growing in its native environment. Instead, he had learned about tea whilst working for the British East India Company in Gujarat, in western India. The merchants in Gujarat had been trading tea, spices, and silk with the Chinese for centuries. According to

Ovington, tea was 'a common Drink with all the Inhabitants of India', where it was mixed with sugar and lemon. Whilst in Gujarat, Ovington also met a Chinese envoy at the local court. The envoy had apparently 'brought with him several kinds of tea'. From his conversations in India, Ovington was able to piece together some of the basic facts about tea, including how it was processed. 'The Leaf is first green, but is made crisp and dry by frying twice . . . and as often as it is taken off the Fire it is roll'd with the Hand upon the Table, till it curls,' explained Ovington. He also suggested it might be possible to grow tea in Europe, if only a specimen could be acquired, writing 'the Shrub itself is of a strong and hardy Constitution . . . the winter in *England*, in some places where it grows, is not more cold'.[39]

Ovington got a lot right, but there was only so much Europeans could learn about tea without actually going to China. Shortly after the publication of this book, however, the problem was solved. Landing on Chusan Island in 1700, James Cuninghame became one of the first Europeans to observe tea growing in its native Chinese environment. A surgeon in the employment of the British East India Company, Cuninghame had been sent to Chusan – a small island just off the coast of eastern China – in order to help set up an early trading station. The station was ultimately a failure, and the British East India Company quickly abandoned the project. Cuninghame, however, decided to stay behind, hoping to learn something about Chinese natural history. During his stay, Cuninghame corresponded with the influential British naturalist James Petiver, promising to acquire a specimen of the tea tree. Petiver also asked Cuninghame to 'inquire what variety theirs is of it & wherein the Bohea Tea differs from the common'. In short, Petiver wanted to know if black and green tea came from the same plant. Cuninghame did his best to answer Petiver's questions. He visited multiple tea plantations, recalling the sweeping hills lined with neat rows of green shrubs, Chinese men and women picking the leaves by hand. Tea was 'a flowering plant, with leaves serrated like nettles and whiteish underneath', Cuninghame explained. Spending over a year on Chusan, he was ultimately able to observe the complete life cycle of the tea tree, including its harvest and processing. This allowed him to produce the first accurate description of the tea tree outside of China.[40]

Cuninghame's account of the tea tree was published in the

Philosophical Transactions, the prestigious journal of the Royal Society in London. In the article, Cuninghame made a crucial observation. 'The 3 sorts of Tea commonly carry'd to *England* are all of the same Plant,' he explained. What mattered was when the tea leaves were picked, and then how they were processed. In the article, Cuninghame went on to explain how 'Bohe', or black tea, 'is the very first bud gather'd, in the beginning of *March*, and dry'd in the *shade*'. In contrast, 'Bing', a variety of imperial tea, 'is the second growth in *April* . . . dry'd a little in *tatches* or Pans over the Fire'. Alongside this article, Cuninghame sent hundreds of specimens of different Chinese plants back to Britain. In fact, the oldest surviving specimen of tea held outside of China, kept today at the Natural History Museum in London, was collected by Cuninghame. It sits in a tiny wooden box along with a label, dating from the eighteenth century, on which the words 'A Sort of Tea from China' are written.[41]

Carl Linnaeus was not the only one developing a new way of classifying the natural world in this period. In China, there was already a well-established tradition in the study of natural history, dating back thousands of years. The Chinese even developed a specific literature dedicated to the scientific study of tea. The most famous of these texts, titled *The Classic of Tea*, was written by a scholar named Lu Yu in the eighth century. Lu's book set out everything you could possibly want to know about tea: where it was cultivated, how different varieties were processed, its medicinal properties, even how it should be served. According to Lu, tea was 'the common drink of every household'. And, unlike alcohol, it did 'not lend itself to extravagance'. *The Classic of Tea* was the first of over a hundred 'tea books' published in China, many of which were written in the seventeenth and eighteenth centuries, just at the time when Europeans started to engage in the tea trade.[42]

As in Europe, the development of trading links with the wider world from the fifteenth century onwards sparked a revolution in the study of natural history in China. Merchants imported maize from the Americas, spices from India, and fruit from East Africa. All this increased demand for new works of natural history. The most significant of these was published in Nanjing at the end of the sixteenth century. Titled *The Compendium of Materia Medica* (1596), this monumental book – which ran to over two million Chinese characters – contained 1,892 entries of different plants, animals, and minerals, many of which had never been classified before. Its author, Li Shizhen, was born in 1518 to a family of doctors in central China. Li wanted to enter the prestigious Chinese civil service, but failed the competitive examinations. However, thanks to his background in medicine, Li was able to obtain a post at the Imperial Medical Office in Beijing.[43]

Li's job was to help regulate medicine across China, setting examinations, awarding licences, and assessing new drugs. Working at the

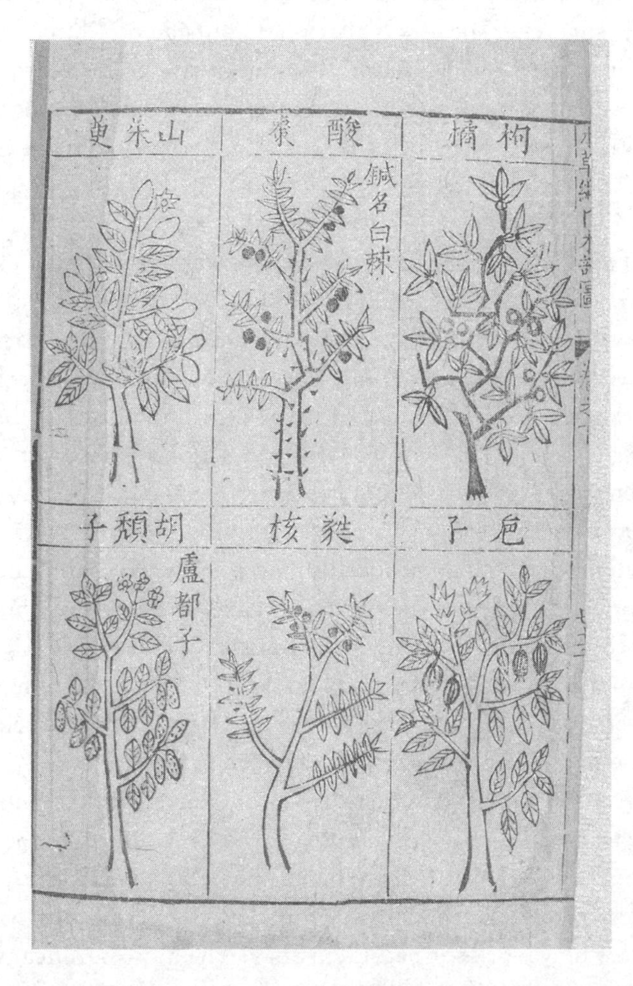

23. An illustration of different plants, including the Chinese bitter orange and cape jasmine, from Li Shizhen, *The Compendium of Materia Medica* (1596).

Imperial Medical Office, Li had access to a vast collection of medicines. He was also able to read many of the ancient Chinese works on natural history, such as *The Classic of Tea*. However, he soon realized that regional diversity in the naming of plants made it exceptionally difficult for the Imperial Medical Office to do its job. How could the Chinese state assess new drugs or collect taxes on medicines if the same plants went by different names? Tea is a good example of this. In Canton, it was referred to as *ch'a* (from which we get 'chai'), whereas in Amoy it was called *te* (from which we get 'tea'). This was all made harder still by the arrival of various foreign plants, as China traded with the rest of the world. Li decided that there needed to be a standardized way to describe all the different plants, animals, and minerals found within the Chinese Empire.[44]

Li then spent the next thirty years of his life travelling across China, collecting specimens as well as interviewing local doctors and farmers, gathering information for *The Compendium of Materia Medica*. In the introduction to his book, Li explained his taxonomic system, writing, 'my overall system of classification consists of sixteen sections (*bu*) that form the upper level (*gang*) and sixty categories (*lei*) which form the lower level (*mu*)'. The upper level was organized around the 'five phases', a traditional Chinese division of the world, much like the four elements in ancient Greek philosophy. The five phases were wood, fire, earth, metal, and water. These then corresponded to particular qualities (such as warm or cold) as well as particular tastes (such as acidic or sweet). Below this, there was then a further subdivision, often based on the environment in which a particular plant or animal could be located, for example, 'mountain herbs' or 'aquatic birds'. Li also needed to classify various foreign plants, such as maize. The tea tree featured as well, correctly identified as a single species, and said to act as an effective anti-inflammatory. Indeed, as a doctor Li spent a lot of time detailing the medicinal properties of all the plants and minerals listed in the book. He even wrote a separate chapter cross-referencing hundreds of specific illnesses with different medicines.[45]

Li ultimately provided a standardized way to classify the natural world, one that could be used by doctors and bureaucrats across the Chinese Empire. His book proved a phenomenal success. The final publication was accompanied by two volumes of detailed illustrations, depicting many of the plants and animals described in the book. The Chinese emperor was presented with a copy, whilst multiple updated editions were printed throughout the seventeenth century. It became even more popular in China following the rise of the Qing dynasty in 1644. By the middle of the eighteenth century, the Qing controlled an area twice as large as the Ming dynasty, mainly thanks to a series of military conquests to the west. This territorial expansion brought Chinese naturalists into contact with even more new plants and animals, as well as new systems of taxonomy. There was once again an explosion in scientific publishing, as eighteenth-century Chinese naturalists looked to update Li's work.[46]

During the same period, copies of Chinese works of natural history started to reach Europe. In 1742, a French naturalist named Pierre Le Chéron d'Incarville wrote a letter from Beijing describing how he had 'found a book containing drawings of Chinese medicinal plants, a few animals and insects: really a book of natural history'. This was none other than Li's *Compendium of Materia Medica*. D'Incarville quickly purchased two volumes, forwarding them to the King's Garden in Paris. Translated extracts soon appeared in French and English. Joseph Banks, President of the Royal Society, even purchased a copy of Li's book, hoping it would help him identify different Chinese plants sent to London by British merchants. It continued to be consulted by European naturalists well into the nineteenth century, something we'll explore in more detail in the following chapter.[47]

The Compendium of Materia Medica is an important reminder of just how closely the development of natural history in Europe and China mirrored one another. Li, after all, was not so different from Carl Linnaeus. He was a trained doctor who, in the context of a growing world of trade and empire, saw the need for a standardized system of classifying the natural world. Li's taxonomy, again like Linnaeus's, was based on a mixture of physical characteristics and environmental considerations. And it was prompted by economic and bureaucratic demands. Yes, there were differences in some of the specifics, particularly Li's use of the five phases. But ultimately, when we think on a global scale, it is clear that the growth of natural history in Europe was not unique. Scientific thinkers in Asia were also developing new ways to classify nature in order to make sense of an increasingly connected world, and as we'll see in the following section, this was also the case in early modern Japan.

IV. Studying Nature in Tokugawa Japan

The shogun wanted an elephant. In 1717, Tokugawa Yoshimune, the ruler of Japan, was browsing the castle library in Edo, modern-day Tokyo. He came across a book his uncle had been given by a Dutch merchant: Johann Jonston's *Natural History of Quadrupeds* (1660). Originally published in Leiden, this lavishly illustrated work featured engravings of many animals the shogun had never seen before: camels, lions, and reindeer. But what most fascinated Yoshimune was the image of an elephant. Yoshimune quickly ordered his personal physician, Noro Genjo, to begin translating Jonston's book from Dutch into Japanese. Yoshimune particularly wanted to know where elephants came from, and what they might be used for. 'These animals exist in great numbers in countries visited by the Dutch . . . the tusks are used for medical purposes,' reported Noro.[48]

Books were all well and good. But what Yoshimune really wanted was an elephant of his own. In 1729, he got his chance. The Dutch East India Company, which was keen to secure favourable trading relations with Japan, agreed to import two Asian elephants from Vietnam, one male and one female. In April, the elephants arrived in Nagasaki, where the Dutch East India Company occupied a small trading station. Crowds lined the streets cheering as the elephants were paraded across Japan. They were first brought from Nagasaki to Kyoto, before finally being delivered to Yoshimune in Edo. Unfortunately, the male elephant died shortly after arriving. But the female elephant survived for another thirteen years, and she was kept on display in the beautiful gardens surrounding Edo Castle. The elephants were just the start. Over the following decades, Yoshimune and his successors acquired a range of different exotic animals, many of which were previously unknown in Japan. By the end of the eighteenth century, Edo Castle housed a porcupine from North Africa, two orangutans from Borneo, horses from Persia, and an entire flock of sheep imported from Europe.[49]

The Enlightenment was a period of major transformation in the study of natural history, not just in Europe, but also in Asia. This was especially true in Japan. In the ancient and medieval periods, most Japanese studies of natural history were undertaken by Buddhist monks or Shinto priests. Natural history served an important religious function. Shinto shrines were often dedicated to sacred animals, whilst Buddhists thought that natural history might help them better understand the cycle of reincarnation. However, by the beginning of the eighteenth century, things had changed considerably. With the growth of global trade, natural history in Japan, just as in Europe, started to take on a much more commercial dimension. This was particularly the case following the foundation of the Tokugawa shogunate in 1600, which brought together Japan's different warring states under one ruler. Although the Tokugawa shogunate followed a policy of *sakoku*, meaning 'closed country', in which foreign access to Japan was restricted, this did not mean a complete cessation of trade. Somewhat counterintuitively, the 'closed country' policy in fact led to an intensification of trade, with only a small number of European, Chinese, and Japanese merchants – sanctioned by the shogun – controlling the flow of valuable goods in and out of the country.[50]

Yoshimune's interest in exotic animals was therefore not simply down to curiosity. He was deeply concerned about the economic and political future of Japan, and believed that studying the natural world might help unlock the key to prosperity. This was all the more important given that Japan also suffered in the balance of trade, importing far more than it exported, in part as a consequence of the 'closed country' policy. With this in mind, Yoshimune commissioned a series of surveys of Japanese natural history, hoping to identify home-grown alternatives to expensive imports. The largest of these surveys was conducted in the 1730s by Niwa Shohaku, another of Yoshimune's court physicians. Niwa travelled right across Japan, much as Li Shizhen had done in China. He distributed questionnaires in every domain, requesting that local lords report 'all produces generating from the earth' along with 'all species in that region without exception'. The questionnaires were accompanied by a letter, signed by Yoshimune himself, reminding Japanese lords of their obligation to the shogun in Edo. The final survey, titled *A Classification of All Things*, included 3,590 entries, encompassing not just plants and animals, but also metals, minerals, and gemstones.

Niwa's survey confirmed what Yoshimune had suspected. Japan possessed incredible natural wealth, particularly in copper and camphor oil, two commodities that European trading companies were keen to purchase.[51]

Yoshimune also supported the expansion of botanical gardens in Japan, particularly the Koishikawa Botanical Garden on the outskirts of Edo. Originally founded in the seventeenth century, it transformed into a site of commercial botanical research during the eighteenth century. There is a remarkable parallel here with what was happening in Europe. At just the same time that Carl Linnaeus was trying to grow exotic plants in Uppsala, Japanese naturalists were doing the same thing in Edo. By the 1730s, the Koishikawa Botanical Garden housed thousands of foreign plants, many of which had previously been imported at great expense, including ginseng from China, sugar cane from Southeast Asia, and sweet potatoes from the Americas. The garden was so successful that, by the 1780s, Japan actually went from importing ginseng to exporting it.[52]

Trading links brought Japan into contact with, not just exotic goods, but also a variety of scientific cultures. Initially, the most significant of these links was with China. The two countries shared a long history of intellectual and commercial exchange, dating back well over 1,000 years. The Japanese language, along with much of Japanese philosophy, borrowed heavily from China. This flow of goods and ideas intensified in the seventeenth century, particularly following the foundation of the Tokugawa shogunate in 1600. Alongside silk and tea, Chinese merchants started selling more and more books. They brought with them copies of Chinese works on astronomy, medicine, and natural history. In 1604, just a few years after its publication in Nanjing, Li Shizhen's *Compendium of Materia Medica* was being sold in Nagasaki. The shogun himself purchased a copy, adding it to the castle library in Edo. By 1637, Li's book had been reprinted in its entirety in Japan. It proved incredibly influential, forming the basis of the majority of studies of natural history in seventeenth-century Japan.[53]

At the beginning of the eighteenth century, a Japanese naturalist decided to write a new book, one that would combine the best bits of Chinese natural history with an updated survey of the plants of Japan.

Kaibara Ekiken came from a humble background. He was born in 1630 on the southern island of Kyushu, the son of a village doctor. However, Kaibara went on to become one of the most influential naturalists in Tokugawa Japan. Unlike other Japanese naturalists at the time, he was not content simply to follow the teachings of Chinese scholars. Kaibara complained that *The Compendium of Materia Medica* 'treats many exotic species that do not live or grow in Japan'. He therefore decided to leave Kyushu and travel across Japan in order to 'record in one single text all those species that people can actually see in our country'. Kaibara's approach represented an important shift in Japanese natural history. Rather than basing his knowledge solely on existing Chinese books, Kaibara emphasized the importance of personal experience. 'I climbed tall mountains. I penetrated into deep valleys. I followed steep paths and walked through dangerous grounds. I have been drenched by rains and lost my way in the fog. I endured the coldest winds and the hottest sun. But I was able to observe the natural environment of more than eight hundred villages,' explained Kaibara.[54]

On returning from his travels, Kaibara published *Japanese Materia Medica* (1709–15). This was a classic fusion of different scientific traditions. Kaibara still borrowed a lot from Li Shizhen. The organization of *Japanese Materia Medica* mirrored that of *The Compendium of Materia Medica*, particularly the use of the five phases. Many of the species common to both Japan and China were also copied straight from Li's book. However, even in these cases, Kaibara listed Japanese names, as well as regional varieties, rather than simply relying on the Chinese text alone. On top of this, Kaibara added a further 358 species of plant that could only be found in Japan. These included the famous Japanese cherry tree, known as the *sakura*, with its beautiful pink and white blossom. 'The Japanese cherry-tree, however, does not exist in China, as Chinese merchants testified when I interviewed them in Nagasaki,' explained Kaibara; 'if such a tree did exist, it would have been mentioned in Chinese books.'[55]

As it happens, Kaibara was only half right. Although very few Chinese works of natural history specifically identified the Japanese cherry tree, it did in fact grow in some areas of China, and around the Korean Peninsula. Nonetheless, what really mattered was the idea that Kaibara started to promote. It wasn't enough to simply rely on existing Chinese texts. Japanese naturalists needed to travel, to observe, and collect. Only

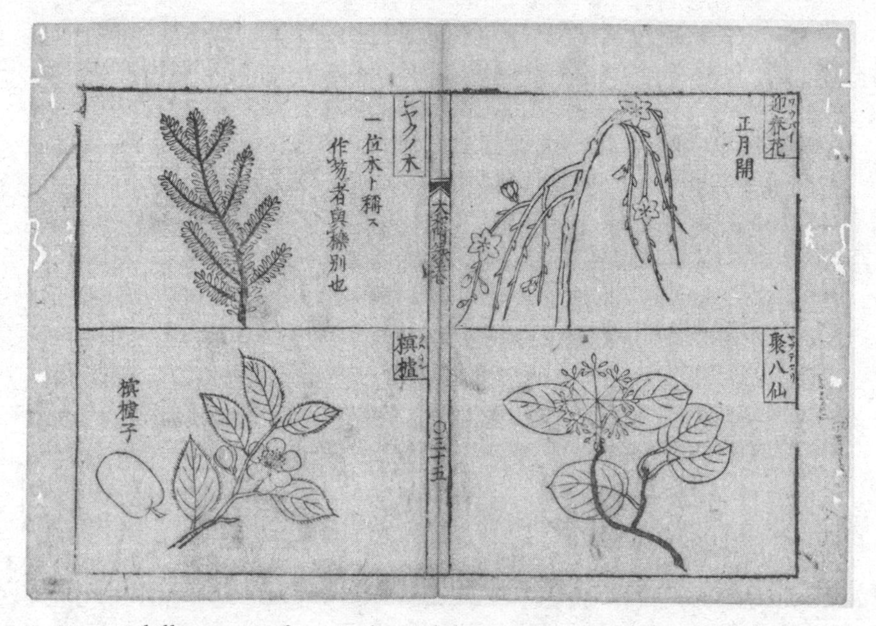

24. Botanical illustrations from Kaibara Ekiken, *Japanese Materia Medica* (1709–15).

then, Kaibara argued, would natural history 'be of concrete help to the people of *this* country'.[56]

Alongside China, the other major source of scientific knowledge in Japan was the Dutch East India Company. As discussed earlier, the Tokugawa shogunate had operated a policy known as *sakoku*, or 'closed country', since the early seventeenth century. Under this policy, European access to Japan was severely restricted. Christian missionaries were excluded entirely, as were most European merchants. Only the Dutch East India Company was granted permission to trade with Japan, and even then, the Dutch were confined to Deshima, a tiny island off the coast of Nagasaki. Over time, however, Japanese and European scientific cultures began to come into contact with one another. Dutch merchants not only presented scientific books at the court in Edo, but Japanese naturalists started to learn Dutch, hoping they might be able to learn something about distant lands. Indeed, the 'closed country' policy ultimately created a particularly intense form of cultural exchange, with a small number of Japanese and Dutch thinkers working very closely together.

Yoshimune, as we've seen, was impressed by what the Dutch knew. He owned, not only a copy of Jonston's *Natural History of Quadrupeds*, but many other Dutch works of natural history, including Rumphius's *Ambonese Curiosity Cabinet*. With this in mind, Yoshimune decided to relax an old law forbidding the import of European books. (This law had originally been put in place in the seventeenth century in order to stop the spread of Christianity.) Yoshimune gave a select group of scholars permission to purchase Dutch books with a view to translating them into Japanese. Before long, there were even specialist schools dedicated to *rangaku*, or 'Dutch learning'. Crucially, however, this was not simply a one-way relationship. At the same time as Japanese naturalists were learning from Europe, European naturalists were learning from Japan.[57]

Carl Peter Thunberg longed to escape Deshima. Employed as a surgeon, Thunberg had arrived in Japan aboard a Dutch East India Company ship in August 1775. He intended to start collecting exotic plants immediately. However, Thunberg soon found his movements severely restricted. 'It grieves me in my heart to see these rare and beautiful hills, cultivated by the industrious Japanese . . . without having the liberty to go there,' Thunberg grumbled in a letter to a friend. The island of Deshima consisted of just two streets, lined with wooden houses and storerooms. There was also a building for the Japanese interpreters, employed to translate for the Dutch, as well as a single bridge connecting Deshima to the city of Nagasaki. Thunberg started to despair. He had studied under Carl Linnaeus at the University of Uppsala and hoped to be the first to apply the binomial system of classification to Japanese plants. But Thunberg could achieve nothing if he couldn't explore beyond Nagasaki, or even the island of Deshima. 'I have indeed never been circumscribed within such narrow limits, never less free, never more secluded from my beloved flora,' he complained.[58]

Thunberg soon realized that he would need to make some friends. Each day, he would pop into the Japanese interpreters' building. As luck would have it, many of the Japanese interpreters, who were officially there to assist with trade, were also trained in medicine. In Japan, some doctors chose to learn Dutch in order to read European works of natural history and medicine. The interpreters were impressed by what Thunberg knew, as he advised on various new medical treatments, including the use of mercury to treat syphilis, at that point endemic in Japan. (Unfortunately, this treatment probably did more harm than good.) Thunberg had also brought with him some exotic specimens he hoped to trade, including a rhino horn purchased in Java.[59]

One of the Japanese interpreters, Shige Setsuemon, finally agreed to help Thunberg. In exchange for books and medical advice, he promised to supply Thunberg with specimens from the Japanese mainland. This was an incredibly risky thing to do, given the harsh punishments enacted against anyone caught smuggling. Every day, Shige crossed the bridge onto Deshima, hoping the guards would not search his bags, stuffed as they were with seeds and dried plants. The two men would meet in the evenings at the interpreters' building, quickly swapping bundles under the table, hoping no one would notice. Thunberg was delighted, writing that Shige brought him 'various beautiful and rare plants, previously unknown and peculiar to the country'. These included seeds of a Japanese chestnut as well as prints from various Japanese books of natural history.[60]

Still, there was only so much Thunberg could learn from one person. He still wanted to explore Japan for himself, and luckily for Thunberg, his chance came in March 1776. Every year, the Dutch East India Company sent a delegation to visit the shogun in Edo. For the first time, Thunberg was allowed to leave his island home. Perhaps, Thunberg hoped, he would finally get to explore Japan properly? As it turned out, things weren't quite that straightforward. Thunberg was forced to travel to Edo aboard a *norimono*. This was a large cabin in which the occupant would be carried by servants, similar to an Indian palanquin. Thunberg was not allowed to exit the cabin without the permission of his Japanese guards, and certainly wasn't free to roam as he pleased. Over the next few months, he was carried aboard the *norimono* over 700 miles from Nagasaki to Edo. It must have been exceptionally frustrating, rocking from side to side, as the beautiful scenery passed him by. Whenever he could, Thunberg did try and hop out, and collect a few plants. Crossing Mount Hakone, close to Edo, he even managed to lose his Japanese guards for a bit, trekking around in the undergrowth before he was summoned back to his *norimono*. In the end, Thunberg managed to collect sixty-two species of plant, all previously unknown in Europe, including a Japanese maple.[61]

On arriving in Edo, Thunberg presented himself at court. For the occasion, he wore a black silk cloak with gold trim, much like a traditional Japanese kimono. Although he enjoyed the ceremony, and was pleased to have escaped Deshima, Thunberg once again found himself confined. Along with the rest of the Dutch East India Company delegation, he was forced to stay in a small house on the outskirts of Edo Castle. He was not allowed to roam the city, nor the surrounding countryside. Still, he made the best of the situation. At court, Thunberg befriended two influential Japanese physicians: Nakagawa Jun'an and Katsuragawa Hoshu. Both Nakagawa and Katsuragawa were fluent in Dutch. They had also been part of a team responsible for translating one of the first European anatomy textbooks into Japanese. Nakagawa and Katsuragawa visited Thunberg every day for nearly a month, discussing the latest European medical theories, as well as sharing their knowledge of Japanese natural history. Nakagawa brought Thunberg a 'small collection of drugs, minerals, and numerous fresh plants', identifying each with its Japanese name. He also gave Thunberg a copy of a Japanese book titled *Splendours of the Earth*. This book, published in Edo in the early eighteenth century, included illustrations of hundreds of Japanese plants, as well as advice on their proper cultivation.

After a month in Edo, Thunberg returned to Nagasaki. On the way back, he visited a botanical garden in Osaka, full of 'the rarest shrubs and trees, planted in pots'. He even managed to convince the director of the garden to sell him some specimens, including a sago palm. (Technically, this too was illegal, Thunberg noting that 'the exportation of it is strictly prohibited'.) In November 1776, Thunberg finally left Japan, boarding a Dutch East India Company ship bound for Europe. Despite all the obstacles, he had managed to amass a vast collection of Japanese plants and books. In total, Thunberg returned to Europe with over 600 specimens. These formed the basis of his *Flora of Japan* (1784). This was the first work to apply the Linnaean system of classification to Japanese plants. It made Thunberg's name. Shortly after the publication of the book, Thunberg was appointed to Linnaeus's old position as Professor of Medicine and Botany at the University of Uppsala.[62]

At first glance, *The Flora of Japan* looks like a typical work of European natural history. But look a little closer, and we can see the traces of Thunberg's time in Japan. Many of the plants are cross-referenced against their traditional Japanese names. The sago palm is a good example. Thunberg gave the sago palm its Latin name, *Cycas revoluta*, indicating it was a member of the same genus as various other palms growing across Asia. However, Thunberg also noted that the sago palm was called 'sotits', or *sotetsu*, in Japanese. This, of course, is something he could only have learned from the Japanese naturalists he met in Nagasaki and Edo. Thunberg's *Flora of Japan* is therefore a perfect example of how eighteenth-century science relied on the exchange of knowledge between different cultures. On the one hand, it was a work of European natural history, extending the reach of Linnaeus's system of classification as far east as Japan. But on the other hand, Thunberg could not have written this book without the help of the many Japanese naturalists he met during his travels.[63]

V. Conclusion

As shown throughout this book, the best way to understand the history of modern science is to examine key moments in global history. In the case of natural history, we need to look to the expansion of global trade during the seventeenth and eighteenth centuries. This expansion was fuelled by the growth of European empires. Agents working for trading companies, such as the Royal African Company and the British East India Company, returned to Europe with specimens from far-off lands. At the same time, naturalists across the colonial world helped establish botanical gardens, with the hope of growing exotic plants for export. The expansion of European empires also brought a variety of different scientific cultures into contact with one another. People from across Africa and Asia possessed a sophisticated understanding of the natural world that is often forgotten today. African healers were responsible for identifying many of the plants in Hans Sloane's *Natural History of Jamaica*, whilst Hendrik van Rheede relied on Brahmin priests to write his *Garden of Malabar*. In China and Japan, knowledge of natural history was particularly advanced, part of a long tradition of scientific texts dating back well over 1,000 years. By the end of the eighteenth century, European naturalists were collecting, not just exotic plants, but also foreign books. Joseph Banks, President of the Royal Society in London, even owned a copy of Li Shizhen's *Compendium of Materia Medica*. Crucially, this was all happening at exactly the same time that the scientific cultures of China and Japan were themselves being transformed by connections to the wider world.

How then should we characterize the history of science during the Enlightenment? Traditionally, the Enlightenment was understood as the 'age of reason'. However, as the previous two chapters demonstrated, we need to remember that the Enlightenment was also the age of empire. In my view, it is that connection to empire – along with the violence and appropriation that went with it – which best explains the development of Enlightenment science. That's certainly true of the two most important sciences of the eighteenth century: astronomy and natural history. Without empire, Isaac Newton could not have discovered the laws of motion, relying as he did on observations made during the voyages of slave traders. And without empire, Carl Linnaeus could not have developed his system of biological classification, as this too depended upon botanical information collected during the expansion of European trading empires in Asia and the Americas. In the next two chapters, we follow the history of science into the nineteenth century, a period in which the link between science and empire only grew stronger. This was a world of factories and machines. A world of nationalism and revolution. And a world of capitalism and conflict. Science was about to enter the industrial age.

PART THREE

Capitalism and Conflict, *c.* 1790–1914

5. Struggle for Existence

Climbing down a makeshift ladder, Étienne Geoffroy Saint-Hilaire descended into an ancient Egyptian tomb. The sun was beating down outside, but at the bottom of the shaft it was almost completely dark. When he reached the foot of the ladder, Geoffroy lit a torch and held it up to the walls of the chamber. He could hardly believe his eyes. The walls were covered in hieroglyphics, the majority of which seemed to depict animals: a bird, a monkey, a beetle, and a crocodile. This was promising. Geoffroy, a young French naturalist, had learned about the existence of a 'sacred animal necropolis' from some locals. Perhaps this was it? He noticed a small opening on one side of the chamber. Crawling through, he emerged on the other side to find a room packed with earthenware jars. This was exactly what he had been looking for. He took one of the jars and cracked it against the floor. Just as he hoped, the jar contained the mummified remains of some kind of small bird. Geoffroy then called out to a group of French soldiers who were guarding the entrance of the tomb. Somewhat reluctantly, one of the soldiers climbed down, and began carrying out a number of the jars, which would later be sent to the Museum of Natural History in Paris. He didn't know it at the time, but Geoffroy had in fact just made a discovery which would spark one of the most important scientific debates of the nineteenth century.[1]

In the scorching summer of 1798, the French army, led by Napoleon Bonaparte, invaded Egypt. The French hoped that by securing Egypt they could better control both the Mediterranean and the overland route to India, challenging the British dominance of the region. This, however, wasn't simply a military operation. It was also a scientific one. Along with 36,000 soldiers, Napoleon recruited a group of scholars – mathematicians, engineers, chemists, and naturalists – who formed the Commission of Science and Arts. Geoffroy, aged just twenty-six, jumped at the chance to travel to Egypt. Over the next three years, the Commission followed the military, surveying the land and identifying valuable natural resources, all with the hope of turning Egypt into a profitable colony. The French also established the Institute of Egypt in Cairo, located in a lavish palace that had been captured by Napoleon's troops, where the Commission of Science and Arts held weekly meetings and even published a scientific journal.[2]

Today, Geoffroy is best remembered as one of the first European naturalists to advance a theory of evolution. Writing in his *Anatomical Philosophy* (1818), he argued that species are not fixed entities, but rather undergo change in response to the environment. He even claimed that it was possible to identify evidence of evolution in living species, particularly by studying the growth of embryos or by comparing the anatomy of seemingly distinct animals. All this set the stage for the development of modern evolutionary thought over the course of the nineteenth century. Traditionally, this is a story which focuses on Europe alone. We are told that, working amongst the vast anatomical collections held at the Museum of Natural History in Paris, Geoffroy, alongside a number of other early nineteenth-century French naturalists, started to make the case for evolution. What is less often recognized, however, is that Geoffroy first started thinking about evolution when he was in Egypt, not France. In fact, some of his most important early work was published by the Institute of Egypt in Cairo. In order to understand the history of evolution, we therefore need to begin with Geoffroy and the French army in North Africa.[3]

Less than a year after arriving in Cairo, Geoffroy set out on an expedition up the River Nile. He was accompanied by a group of soldiers, as Napoleon wanted to secure more of the territory to the south. Geoffroy, however, wasn't particularly interested in Napoleon's military objectives. Instead, he wanted to explore the ruins at Saqqara, one of the oldest ancient Egyptian sites on the Nile. Geoffroy had heard a rumour that the tombs there contained mummified animals, particularly those that the ancient Egyptians associated with their gods. This rumour turned out to be true. At Saqqara, Geoffroy uncovered hundreds of pots containing mummified birds, cats, and even monkeys.[4]

Geoffroy collected his most important specimen there too – the mummified remains of a bird known as the 'sacred ibis'. This specimen, which is still on display at the Museum of Natural History in Paris, ended up at the heart of early European debates over evolution. For his part, Geoffroy believed that the mummified ibis might provide the crucial piece of evidence he needed to support his new theory. After all, these specimens were known to be well over 3,000 years old. And thanks to the process of mummification, the bodies were perfectly preserved, sometimes even down to the level of the feathers and skin. With this in mind, Geoffroy suggested that the mummified ibis be compared to the contemporary ibis found in modern Egypt. Perhaps, he thought, it might be possible to identify some kind of anatomical

25. The skeleton of the 'sacred ibis' collected by Étienne Geoffroy Saint-Hilaire in Egypt in 1799.

difference between the two? This would then prove that species really did evolve.[5]

Unfortunately for Geoffroy, things were not so simple. Back in Paris, another leading French naturalist, Georges Cuvier, began examining the mummified specimens collected at Saqqara. 'For a long time, it was desired to know if any species changed form with the passage of time,' noted Cuvier. He then set to work measuring the mummified ibis and comparing it to a modern specimen. Cuvier even compared the two specimens to an image of an ibis carved on an ancient Egyptian temple. The results, however, were not what Geoffroy had hoped for. 'These animals are perfectly similar to those of today,' wrote Cuvier. The modern ibis was 'still the same as in the time of the pharaohs', he concluded. As it turned out, 3,000 years was simply not enough time to detect any meaningful anatomical difference between the two specimens. Evolution requires far longer timescales to operate, a fact that was only recognized much later in the nineteenth century. And yet, as we will see, this debate over a mummified ibis represents an important moment in the history of science. The world of eighteenth-century natural history was about to give way to a new era of evolutionary thought.[6]

Napoleon's invasion of Egypt in 1798 marks the beginning of our next key period of global history. The French campaign in Egypt was the first in a series of devastating wars which continued throughout the nineteenth century, culminating in the outbreak of the First World War in 1914. Nationalism too was on the rise, as states competed against one another for resources and territory. The nineteenth century was also a period of industrialization. This began in Northern Europe, particularly in Britain, but soon extended to Asia and the Americas. From the cotton mills of Bombay to the railways of Argentina, industrialization transformed the way people lived and worked, right across the globe.

The growth of war, nationalism, and industry had a profound effect on the way people understood the natural world in the nineteenth century. Étienne Geoffroy Saint-Hilaire even went as far as to describe nature as 'at war with itself', a striking choice of words given his experience with the French army in Egypt. This reflected Geoffroy's belief that everything, from chemical atoms through to living species, went through a cycle of destruction and renewal. Later in the nineteenth century, the British naturalist Charles Darwin used similar language, famously describing evolution in terms of a 'struggle for existence'. These ideas were then taken up by a range of other scientific thinkers, both in Europe and beyond, merging into a broad philosophy of Darwinism. One of the most influential of these later thinkers was the British Social Darwinist Herbert Spencer, who captured the mood of the age when he coined the phrase 'the survival of the fittest' in his *Principles of Biology* (1864), a book that was later read in Japan and Egypt. More often than not, this kind of Social Darwinism was used to reinforce existing forms of discrimination, particularly racial, with damaging consequences that extended well into the twentieth century, a theme we pick up on again later in this book. In the next two chapters, however, we uncover how the world of capitalism and conflict shaped the development of science during the nineteenth century. In this chapter, we follow the history of evolution onto the battlefield, whilst in the next, we explore the close connection between the growth of industry and the development of the modern physical sciences.[7]

When we think of the history of evolution, we often think of Darwin and his voyage aboard HMS *Beagle*. Between 1831 and 1836, the young British naturalist circumnavigated the globe, spending most of his time in South America before crossing the Pacific Ocean on his way back to Britain. On visiting the Galápagos Islands, located around 600 miles off the western coast of modern-day Ecuador, Darwin started to notice subtle differences between those species he had already encountered on mainland South America, such as the mockingbird. Over the next twenty-five years, Darwin transformed these early observations into his theory of evolution by natural selection, culminating in the publication of *On the Origin of Species* (1859). In this famous work, Darwin argued that members of the same species compete for survival, and ultimately to reproduce. Those with characteristics which aid in survival are more likely to pass on those traits to future generations. Given enough time, and especially in the context of geographical separation, this would result in the formation of new species, hence the differences Darwin observed during his voyage. *On the Origin of Species*, we are told, was the starting point for modern evolutionary thought.[8]

Darwin was certainly important, but in this chapter I want to suggest an alternative way of thinking about the history of evolution, one that begins prior to the publication of *On the Origin of Species* and extends well beyond Darwin and his voyage aboard HMS *Beagle*. After all, Darwin was not the first evolutionary thinker, not even in Europe. Amidst ancient Egyptian monuments and mummified remains, the French naturalist Geoffroy began to develop his own evolutionary theory, almost a decade before Darwin was born. Evolutionary thinking was in fact remarkably common in the early decades of the nineteenth century, right across the world. In Moscow, a Russian botanist proposed a theory of evolution as early as the 1820s, describing the natural world as one of 'constant change'. Similarly, in early nineteenth-century Kyoto, a Japanese philosopher published an evolutionary account, not just of new species, but of the entire planet. Inspired by Buddhist teachings, the philosopher described how the Earth itself had evolved from a combination of fire and water. This was then followed by the development of plant and animal life. 'One species of plant changes and becomes the manifold of plants. One species of animal, insect, fish, changes and becomes the manifold of animals, insects, and fish,' he explained.[9]

In order to properly understand the history of evolution, we therefore need to recognize that, before Darwin had even boarded HMS *Beagle*, people were already discussing the possibility that species might undergo transformation. At the same time, it is worth remembering that *On the Origin of Species* did not represent the final word on evolution. Darwin in fact left many unanswered questions, not least the actual mechanism of inheritance as well as the evolutionary origins of humans. These were questions Darwin himself later tried to address, with varying levels of success, but they were also taken up by scientists around the world. Moving from Latin America to East Asia, this chapter uncovers a forgotten history of evolutionary thought, a history which highlights the global origins of the modern biological sciences in an age of conflict.

I. Fossil Hunters in Argentina

Trekking across the Argentine plains, Francisco Muñiz came across a ferocious beast. With 'enormous fangs' this 'brutal king of the jungle' would make short work of anyone foolish enough to get too close. 'The strength of the entire structure knows no peer,' warned Muñiz in a local newspaper in 1845. Even the 'African lion . . . would see its throat sliced and its deepest entrails spilled with a single swipe of those fangs'. Thankfully, the population of Argentina had little to fear. The beast in question – a sabretooth cat – was long extinct, having died out over 10,000 years ago. Muñiz was in fact describing a collection of fossils he had recently discovered near the town of Luján in the Province of Buenos Aires. The region was already well known to fossil hunters. In 1788, Spanish labourers working near the River Luján unearthed the fossilized remains of an immense land mammal, a kind of giant sloth subsequently named the *Megatherium* (literally, 'great beast'). The bones of the *Megatherium* were shipped across the Atlantic to the Royal Cabinet in Madrid, where they generated much excitement amongst European naturalists. The French naturalist Georges Cuvier identified the fossils as conclusive evidence that 'the animals of the ancient world were entirely different from those seen on Earth today'. Cuvier himself didn't actually go as far as to say that new species might have evolved from earlier ones. But by the beginning of the nineteenth century, many people were starting to believe that the natural world wasn't quite as static as earlier scientific thinkers had tended to assume.[10]

Muñiz certainly believed in the possibility of evolution. After all, what else could explain the striking similarity between living species and those that were now extinct? 'I recognise the skeleton in question as belonging to an individual of the genus *Felis*, and resembling the *lion* in many particulars of its structure,' explained Muñiz in his article on the sabretooth cat. This 'new species' was nothing less than 'the first monster of the feline tribe', perhaps an ancient ancestor of the modern lion or tiger. Unlike Darwin, Muñiz did not have a fully worked out theory of natural selection. Nonetheless, he is another good example of how, even before the publication of *On the Origin of Species* in 1859, naturalists around the world were starting to grapple with the idea that species might evolve.[11]

In February 1847, Muñiz's article on the sabretooth cat reached Darwin himself. At this time, prior to the publication of *On the Origin of Species*, Darwin was known mainly as the author of the *Journal of Researches* (1839), his account of the voyage of HMS *Beagle*. Muñiz was aware that Darwin had collected a number of fossils during his time in

South America in the 1830s. However, the sabretooth cat was 'not among those described by the estimable *Mr. Darwin*', noted Muñiz. With this in mind, he decided to forward a copy of his article to Darwin, asking whether it might be translated from the original Spanish and published in a British scientific journal. As expected, Darwin was delighted to learn about the discovery of a new species of extinct mammal. In a letter to Richard Owen, Curator of the Hunterian Museum at the Royal College of Surgeons, Darwin described the 'wonderful collections of Fossil Bones' discovered by Muñiz. He even suggested that Owen try and purchase the fossils for the Royal College of Surgeons, or at least plaster casts of them, so that they could be compared to the ones Darwin had collected a few years earlier. Darwin also arranged for Muñiz's article to be translated into English. And although it was never published in Britain, a copy of the translation was held at the library of the Royal College of Surgeons, where it was consulted by Owen and other British naturalists. Darwin and Muñiz continued to exchange letters throughout the 1840s, discussing everything from the origins of Argentine cattle to the breeding habits of wild dogs. Darwin even made reference to Muñiz's work in a number of his books, including in a later edition of *On the Origin of Species*.[12]

Born in Argentina at the end of the eighteenth century, Muñiz was the first of a new generation of Latin American naturalists, many of whom contributed to the development of evolutionary thought. He entered the Military Medical Institute in Buenos Aires in 1814, right in the middle of the Argentine War of Independence. By the time he graduated in 1821, Argentina, Chile, Peru, and Mexico – along with many other former colonies – had declared independence from Spain. The Portuguese Empire in the Americas also collapsed around the same time, culminating in the Brazilian War of Independence of 1822–4. This was all part of the broader age of revolution which swept across the Atlantic world during the late eighteenth and early nineteenth centuries. Muñiz himself served as an army doctor in a number of military campaigns, and it was during these early postings that he first began to collect fossils.[13]

Muñiz donated most of his specimens to the Public Museum of Buenos Aires, which had been founded in 1825 with the support of the provincial government. In the past, an impressive fossil specimen like that of the *Megatherium* might have been shipped back to Spain or Portugal. But now, with independence, Latin American naturalists were encouraged to see their work as part of the process of building new national collections. 'Our sentiments ran high indeed when we learned that an object of such great worth, to be found only in our soil, was on display in the museum of a foreign country,' complained a newspaper in Buenos Aires when it was revealed that a valuable collection of fossils had recently been exported to Britain. This was all part of a wider effort to establish new scientific institutions, which were widely seen as the key to both economic prosperity and military strength. 'Without its own science, there is not a strong nation,' declared one of the members of the Argentine Scientific Society in Buenos Aires.[14]

Francisco Moreno was one of the most prolific fossil hunters in nineteenth-century Argentina. Born into a wealthy family in 1852, Moreno began collecting fossils as a young boy, digging in the soil on the banks of the River Luján, just as Francisco Muñiz had done a decade or so earlier. By the age of fourteen, Moreno had already assembled a small private museum at his parents' home in Buenos Aires. A glass cabinet housed specimens of fossilized teeth, precious stones, and sparkling shells. In 1873, inspired by reading Charles Darwin's *Journal of Researches*, Moreno decided to set out on his own scientific expedition. Sponsored by the Argentine Scientific Society, Moreno intended to head south to Patagonia, where Darwin had collected many valuable specimens, including the remains of a *Megatherium*. Moreno spent the next five years retracing Darwin's footsteps, paddling up the Río Santa Cruz and reaching as far south as the Strait of Magellan. He amassed a vast collection, including the carapace of an extinct species of giant armadillo as well as the fossilized remains of various marine animals.[15]

Much like the French naturalists in Egypt, Moreno's exploration of Patagonia would not have been possible without military support. Throughout the 1870s, Argentina sought to expand its territory to the south. Political leaders worried that if Argentina did not claim Patagonia, then another foreign power might do so. Chile had already secured much of the Pacific coastline, whilst the British had claimed a number of smaller territories in the South Atlantic, including the Falkland Islands. Darwin himself had stopped off on the Falklands during the voyage of HMS *Beagle*, which had been sent to survey the coast of South America for this very reason: to secure British interests in the region. With potential competitors in mind, the Argentine government authorized military action in 1875. In a series of brutal wars, the Argentine army pushed south into Patagonia, killing thousands of Indigenous people. Those who were not killed became forced labourers. Moreno himself assisted the military, undertaking geographical surveys and scouting the terrain in advance. In return, he was provided with men, weapons, and supplies. The Argentine army even allocated a steamship to Moreno, allowing him to explore the Patagonian coastline more closely.[16]

It was during this time that Moreno started to become interested in human evolution. 'When, in 1873, I visited for the first time the Patagonian lands, I was struck by the number of human types in the graves of the old Indian encampments,' Moreno later recalled. Seemingly unconcerned with the cultural and physical violence he was about to inflict, Moreno decided to start collecting Indigenous skulls. 'I reaped an abundant harvest of crania and skeletons,' Moreno noted enthusiastically in his diary. To be clear, these were not human fossils, but the bodies of the recently deceased. Most of the skulls were raided from the graves of Indigenous people, whilst others were collected from the battlefield in the wake of the Argentine military campaign. In one instance, Moreno even dug up the body of an Indigenous guide who had been escorting him through Patagonia. 'On a moonlit night, [I] exhumed his cadaver, whose skeleton is now preserved in the Buenos Aires Anthropological Museum,' Moreno recorded coldly.[17]

Moreno's skull-collecting activities are an important reminder of the darker side of the history of evolution. In the second half of the nineteenth century, and particularly following the publication of Darwin's *Descent of Man* (1871), naturalists around the world started to discuss the evolutionary origins of humans. In doing so, they often reinforced existing hierarchies of racial difference, wrongly assuming that Indigenous people represented evolutionary 'relics' of earlier humans. That is partly why Moreno was so obsessed with acquiring the skulls of Indigenous people. He believed that the skulls he collected in Patagonia might tell him something about the origins of the 'prehistoric Indians' who lived in the Americas. 'It seemed to me that here was a general burial place of all the American races during their forced migrations to the extreme south,' explained Moreno.[18]

All this strengthened a narrative, common across Latin America in the late nineteenth century, that presented Indigenous people as little more than the remnants of a dying civilization, soon to be displaced by modern nation states. Domingo Sarmiento, who served for a time as President of Argentina, even deployed Darwin's famous metaphor to make this very point. 'Once they have come into contact with civilised peoples, they are condemned to final extinction,' declared Sarmiento in 1879, right at the peak of the Argentine conquest of Patagonia. This was 'the struggle for existence in full force', he concluded. Here, Sarmiento wilfully confused natural selection with military aggression. The Indigenous people of Patagonia were not simply dying out. Rather, the Argentine army was trying to exterminate them.[19]

Following his return to Buenos Aires, Francisco Moreno was appointed as the first director of the new Museum of La Plata, established by the Argentine government in 1884. His extensive private collection then formed the basis of this new public museum dedicated to evolution. The fossils he had collected in Patagonia were arranged in order to demonstrate the progression of different species over time. Moreno did the same with the human remains, which were displayed in a glass cabinet alongside the caption 'Argentine man, both modern and prehistorical'. Like many Argentine naturalists, Moreno presented the region as a particularly rich site for the study of evolution. 'These animals, whose remains roiled the oceans and rivers before being deposited beneath the surface of Patagonia, display the great richness and variety of beings that once paraded their curious forms across the tertiary landscapes,' he wrote.[20]

In 1886, Moreno was joined at the Museum of La Plata by Florentino Ameghino, another ambitious Argentine fossil hunter. Born in 1854 in Luján, Ameghino grew up in the town made famous by the discovery of the *Megatherium*. Like Moreno, he collected fossils as a young boy. But unlike Moreno, Ameghino came from a much more humble background. His father was a shoemaker, and Ameghino had to support himself as a young man selling fossils to more wealthy collectors in Buenos Aires. However, he made sure to keep the best specimens for his

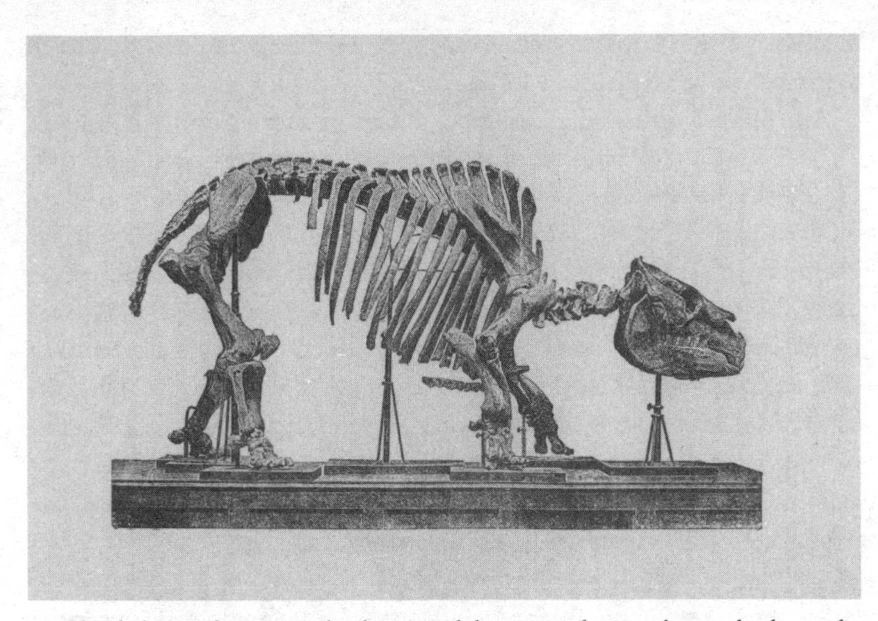

26. The skeleton of an extinct land mammal, known as the *Toxodon*, on display in the 1890s at the Museum of La Plata, Argentina.

own private collection, including a complete skeleton of an extinct armadillo. In 1882, Ameghino presented his fossils at the South American Continental Exhibition in Buenos Aires, an immense celebration of science and arts attended by over 50,000 people.[21]

Working at the Museum of La Plata, Ameghino began to reflect on how best to organize the collection. He often got into arguments with Moreno, as the two debated the exact evolutionary relationship between different fossils. Where should the giant armadillo sit in relation to the sabretooth cat? And what was the correct arrangement of the bones of the *Megatherium*? Ameghino started to believe that such arguments were futile. It was just one person's opinion against the other. 'I soon came to the conclusion that it was not the specimens that were unclassifiable, but the classifications that were defective,' he wrote. With this in mind, Ameghino decided to develop 'a new classification, built on new foundations'. The basic idea was to adopt a much more mathematical approach to evolution. Naturalists, Ameghino argued, should use precise mathematical formulae to compare the different dimensions of particular fossils. This would allow 'for the determination of the relationships among them with the same exactitude with which astronomers determine the relationship between the stars: an exactitude grounded in numbers', he explained.[22]

Ameghino's ideas were all set out in a book titled *Phylogeny* (1884), a major work of evolutionary theory. Whereas Darwin presented evolution as a broad theory to explain the origin of different species, Ameghino understood evolution as a mathematical law of nature – no different from gravity. This led Ameghino to an even more radical conclusion. If it was possible to mathematically classify extinct species of animal, then might it not also be possible to mathematically predict the existence of species that had yet to be discovered? 'Catalogued fossil animals compromise the set of known terms with which we ought to be able to determine the unknowns,' argued Ameghino. This was a bold proposal. Even today, few biologists think of evolution as a predictive science. But Ameghino, writing in the 1880s, ultimately believed that if Darwin was right, then there was no reason why naturalists could not predict the existence of other species that were still waiting to be unearthed.[23]

Over the course of the nineteenth century, Charles Darwin's ideas reached a wide audience right across Latin America. Cheap Spanish translations of *On the Origin of Species* could be purchased in Mexico City, whilst medical students in Uruguay were taught evolution as part of their degree course. Evolution even reached Cuba, where Darwinism was taught at the University of Havana in the 1870s. Some Catholic leaders did worry about the religious implications of Darwin's theory. But by and large evolution was taken up with great enthusiasm in Latin America. Argentina in particular housed a thriving community of evolutionary thinkers. As elsewhere, evolutionary thought and armed conflict seemed to go together, with Argentine collectors following the military into Patagonia in search of fossils. These collections then formed the basis of new scientific institutions dedicated to evolution, like the Public Museum of Buenos Aires and the Museum of La Plata.[24]

For many in Argentina, such impressive collections of fossils suggested that the country had much more to offer the world of science. After all, Darwin himself had first started to think about evolution when collecting fossils in Patagonia. This point was not lost on the nation's political leaders. 'With our Argentine fossils and breeds, we gave Darwin science and fame,' declared the former President of Argentina, Domingo Sarmiento, at a public lecture in Buenos Aires. Some pushed this even further, claiming that the origins of life itself might be located in Patagonia. 'There lived in the Argentine territory, not only the ancestors of mammals which inhabit it now, but also those which live in all parts and all climates of the world,' argued Florentino Ameghino during a speech celebrating the opening of the new University of La Plata in 1897. This was a radical suggestion, one that in fact turned out to be incorrect. Still, in an age in which Latin American states were searching for new national identities, such evolutionary narratives proved incredibly attractive. Ameghino wanted the world to know that Argentina was no longer just some Spanish colony, sitting on the periphery of the Atlantic Ocean. Rather, it was at the centre of a new evolutionary history of the planet.[25]

II. Evolution in Tsarist Russia

Nikolai Severtzov watched as a large brown bear made its way down the steep cliff. The air was ice cold, the ground covered in snow. Severtzov could feel his heart beating more quickly. The bear was getting close now. Slowly, he raised his rifle and pulled the trigger. The sound of the shot echoed around the mountains, as the bear fell on its side, bleeding into the snow. A few minutes later, the animal was dead. Pleased with his kill, Severtzov approached the bear and began to examine its claws. As he suspected, they were white. Making a note in his journal, Severtzov ordered his local Kyrgyz guide to skin the bear, before the two continued on their path through the Tian Shan Mountains, in what is today modern Kyrgyzstan. Throughout the 1860s, Severtzov – an influential Russian naturalist – travelled far and wide across Central Asia. He collected hundreds of animal specimens, many of which he shot himself, including bears, bats, and eagles. Severtzov later donated most of his collection to the Museum of Zoology at Moscow University. These specimens also formed the basis of Severtzov's major work of natural history, *The Vertical and Horizontal Distribution of Turkestan Animals* (1872).[26]

Much like the other naturalists we've encountered, Severtzov's scientific expeditions were part of a broader military campaign. From the late 1840s onwards, the Russian Empire expanded into Central Asia, beginning with an attack on the Khanate of Kokand in 1847 and culminating with the conquest of Turkestan in 1865. This was all part of a wider struggle, which came to be known as the 'Great Game', in which the British and Russian empires battled for control of Central Asia. Severtzov himself came from a military background. His father had fought against Napoleon during the French invasion of Russia, leading a company of troops at the Battle of Borodino in 1812. Severtzov too entered the military, although he served in quite a different capacity. After graduating from Moscow University, where he studied zoology, Severtzov joined the army as a naturalist. Sponsored by the Saint Petersburg Academy of Sciences, he spent over a decade collecting specimens and documenting the landscape during the Russian conquest of Central Asia. At one point, he was even taken captive by a group of Kokand rebels, spending a month in prison, chained to the wall. Severtzov was eventually liberated during a Russian counteroffensive. However, during the battle he was struck across the face, leaving him with an impressive scar for the rest of his life.[27]

On returning from Central Asia, Severtzov presented his findings at a meeting of the Moscow Society of Naturalists in 1872. He must have looked somewhat out of place amongst all the well-to-do gentlemen in their frock coats. Severtzov seemed more like a rugged explorer than a traditional scientist, with his long tangled hair, scruffy grey beard, and old fur coat. But whatever his appearance, Severtzov had a keen scientific mind. For him, the natural history of Turkestan provided striking evidence of the effects of the environment on the evolution of animals. For example, in the case of the brown bear, Severtzov documented how the colouration of the claws and fur seemed to vary with altitude. Bears living at higher altitudes – like the one he shot in Tian Shan – tended to have white claws and light fur. In contrast, those living at lower altitudes tended to have black claws and darker fur. Severtzov reasoned that this must be an evolutionary adaptation to the environment, as lighter fur and white claws would clearly provide better camouflage in snowy conditions. He also documented the 'origins of the wild and tame sheep of Central Asia', noting that the wild sheep – with their large horns and strong muscles – 'were obliged to modify in order to exist, and to avoid being driven altogether away by the tame flocks'.

The suggestion was that the introduction of domesticated sheep had caused the wild sheep to adapt and become more hardy in the face of competition. All this was evidence of 'the laws of the variation of species', argued Severtzov.[28]

Although his major work was published in the 1870s, Severtzov had in fact been writing about evolution since the early 1850s. In 1855, he completed an MA dissertation at Moscow University on the relationship between the environment and the variation of species around the city of Voronezh, in southwest Russia. In the dissertation, Severtzov, much like Charles Darwin, described the development of species in terms of a 'tree' of life. Later, in a talk at the Saint Petersburg Academy of Sciences in 1857, Severtzov expanded upon his earlier ideas. He told the members of the Academy that there was 'a principle of evolution, of modification, inherent to the organism'. Like many Russian naturalists, Severtzov placed particular emphasis on the power of the environment to direct evolution. 'Under the influence of the environment, the species type is modified,' he told the Academy. It was this talk that persuaded the Saint Petersburg Academy of Sciences to sponsor Severtzov's trip across Central Asia with the Russian army. Many of the members of the Academy were already convinced that species could undergo transformation, and were therefore sympathetic to Severtzov's ideas. They hoped that the specimens he collected in Turkestan would provide the evidence needed to substantiate such a theory.[29]

Following the publication of *On the Origin of Species* in 1859, Charles Darwin's ideas found a receptive audience in nineteenth-century Russia. This was in part because, much as we've seen elsewhere, Russian naturalists were already thinking about evolution. Members of the Saint Petersburg Academy of Sciences had been debating the theory of 'progressive metamorphosis' since the early 1820s, whilst students at Moscow University were taught about evolution from the 1840s onwards. Darwin himself even acknowledged an earlier Russian embryologist, named Karl von Baer, as having made an important contribution to the development of evolutionary theory in the 1820s.[30]

The timing of the publication of *On the Origin of Species* was important, as Darwin's ideas entered Russia during a period of renewed scientific growth. Following defeat in the Crimean War of 1853–6, Tsar

Alexander II authorized a series of sweeping educational and political reforms. At the time, it was widely believed that Russia had fallen behind other European nations, and needed to modernize once again, much as it had under Peter the Great at the end of the seventeenth century. 'If our enemies are superior to us, it is only because of the power of their knowledge,' declared the Minister of National Education in 1855. In response, the government introduced scientific education into all Russian schools for the first time. As part of the same reforms, universities were given greater autonomy in the appointment of professors and the allocation of funds. This led to the creation of new museums and laboratories dedicated to science, including the aforementioned Museum of Zoology at Moscow University, established in 1861, as well as the Sevastopol Biological Station, established in 1869.[31]

As in many other countries, enthusiasm for Darwin's ideas was closely associated with this wave of modernization. The *Russian Herald*, a new liberal magazine published in Moscow, described *On the Origin of Species* as 'one of the most brilliant books ever to be written in the natural sciences', whilst the secretary of the Saint Petersburg Society of Naturalists noted that 'almost all leading contemporary biologists are followers of Darwin'. *On the Origin of Species* quickly appeared in a Russian translation in 1864, as did many other works by leading British evolutionary thinkers. Thomas Henry Huxley's *Evidence as to Man's Plan in Nature* (1863) was translated into Russian, as was Alfred Russel Wallace's *Contributions to the Theory of Natural Selection* (1870). Evolution also seeped into Russian literary culture. Leo Tolstoy's *Anna Karenina* (1878) even featured a passage in which one of the characters starts explaining the 'struggle for existence' and 'natural selection' to Anna herself. Russia's other great nineteenth-century novelist, Fyodor Dostoevsky, was similarly enthusiastic about Darwin, going as far as to describe the British naturalist as 'the leader of European progressive thought'. There was, as elsewhere, some resistance from religious authorities, particularly following the publication of Darwin's *Descent of Man* (1871), which was briefly banned by a government censor. But on the whole, belief in evolution was considered perfectly respectable in nineteenth-century Russia.[32]

Enthusiasm for evolution, however, did not mean that Darwin's ideas were simply accepted at face value. Even the most committed Darwinists recognized that *On the Origin of Species* left a lot of unanswered questions. Russian naturalists in particular often criticized Darwin for placing too much emphasis on competition between individuals as the main driving force behind evolution. Many instead chose to focus on the importance of the environment or disease in natural selection. Darwin's emphasis on competition also seemed to ignore the role of cooperation in both human and animal societies. This was something Darwin himself was aware of and tried to address in his later work, particularly in *The Descent of Man*. But even then, Darwin found it difficult to explain how a world governed by brutal competition could give rise to such complex forms of cooperation, whether that was a colony of bees working together to build a hive or wolves hunting in a pack. With this in mind, a number of Russian naturalists took up, extended, and sometimes even challenged Darwin's earlier ideas. In doing so, they made significant contributions to the development of evolutionary thought.[33]

Ilya Mechnikov peered through the microscope, carefully watching a starfish embryo that he had collected the previous day. It was a beautiful if surreal sight. Mechnikov could see all the cells moving inside the developing animal, which at this stage was still translucent. He then did something rather cruel. Mechnikov took a sharp thorn and pushed it into the embryo. Then, he waited. Just as Mechnikov hoped, the starfish embryo began to react. Under the microscope, Mechnikov could see a group of cells start to migrate to the site of the puncture, clustering around the thorn. Over the following hours, this group of cells was then able to push the thorn out of the embryo. Mechnikov realized that he was observing something incredibly important. This was the first direct evidence that animal cells could coordinate in order to produce an immune response. Reporting his findings at a meeting of the Congress of Russian Naturalists and Physicians in 1883, Mechnikov described what he called the 'phagocyte theory'. Whilst scientists had known about white blood cells since the middle of the nineteenth century, no one really knew what they were for. Most doctors at the time simply believed that inflammation was a symptom of disease, something that needed to be controlled. Mechnikov, however, realized that this was wrong. Inflammation was not simply a symptom of disease, rather it was the coordinated response of various cells – such as the phagocytes

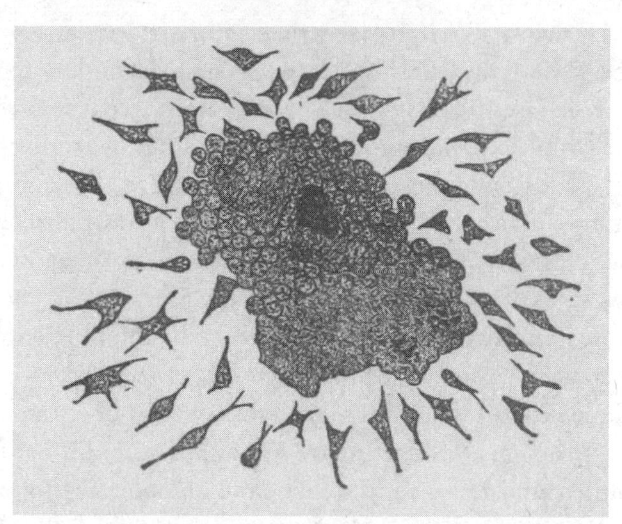

27. The formation of phagocytes around the site of a puncture in a starfish embryo, as seen under the microscope by Ilya Mechnikov.

he had observed in the starfish – to combat infection. This was a major breakthrough in the scientific understanding of disease for which Mechnikov shared the 1908 Nobel Prize in Physiology or Medicine.[34]

Today, Mechnikov is remembered as a pioneer of medical science. However, he was also an important evolutionary thinker. Born in Kharkov in the middle of the nineteenth century, Mechnikov first learned about Charles Darwin whilst studying in Germany in the 1860s. In Leipzig, he bought a German translation of *On the Origin of Species*, which he read with great excitement. Mechnikov was sympathetic to the idea of evolution, although like many other Russian naturalists he thought that Darwin placed too much emphasis on competition for resources between members of the same species. 'The view that every corner of the earth is teeming with life is positively untrue,' he wrote. Nonetheless, it was this interest in evolution which brought Mechnikov to the study of the immune system. After completing a PhD in embryology at Saint Petersburg University, Mechnikov joined the University of Odessa in 1870. Established in 1865, this was one of the new universities created as part of Alexander II's educational reforms. And it was here, next to the Black Sea, that Mechnikov began to work on the evolution of immunity in marine animals.[35]

Whilst Darwin had emphasized that natural selection involved the struggle between individuals within a species, Mechnikov highlighted the role of disease. Throughout the nineteenth century, the world experienced multiple waves of pandemics, ranging from cholera to influenza. These grew in severity over the course of the century, as the world became increasingly connected via new industrial technologies like railways and steamships, allowing disease to spread at a faster rate. Mechnikov himself lived through one of the most deadly cholera outbreaks of the nineteenth century, in which over a million people died in Russia between 1846 and 1860. And in 1873, Mechnikov's first wife, Ludmilla, died from tuberculosis, aged just twenty-one. Life really did seem like a struggle, and Mechnikov sometimes found it hard to cope. (He attempted suicide twice, the first time immediately following his wife's death.) Ultimately, for Mechnikov, as for many other Russians, the greatest threat to life was not competition for resources. Rather, the struggle for existence was a struggle to survive in the face of deadly diseases. It was this idea that shaped how Mechnikov thought about evolution.[36]

In the nineteenth century, most evolutionary thinkers believed that anatomical similarities between humans and apes provided the best evidence of a shared ancestry. Mechnikov, however, took a different approach. He argued that the existence of immune cells was direct evidence for the shared ancestry of all living things. A single-celled organism, like a bacterium, typically survived by engulfing other smaller organisms and digesting them within the cell. Mechnikov pointed out that this is exactly what immune cells do. A white blood cell, like a macrophage, engulfs the bacteria, and digests it within the cell in order to combat the spread of disease. Mechnikov reasoned that white blood cells must be the evolutionary relics from when single-celled organisms evolved into multicellular organisms, probably through the very process of engulfing another cell. He also noted that many different kinds of animals, from humans to starfish, had similar kinds of immune cells, indicating a shared evolutionary history. This was the best evidence that 'man is a blood relative of the animal', as Mechnikov put it.[37]

Mechnikov believed that the existence of immune cells provided direct evidence for evolution, and at the same time, he understood inflammation itself as a form of natural selection taking place within the body. The job of the different immune cells was to overcome bacteria and other foreign bodies. Often this was achieved by engulfing the foreign cell, and destroying it before it could spread further and reproduce. 'It is a veritable battle that rages in the innermost recesses of our beings,' wrote Mechnikov in 1903. At other times, he pushed the military metaphor even further. The immune system was 'like a highly organized state . . . fighting against the savage tribes', declared Mechnikov at a lecture in Odessa, right at the time when the Russian Empire was expanding into Central Asia. 'Against the bacteria it sends an army of amoeboid cells,' he concluded. Here, we can again see how the growth of nationalism and war in the nineteenth century conditioned how scientists thought about nature itself. For Mechnikov, the body was just another battlefield.[38]

Whilst Ilya Mechnikov was working in Odessa, another group of Russian naturalists was conducting important research on the other side of the Black Sea. Based at the Sevastopol Biological Station, they were led by Sofia Pereiaslavtseva, a pioneering embryologist and one of the first women in the world to direct a scientific laboratory. Getting this far had not been easy. Whilst the educational reforms introduced by Tsar Alexander II had resulted in an increasing number of Russian men attending university, the same was not true for women. In 1861, a group of students marched on Saint Petersburg University, demanding greater access to higher education for women. Alexander II not only ignored the demands, but in fact decided to formally ban all women from attending Russian universities. (Prior to this, some women had been able to study informally at Russian universities, although they were not awarded degrees.) Undeterred, a number of women decided to take things into their own hands. Rather than waiting for things to change in Russia, they instead chose to study abroad. That is exactly what Pereiaslavtseva did in 1872, when she travelled to Switzerland to enrol at the University of Zurich. This was a popular destination for many Russian women at the time, as the University of Zurich not only allowed women to study, but also awarded formal degrees. The same was not true in Russia until after the Bolshevik Revolution of 1917.[39]

Pereiaslavtseva had always loved natural history. The daughter of an army colonel, she collected butterflies as a young girl around her hometown of Voronezh. Her dream was to become a professional naturalist. Pereiaslavtseva must therefore have felt a deep sense of frustration when she learned that Alexander II had barred women from entering university in Russia. Nonetheless, she convinced her father – who was sympathetic to the cause of women's education – to allow her to study in Switzerland. In 1876, after four years of intense study, Pereiaslavtseva graduated with a PhD in zoology, one of the first Russian women to do so. She returned to Russia in 1878, and was quickly appointed as director of the Sevastopol Biological Station.

Over the next ten years, Pereiaslavtseva conducted research on evolutionary embryology. Working next to the Black Sea, she collected the embryos of different marine animals, examining them under the microscope back in the laboratory. This kind of work required immense skill and patience. Pereiaslavtseva needed to compare the stages of embryological development between different species. She believed, much like the French naturalist Étienne Geoffroy Saint-Hilaire, that this would reveal something about the evolutionary history of different animals. However, to make these kinds of comparisons, Pereiaslavtseva needed to watch the development of the embryo under the microscope for hours on end. In some cases, she would stay at the lab bench for up to thirty hours, with only a few short breaks to rest.[40]

As a champion of women's education, Pereiaslavtseva used her position at the Sevastopol Biological Station to promote the careers of other female scientists. She was soon joined by two other early pioneers of evolutionary embryology: Maria Rossiiskaia and Ekaterina Wagner. Together, these three women worked on the embryology of different marine animals, comparing their results. Pereiaslavtseva studied flat worms, whilst Rossiiskaia and Wagner studied shrimps. In a series of papers published in the *Bulletin of the Moscow Society of Naturalists*, Pereiaslavtseva and her team were able to establish the evolutionary links between different species of marine animal based on their embryological development. In recognition of her work, Pereiaslavtseva was awarded a major prize by the Congress of Russian Naturalists and Physicians in 1883, a rare acknowledgement of the contribution of women to science in a period in which the profession was still dominated by men.[41]

*

The nineteenth century was an age of capitalism and conflict. But it was also an age in which people began to articulate a range of political alternatives, whether that was socialism, communism, or anarchism. In Russia, there was often considerable overlap between those who were interested in evolution and those who were interested in left-wing politics. Leon Trotsky, who went on to lead the Bolshevik Revolution in 1917, read a number of Charles Darwin's books during a spell in prison in the 1890s, later telling a friend that 'the idea of evolution . . . took possession of me completely'. Around the same time, Peter Kropotkin – a leading Russian anarchist – published *Mutual Aid: A Factor of Evolution* (1902). In this book, Kropotkin drew explicit parallels between cooperation in the animal kingdom and the need for humans to work together to survive. 'The unsociable species . . . are doomed to decay,' argued Kropotkin, who at the time was living in London, having fled Russia following persecution for his political views.[42]

Andrei Beketov was another one of those who saw in socialism a different way of thinking about both society and nature. Born in Central Russia in 1825, Beketov was somewhat of a rebel. He had first been sent to study at a military academy, but was soon discharged due to his lack of discipline. The young Beketov then hung around Saint Petersburg for a bit, where he attended a socialist study circle, reading the works of the early French socialist Charles Fourier. This was a dangerous thing to be involved in, as the Tsars did not take kindly to political dissent. Nonetheless, Beketov managed to stay out of trouble, and eventually moved to study botany at Kazan University, where he graduated with a PhD in 1858. Soon afterwards, he was appointed as Professor of Botany at Saint Petersburg University.[43]

Throughout his career, Beketov highlighted the role of the environment in shaping evolution. The 'actual struggle for existence', Beketov told his students at Saint Petersburg, was not between individuals, competing for limited resources. Rather, it was between the individual and the environment. Plants provided a good example of this. Beketov asked his students to imagine the frozen landscape of Siberia, or the exposed terrain of the Russian steppes. In both cases, plants did not really compete against one another for survival. Instead, the threat to life was the cold and the wind. According to Beketov, there was 'a constant and stubborn struggle with the elemental forces of nature'. It was this struggle with the environment that explained the evolution of particular adaptations observed in these regions, argued Beketov. He noted how plants growing in Siberia tended to be resistant to cold, with shallow roots that spread out through the rocky landscape, whilst those growing in the Russian steppes tended to be low-lying, protecting them from the wind.[44]

Beketov also believed that the pressures of a harsh environment could help explain the evolution of cooperation. He gave examples of how, both in Siberia and the Russian steppes, plants of the same species often grow close together, in order to shield each other from the wind. In making this point, Beketov developed an early understanding of what we would now call 'ecology', noting how plants in a forest often rely on one another for support. The 'mutual aid that plants provide to one another' was the key to their survival, he argued. Ultimately, Beketov – the committed socialist – believed that Darwin had been wrong to assume that individual struggle was a necessary part of life. By working together, humans, animals, and even plants could better survive in the face of harsh conditions. 'Sociability' was 'a powerful means of self-defence', he concluded.[45]

When Darwin died in 1882, the Congress of Russian Naturalists and Physicians organized a special conference to celebrate his life and work. Almost everyone agreed that he was one of the most important scientific thinkers of the nineteenth century. However, many also pointed out that Darwin had left a lot of unanswered questions. 'Darwin died before he had completed his work,' declared one of the attendees, echoing a widely shared sentiment amongst Russian naturalists at the time. True, some found the idea of a 'struggle for existence' perfectly reflected the world in which they lived, particularly in the aftermath of the Crimean War. But many Russian naturalists also felt that *On the Origin of Species* could not explain everything. In particular, Darwin's emphasis on the struggle between individuals seemed to ignore the role of the environment and disease in natural selection. Interest in evolution was therefore paired with a sense that Darwin's legacy was incomplete. From Ilya Mechnikov's study of the immune system to Andrei Beketov's research on 'mutual aid', Russian naturalists pushed Darwin's ideas in new directions. In doing so, they helped establish evolution as a fundamental component of the modern biological sciences.[46]

III. Darwinism in Meiji Japan

Edward Morse took to the stage, ready to deliver his first of three lectures on evolution. He had only arrived in Japan a few months earlier, having travelled from the United States in order to study a local species of brachiopod – a kind of ancient marine animal with a deep evolutionary history. However, Morse now found himself preparing to speak in front of over 800 people at the University of Tokyo. On 6 October 1877, he began his first lecture with a striking description of the principle of natural selection. Morse asked his audience to imagine the following scenario:

> If I were to lock tight the doors of this lecture hall, in just a few days the list of the dead would include those in the audience with weak bodies. Those of you in good health would probably die in a week or maybe two or three.

Morse paused for a moment, letting the audience reflect on what he had just said. Some looked around, checking that the doors at the back of the lecture hall were still open. Others made a mental note of those who they thought would be most likely to perish. Morse then continued, suggesting that the natural world was just like the lecture hall, 'a closed space with insufficient food'. In such a scenario, only the strongest would survive, and pass on their physical characteristics. 'If this situation were to continue for a number of years . . . the people of the future would be entirely different from the people of the present,' explained Morse. 'A powerful and terrible kind of human would be born,' he concluded.[47]

Over the following weeks, Morse continued his lecture series on evolution. In the second lecture, he pushed the concept of the 'struggle for existence' even further. The audience at the University of Tokyo listened as Morse claimed that 'groups with traits useful in war tend to survive'. He also explained the importance of technological advances in the survival of the fittest. 'It is obvious that groups capable of creating metal weapons will defeat those fighting with bows and arrows,' argued Morse. Natural selection, he explained, was nothing more than the principle that 'the advanced race survives and the less advanced is destroyed'. These kinds of military metaphors, as we've seen throughout this chapter, were a common part of evolutionary thought in the nineteenth century. And in Japan, such talk was particularly powerful. Less than a decade earlier, the people of Japan had themselves been engaged in a brutal civil war. In 1868, a group of samurai formed an alliance and fought to overthrow the Tokugawa shogunate. They believed that the shogun was holding back the modernization of Japan, and had displayed weakness in the face of foreign military aggression. The samurai fought all the way to the capital city, Edo, where they defeated the shogun's forces, and placed the young Emperor Meiji on the imperial throne. This marked the beginning of the period known as the Meiji Restoration.[48]

Amongst those listening to Morse's lectures at the University of Tokyo was a young Japanese biologist who had personal experience of the civil war. Chiyomatsu Ishikawa was born in Edo in 1861. His father, who worked for the shogun, owned a large collection of traditional Japanese works of natural history and medicine. As a young boy, Ishikawa studied many of the books we encountered in the previous chapter, including Kaibara Ekiken's *Japanese Materia Medica* (1709–15). This sparked a love of natural history, and particularly of zoology. Ishikawa would spend the summers collecting butterflies and crabs around Edo Bay. However, the tranquillity did not last. Following the outbreak of the Japanese civil war, Ishikawa and his family had to flee the city, as allies of the shogun were hunted down. By the time they returned in the 1870s, the shogun had been deposed, and Edo had been renamed Tokyo.[49]

Although his father had lost his position in the shogun's court, the Meiji Restoration nonetheless brought new opportunities for Ishikawa. In 1877, the Emperor Meiji authorized the creation of the University of Tokyo. With a dedicated Faculty of Science, this was to be the first modern university in Japan. It was followed by a number of other new universities, including Kyoto University in 1897 and Tohoku University in 1907. This was all part of a wider programme of modernization that took place during the Meiji Restoration, with new laboratories, factories, railways, and shipyards being constructed across the country. At the same time, the Japanese government began employing foreign scientists and engineers to teach at many of these new institutions.

Morse, who previously worked at the Museum of Comparative Zoology at Harvard University, was one of those employed by the University of Tokyo to teach biology. In fact, between 1868 and 1898, the Meiji government employed over 6,000 foreign experts – mostly British, American, French, and German – to teach in Japan. This was a significant shift in policy compared with the earlier period. As we saw in the previous chapter, the Tokugawa shogunate had strictly regulated the entry of foreigners into the country.[50]

Ishikawa was one of the first to benefit from the reforms introduced following the Meiji Restoration. He entered the University of Tokyo in 1877, the year it was established, and became a student of Morse. Every summer, Morse would take his students – including Ishikawa – to Enoshima, a small island south of Yokohama. It was here that Ishikawa learned the basic techniques of modern biological science: collecting different marine animals from the water, observing them under the microscope, and dissecting them. Morse was also an avid follower of Charles Darwin, having read *On the Origin of Species* back at Harvard. During his trips to Enoshima, Morse spent a lot of time explaining the principles of evolution to his students. Fascinated by the idea of natural selection, it was in fact Ishikawa who suggested that Morse give a series of public lectures on the topic back at the University of Tokyo. And it was Ishikawa who later translated Morse's lectures into Japanese, publishing them under the title *Animal Evolution* (1883).[51]

After graduating from the University of Tokyo, Ishikawa was selected in 1885 to go and study in Germany. By this point, the government had decided it was getting far too expensive to keep employing foreign scientists to work in Japanese universities. Instead, the Minister of Education proposed that promising young students be sent abroad to undertake advanced training in scientific subjects. The idea was that they would then return and take up academic positions at the new universities being established across Japan. 'We will not go forward unless we send people to study in the advancing countries,' declared the Minister of Education. As we'll see over the following chapters, many of the most influential Japanese scientists of the late nineteenth and early twentieth centuries spent some time studying abroad, mostly in Britain, Germany, and the United States. Ishikawa was one of the first to go. Between 1885 and 1889, he studied under the leading German biologist

August Weismann at the University of Freiberg. At this time, Weismann was developing his theory of the 'germ plasm', in which he predicted the existence of some kind of hereditary material that was passed on exclusively via the sperm and the egg. In making this claim, Weismann laid the foundations for modern genetics, challenging the older idea – shared by Darwin – that it was possible to pass on characteristics acquired during life.[52]

Ishikawa was studying at the University of Freiberg right at this crucial time. He even collaborated with Weismann, co-authoring six papers that appeared in leading German scientific journals. In one article, Ishikawa described how he had observed the replication of reproductive cells in a tiny translucent marine animal called a water flea. Watching the water flea under the microscope, Ishikawa spotted the formation of two tiny black dots at the edge of one of the eggs as it divided. What he was in fact observing was the process known as 'meiosis', in which an organism produces reproductive cells through replication and division. The black dots that Ishikawa identified were the remnants left over when a cell divides. These 'polar bodies', as they came to be known, proved a crucial piece of evidence in support of Weismann's germ plasm theory. They suggested that Weismann was correct in arguing that the sperm and egg must be produced through a process of cell division separate from the rest of the body.[53]

Ishikawa returned to Japan in 1889 to take up a position at the University of Tokyo. Over the following years, he helped train a new generation of Japanese biologists, many of whom made important contributions to the study of evolution. As in many other countries, Darwinism was closely associated with modernization in Meiji Japan. The idea of a 'struggle for existence' appealed, not just to biologists, but also to political thinkers. It seemed to justify the need for industrialization and military expansion. 'The struggle for survival through natural selection . . . applies not only to the world of animals and plants, but also with the same compelling necessity to the world of human beings,' wrote Hiroyuki Kato, a political philosopher who also attended Morse's lectures at the University of Tokyo. 'The universe is one vast battlefield,' he concluded, just as Japan was preparing for the First Sino-Japanese War of 1894–5.[54]

At the same time, Darwin's ideas proved popular because they seemed to confirm what many Japanese naturalists already believed. This is something Ishikawa would have understood, having studied traditional works of Japanese natural history as a young boy. 'All human beings may be said to owe their birth to their parents, but further inquiry into their origins reveals that human beings come into existence because of nature's law of life,' wrote Kaibara Ekiken, the seventeenth-century Japanese naturalist we met in the previous chapter. Indeed, unlike in Christian Europe, Japanese naturalists were already comfortable with the idea – found in both Buddhism and Shinto – that all life shares some kind of organic origin. Morse himself recognized this, writing that 'it was delightful to explain Darwinian theory without running up against theological prejudice as I often did at home'. One early nineteenth-century Buddhist philosopher, named Kamada Ryuo, even developed his own theory of evolution. 'It must be the case that all plants and animals have split from one species to become the manifold species,' wrote Kamada in 1822, when Darwin was aged just thirteen. The basic idea of evolution was not therefore new in Japan. What was new, however, was the mechanism. It was Darwin's concept of the 'struggle for existence' which really caught the imagination of Japanese biologists.[55]

Asajiro Oka followed a similar career path to Chiyomatsu Ishikawa. Born in 1868, the year of the Meiji Restoration, Oka grew up in Osaka, the son of a successful bureaucrat in the new government. His early life was, however, tinged with tragedy. Oka's younger sister was killed in a horrific accident in which her kimono caught fire. And then the following year, his mother and father both passed away. Finding himself alone in Osaka, Oka moved to Tokyo, where he was raised by his extended family. Like Ishikawa, he went on to study zoology at the University of Tokyo, graduating in 1891. Oka was then selected to undertake further training in Germany, where he also trained under August Weismann at the University of Freiberg. In 1897, Oka returned to Japan to take up a professorship at the Tokyo Higher Normal School. Over the following decades, he played a major role in popularizing evolution in Japan. Oka's *Lectures on Evolutionary Theory* (1904), based on his time teaching at the Tokyo Higher Normal School, proved a hit. The book sold tens of thousands of copies, and turned Charles Darwin into a household name in Japan. At the same time, Oka made a number of important contributions to evolutionary thought in his own right.[56]

Oka's specialism was the biology of Bryozoa, or 'moss animals'. These curious creatures had been studied by the influential German biologist Ernst Haeckel, and Oka himself probably learned about them in Germany. They seemed to blur the line between plants and animals. Each moss animal was formed from a colony of millions of single-celled organisms. When grouped together, the cells started to form structures that looked very much like a plant. Oka himself went hunting for moss animals around Tokyo. He would search in the undergrowth, next to

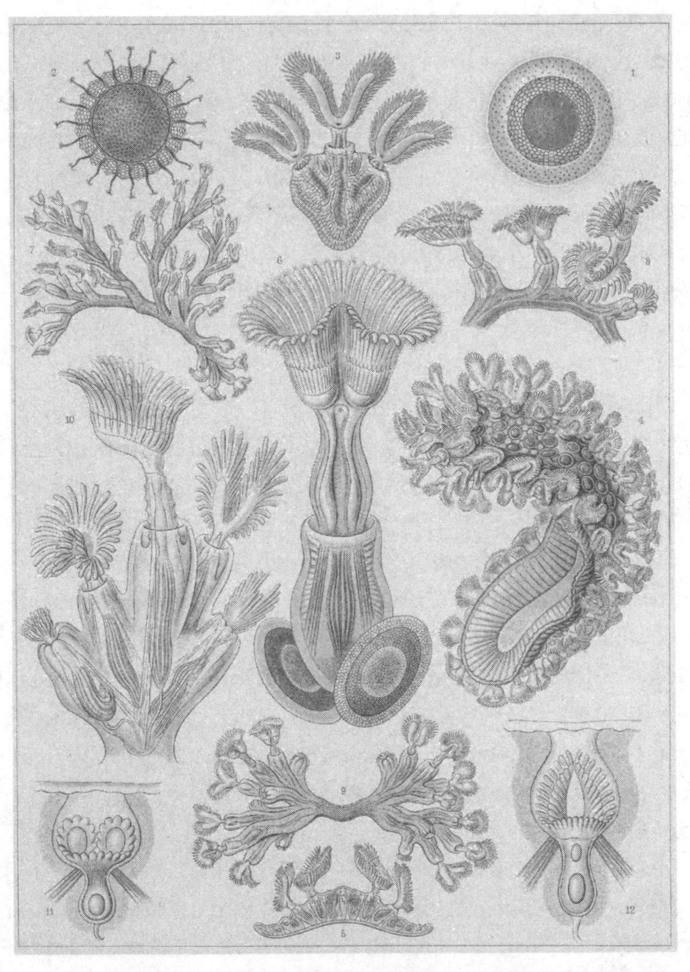

28. Bryozoa, or 'moss animals', as studied by both Ernst Haeckel and Asajiro Oka.

pools of water, collecting specimens in a small glass jar, before examining them under the microscope back at his laboratory. According to Oka, moss animals suggested that the way biologists divided the natural world into different species was wrong. 'It is impossible to establish a clear boundary,' he argued. This was in fact the very same insight that underpinned Darwin's *On the Origin of Species*. After all, what sense did it make to describe something as a particular species if it might evolve into something else? Oka pushed this idea even further, arguing that even some of the most fundamental divisions in the natural world – such as the division between animals and plants – no longer made sense. Animals could sometimes act like plants, and plants like animals. 'Whatever we see in nature is all a continuum of change,' he concluded.[57]

Oka published his *Lectures on Evolutionary Theory* in 1904, the year of the outbreak of the Russo-Japanese War. Over a period of eighteen months, the Japanese and Russian armies fought for control of the region around Korea and Manchuria. 200,000 people lost their lives in what was one of the first industrial wars of the twentieth century. In the end, Japan emerged victorious, although many back home wondered if the fighting had been worth it. Once again, Oka started thinking about moss animals. They seemed to behave rather like human societies, with individual cells coming together to fight as a stronger unit. Within each moss animal, cells shared resources and worked together. Oka even did some experiments, pipetting algae into a Petri dish to feed the colony of cells, or 'nation' as he sometimes referred to it. 'Whichever takes the food, the nutrients will be divided equally,' he reported. The individual cells in a moss animal were clearly capable of cooperation. However, with cooperation came the potential for conflict. Again, Oka did an experiment, placing two separate specimens in the same jar. The two moss animals then battled it out, until only one was left alive. Oka even noted that some moss animals deployed specialized barbed cells, filled with toxins, to attack the enemy. Chemical warfare, it seemed, was just another evolutionary adaptation, a necessary consequence of the struggle for existence. The Japanese had in fact deployed arsenic in the recent war with Russia, a forerunner to the widespread use of chlorine gas in the First World War. 'In this respect, humans do not differ in the least from other organisms,' concluded Oka, a grim reminder of how seemingly innocuous biological ideas could be used to justify the worst acts of violence.[58]

Darwin's ideas entered Japan during a key period of historical change, starting with the Meiji Restoration of 1868. Much as we saw in Argentina and Russia, the idea of a 'struggle for existence' appealed to scientists in Japan because it seemed to reflect the world in which they lived. Both the First Sino-Japanese War of 1894–5 and the Russo-Japanese War of 1904–5 appeared to confirm what Oka called 'the law of life and death'. Humans, Oka argued, were no different from the moss animals he examined in the laboratory, coming together to form larger units and engaging in brutal warfare. As we'll see, it was a very similar attitude to military confrontation that motivated interest in evolution in China, Japan's main imperial rival in the region.[59]

IV. Natural Selection in Qing China

Yan Fu watched in horror as the Chinese flagship was struck by a torpedo. The immense ironclad cruiser was less than a decade old. Now it was stranded, ablaze in the harbour off the coast of Shandong, smoke billowing from the deck. Yan, a Chinese naval engineer, was an eyewitness during the closing stages of the First Sino-Japanese War of 1894–5. A few months earlier, in September 1894, most of the Chinese fleet had been destroyed off the coast of Korea. The remainder of the fleet was then pursued by the Japanese Navy, culminating in the Battle of Weihaiwei in January 1895, in which over 4,000 Chinese sailors lost their lives. By April, the Chinese government had surrendered, signing a peace treaty which guaranteed Japanese control over Korea and Taiwan. This was a humiliating defeat for China, which had long considered itself superior to Japan, and prompted much soul-searching.[60]

Yan was one of those who called for a complete overhaul of the Chinese educational and political system. China, he believed, needed to modernize, or risk being taken over by its rivals. 'Our country is enclosed by hostile powers,' exclaimed Yan in a newspaper article published immediately following the battle. 'We do not have enough time to acquire development before our country follows the deteriorating patterns of India and Poland,' he continued, warning that it might not be long before China found itself turned into a European or Japanese colony. Such calls for reform were common following the Chinese defeat in the First Sino-Japanese War. What made Yan's article unique was that he pitched his argument in terms of evolution. The war with Japan was an example of 'Darwinian principles' at work, argued Yan. 'Natural selection', he claimed, applied just as much to nations and society as it did to individuals. Yan went on, explaining the basics of Charles Darwin's theory to his Chinese audience. 'Men and all other living things are born on the earth in great profusion . . . they band together, and each people and each species struggles to preserve itself,' he explained. China was locked in a 'struggle for existence', concluded Yan. The choice was simple: evolve or die.[61]

As we've seen throughout this chapter, the development of evolutionary thought was closely associated with the rise of war and nationalism in the nineteenth century. This was certainly true in China. Although there were some earlier scattered references to Darwin's ideas, it was Yan Fu who popularized the idea of a 'struggle for existence' in China. Yan himself first learned about Darwin whilst studying engineering at the Royal Naval College in London. He was one of the many Chinese students sent abroad to study science in the second half of the nineteenth century, as the Qing dynasty looked to modernize its army and navy. During his time in Britain in the 1870s, Yan began reading the works of leading Victorian scientific thinkers, including *On the Origin of Species*. Watching the destruction of the Chinese fleet, he was suddenly reminded of Darwin's bleak description of the natural world as one of constant struggle. 'Species struggle with species, group struggles with group, and the weak are devoured by the strong,' recalled Yan, as he witnessed the Japanese Navy bombarding the Chinese coastline.[62]

Yan's article generated widespread interest in Darwin's work in China. Encouraged by the response, he decided to write a longer piece on the subject. Published as *The Theory of Evolution* (1898), this book expanded on many of the themes of Yan's earlier article, extending evolutionary theory into the world of society and nations, with all the dangerous implications that implied. Yan was in fact one of a number of Chinese thinkers who played a significant role in the development of Social Darwinism in the late nineteenth century. 'Living forms progress in natural evolution; hence social evolution is undoubtedly progressive,' he argued. In making this claim, Yan echoed the words of the British evolutionary thinker Herbert Spencer, who described society as a kind of 'social organism'. (Yan later translated Spencer's major work, *The Study of Sociology* (1873), into Chinese.) Like Spencer, Yan believed that society only progressed through competition. 'Men enjoy relaxation and hate labouring; if we stop them from competing, they will not use their mental and physical power; hence . . . they will not progress,' argued Yan. And so, rather than withdrawing from the world of capitalism and conflict, Yan recommended that China double down on industrialization and militarization. The alternative was to face what he called 'racial extinction'.[63]

Yan's *Theory of Evolution* was read by many of the most influential scientific and political thinkers of the day, who saw in the 'struggle for existence' a clear diagnosis of the problems that China faced. Liang Qichao, a prominent Chinese journalist who knew Yan personally, was fascinated by evolution. He too warned that China must reform its educational and political system, or risk colonization. 'The strong flourish; the weak are destroyed,' wrote Liang, before describing the European conquest of Africa and India. Darwin's ideas were equally popular amongst more radical political thinkers. Sun Yat-sen, who went on to lead the Revolution of 1911, first learned about evolution whilst studying at the College of Medicine in Hong Kong. 'I was most fascinated by the Way of Darwin,' he later wrote. Sun drew the same conclusion as many other Chinese thinkers at this time, although he took things a step further. Whilst Liang argued for reform, Sun firmly believed that the only way to save China was to overthrow the Qing dynasty. 'If you do not struggle, there is no way to exist,' he argued.[64]

Sun's remark about 'the Way of Darwin' provides another clue as to why evolution was so popular in late nineteenth-century China. Here, Sun was referring to the ancient Chinese belief in the 'way'. Although interpretations differ, the 'way', or *tao*, was generally believed to be an underlying natural force in the universe, one that humans should try and live in harmony with. Unlike in Christian Europe, there was no religious tradition of a creator god in China, nor was there any concept that humans were somehow separate from the natural world. Instead, Chinese thinkers, going right back to ancient times, believed that all life was connected through some kind of natural force. 'The ten thousand things, the ten thousand forms, all go back to one,' wrote the influential Taoist philosopher Wang Pi in the third century. These ideas developed over the course of the early modern period into more advanced theories of evolution. Li Shizhen's *Compendium of Materia Medica* (1596), which we encountered in the previous chapter, even included passages documenting the adaptation of species to different environments as well as the patterns of inheritance in plants, such as the lotus flower. By the early nineteenth century, Chinese naturalists were perfectly comfortable with the idea that species could undergo change. Writing in *Additions to Materia Medica* (1803), the naturalist Chao Hsueh-Min noted that 'with the passage of time, species and varieties become more abundant . . . these then are new sorts and varieties'.[65]

Darwin himself was actually well aware of this long history of evolutionary thought in China. 'The principle of selection I find distinctly given in an ancient Chinese encyclopaedia,' noted Darwin in *On the Origin of Species*. The 'ancient Chinese encyclopaedia' that Darwin was referring to was none other than Li's *Compendium of Materia Medica*. Intrigued by Chinese natural history, Darwin had asked a friend at the British Museum in London to translate some relevant extracts from Li's book. A number of Darwin's other works contain similar references to Chinese texts. In *The Variation of Animals and Plants under Domestication* (1868), Darwin cited a French translation of an eighteenth-century Chinese agricultural text as his source for the development of different varieties of silkworm. Ultimately, the basic idea of evolution was not new in China and Darwin himself knew this, even if it is rarely acknowledged today. Rather, the bit that was new – and really appealing at the time – was the 'struggle for existence'. In the wake of a humiliating military defeat, and with the fate of the Qing dynasty hanging in the balance, Darwinism seemed to provide answers to many of the questions that Chinese thinkers were grappling with in the closing decades of the nineteenth century.[66]

Following the First Sino-Japanese War, the Qing emperor agreed to a widespread programme of modernization. This included the overhaul of the traditional Chinese civil service exams, which were eventually abolished in 1905, as well as the establishment of a series of new scientific and educational institutions. In 1898, the old Imperial Academy was converted into the Imperial University of Peking. This was China's first modern university, with a curriculum focused on mathematics, physics, and biology, rather than just the Confucian classics. Yan Fu was later appointed president of the university, where he continued to promote Darwinism. Alongside new universities, the Chinese government also established a series of agricultural experimental stations. The largest of these was set up on the outskirts of Beijing in 1906, although hundreds of others were established right across China. The idea was that by applying evolutionary theory, breeders could create improved varieties of staple crops, such as rice and wheat.[67]

During the same period, the Qing government began sending more and more students to train abroad. A number went to Europe and the United States, but many also studied in Japan. This made a lot of sense. After all, the recent war had demonstrated Japan's military and industrial strength. Japan was also a lot easier to travel to, and shared much in common in terms of culture and language with China. By 1907, over 10,000 Chinese students had completed degrees at Japanese universities, mostly in the sciences. Many Japanese textbooks were also translated into Chinese during this period, and a number of Japanese scientists were even invited to teach at the Imperial University of Peking. This marked a significant shift in the scientific relations between the two countries. As we saw in the previous chapter, seventeenth- and eighteenth-century Japanese naturalists tended to base a lot of their work on existing Chinese texts. Some even studied in China. By the end of the nineteenth century, that relationship was reversed, with Japanese science forming the basis of Chinese modernization.[68]

Amongst those who travelled to Japan in this period was the author of the first Chinese translation of *On the Origin of Species*. Ma Junwu was born in 1881 in southern China. He received a traditional education in the Chinese classics until, aged twenty, he was selected to go and study in Japan and train as a scientist. Between 1901 and 1903, Ma studied chemistry at Kyoto University. It was here that he met Sun Yat-sen, who at this point was living in exile. Following this meeting, Ma became increasingly radical. He shared Sun's view that the only way to save China was to overthrow the Qing dynasty. It was also during this period that Ma began translating Darwin's *On the Origin of Species*. He most likely first learned about Darwin from reading the *Journal of the New People*, a Chinese periodical published in Japan by Liang Qichao, who was also living in exile. The *Journal of the New People* regularly featured articles on evolution. One article even included a detailed biography of Darwin, accompanied by a photograph. Ma himself first published some early extracts of his translation of *On the Origin of Species* in the *Journal of the New People*, shortly before he returned to China in 1903.[69]

Back in China, Ma continued to work on his translation. It was, however, taking a while. This was mainly because Ma was spending more and more time on radical politics. He had secretly joined Sun's Chinese Revolutionary Alliance, helping to organize local activists and distribute pamphlets in Shanghai. Realizing the complete translation wouldn't be finished any time soon, Ma decided to publish the first five chapters of *On the Origin of Species* as a separate book in 1903. This contained all the most important stuff, including the chapters on 'The Struggle for Existence' and 'Natural Selection', as well as Darwin's famous tree diagram, in which he illustrated the branching of different species from a single common ancestor. For the first time, Chinese readers could purchase a translation, even if technically incomplete, of Darwin's classic work.[70]

Ma's translation was published by the Guangyi Book Company, which was owned and run by the Chinese Revolutionary Alliance. This was not a coincidence. Like many of his contemporaries, Ma made an explicit connection between Darwinism and the political situation in China. 'People from different countries struggle with each other; the surviving nations must have the equivalent forces to counter foreign invasions,' wrote Ma in 1903. This was in fact a direct reference to the recent occupation of Beijing by the Eight-Nations Alliance during the Boxer Rebellion of 1899–1901. Ma's translation of *On the Origin of Species* was littered with similar allusions to the struggle between nations, going well beyond the wording of Darwin's original text. 'All that want to survive must pay attention to . . . natural selection,' claimed Ma. He then concluded with a thinly veiled reference to revolution, writing that 'the natives must evolve in order to resist the intruders without fear'. Others were more explicit. 'Revolution is the universal principle of evolution,' declared Zhou Rong, another member of the Chinese Revolutionary Alliance and a keen reader of *On the Origin of Species*.[71]

In 1911, Ma got what he wanted. Following a series of local uprisings, the Chinese Revolutionary Alliance seized control of major cities across the country. Four months of intense fighting followed, in which over

200,000 people were either killed or wounded. Finally, following the abdication of the last Qing emperor, Sun Yat-sen was elected as the Provisional President of the Republic of China on 29 December 1911. This marked the end of over 2,000 years of dynastic rule. At the time of the revolution, Ma was studying abroad at the Agricultural University of Berlin. He returned to China to support the new national government, working for a while in a munitions factory manufacturing dynamite, but also found time to complete his long-awaited translation. It had taken nearly two decades, interrupted by war and revolution, but in 1920 Ma finally presented Chinese readers with a complete translation of *On the Origin of Species*.[72]

Much as we've seen elsewhere, Chinese interest in Darwinism was stimulated by the growth of war and nationalism. The Revolution of 1911 ultimately brought about an abrupt end to the Qing dynasty. For the revolutionaries, there was a Darwinian element to this too. 'Our splendid, superior, majority race is under the control of an evil, inferior, minority race,' declared Hu Hanmin, another member of the Chinese Revolutionary Alliance who had read *On the Origin of Species*. Hu was referring to the division between the Han Chinese majority and the Manchu minority who had governed China since the establishment of the Qing dynasty in the middle of the seventeenth century. According to Hu, the Manchus were an 'unfit' race, destined to be eliminated in the struggle for existence. For Hu, the Revolution of 1911 was therefore simply an example of natural selection at work. 'It is all a matter of evolution,' concluded Hu, just as China descended into civil war, another reminder of how Social Darwinism was used to reinforce racial discrimination and conflict.[73]

V. Conclusion

By the outbreak of the First World War, Charles Darwin's *On the Origin of Species* had been translated into at least fifteen different languages, including Russian, Japanese, and Chinese. For many readers, however, the basic concept of evolution was not entirely new. Whether in Tsarist Russia or Qing China, evolution was in fact widely discussed from the late eighteenth century onwards. This was particularly the case in countries like China and Japan, where evolutionary ideas could be found within existing religious and philosophical traditions, such as Taoism and Buddhism. This was something Darwin himself acknowledged, even citing the works of earlier Russian and Chinese authors in *On the Origin of Species*. What made Darwinism so popular, then, was not simply that it was a theory of evolution. There was nothing new about that. What made Darwinism so appealing at the time was the idea of a 'struggle for existence'. At the heart of *On the Origin of Species* was a vision of the natural world as one of constant conflict. There was a 'war of nature', argued Darwin. Evolution was the result of 'the great battle for life'.[74]

It was this metaphor of struggle that caught the imagination of so many different scientific thinkers in the nineteenth century, not just in Europe, but also in Asia and the Americas. It seemed to capture something of the world in which they lived. Towards the end of the nineteenth century, Darwin's theory was increasingly applied, not just to plants and animals, but also to societies and nations. The promotion of the damaging ideas behind Social Darwinism was another product of this period. From the Argentine conquest of Patagonia to the Japanese invasion of Manchuria, evolution was a science born of an age of brutal conflict. In fact, one of the most striking things about the history of evolution is just how many of the key individuals were involved in some way or another with the military. Francisco Muñiz, one of the earliest Latin American evolutionary thinkers, served as an army doctor in the Argentine War of Independence, whilst Yan Fu, the writer responsible for popularizing Darwinism in China, initially trained as a naval engineer. In the following chapter, we move on to explore how this same world – a world of capitalism and conflict – shaped the development of the modern physical sciences.

6. Industrial Experiments

From the top of the Eiffel Tower, Peter Lebedev could see all of Paris. The 'City of Light' certainly lived up to its name, with electric lamps illuminating all the major landmarks. In the distance, Lebedev could make out the glass dome of the Grand Palais, located on the other side of the River Seine, as well as the famous Sacré-Coeur Basilica, way off in Montmartre. Lebedev, however, hadn't come to Paris to see the sights. He was certainly no tourist. Lebedev, a professor at the University of Moscow, was in fact an accomplished physicist, one who had just made a major contribution to the study of light. In August 1900, along with over 500 other scientists from around the world, Lebedev arrived in Paris to attend the First International Congress of Physics. The meeting had been organized to coincide with the Paris Exposition of 1900, one of the vast international exhibitions that proved so popular throughout the late nineteenth and early twentieth centuries. These began with the Great Exhibition of 1851 in London, which was intended as a showcase of Victorian science and industry, but soon spread across the rest of the world. By the end of the nineteenth century, cities ranging from Tokyo to Chicago had hosted similar exhibitions, which were often accompanied by scientific meetings.[1]

The Paris Exposition of 1900 was visited by over fifty million people. The highlight for many was the Palace of Electricity, an Art Nouveau masterpiece built in the shape of a giant peacock feather. Located on the Champs de Mars, right in front of the Eiffel Tower, the Palace of Electricity was covered in over 7,000 multicoloured electric lights. Inside, visitors could inspect all kinds of electrical machines, as well as see enormous steam-powered turbines in action. Nearby, there was also the Palace of Optics, in which visitors could peer through a giant telescope, or watch an early motion picture. Major private companies, including Siemens and General Electric, sent representatives to the Paris Exposition, hoping to sell their industrial machines to countries around the world.[2]

This was an era of internationalism and industrialization. The Paris Exposition of 1900 perfectly captured that mood. Through new communication technologies, such as the electric telegraph, invented in the 1830s, and new transportation technologies, such as the ocean-going steamship, which was invented in the 1810s, the world started to feel much more connected. Many believed that these technological advances had helped accelerate the development of the sciences. 'Ideas . . . fuse and cross all over the world, just like the thin filaments on which human thoughts are transported at lightning speed,' declared one French politician at the opening of the Paris Exposition. This was part of the motivation for bringing physicists together from around the world. The purpose of the First International Congress of Physics was 'to take stock of the knowledge definitively acquired in the field cultivated by these scientists', explained the organizers. This would be 'the first meeting of physicists from all countries'.[3]

When not visiting the Eiffel Tower or the Palace of Electricity, the scientists who attended the First International Congress of Physics discussed their latest research. Much of this concerned the theory of electromagnetism. For hundreds of years, scientists had studied the properties of light, electricity, and magnetism. However, during the second half of the nineteenth century, there was a growing consensus that these seemingly separate phenomena all shared something in common. The initial theoretical contribution came from the British physicist James Clerk Maxwell. In an article published in 1864, Maxwell described how the properties associated with light, electricity, and magnetism could all be explained by the existence of an 'electromagnetic field' in which oscillating waves travelled.

Since the early nineteenth century, scientists had known that, when an electric charge moves through space, it creates a magnetic field. They had also known that, when a magnet moves through space, it creates an electric field. These two principles underpinned the development of early electric motors and generators, as by moving a magnet through a coil of wire you could create an electric current. Maxwell, however, realized that it was possible to combine the concept of an electric field with that of a magnetic field to create a single 'electromagnetic field'. This was the key idea, as it allowed him to explain what light had to do with electricity and magnetism. According to Maxwell, light was simply an 'electromagnetic disturbance' moving through this field, a bit like a wave moving through the sea. He also predicted that there must be other electromagnetic waves, such as radio waves, which behaved in the same way as light. Following Maxwell's publication, physicists around the world set to work investigating the properties of electromagnetic waves. From Moscow to Calcutta, the race was on to prove that Maxwell was right.[4]

Existing histories of modern physics and chemistry tend to focus on a small group of European pioneers. This list typically includes James Clerk Maxwell himself, as well as a number of later scientists who lived and worked in Europe – people like the German physicist Heinrich Hertz, who discovered radio waves in 1887, as well as the Polish physicist Marie Skłodowska Curie, who discovered radioactivity in 1898. And whilst it is true that Europe was very much at the centre of the scientific world in the late nineteenth century, thanks in large part to the economic advantages provided by the expansion of imperialism we saw in the previous chapters, this does not mean that scientists from outside of Europe had nothing to contribute. In fact, by examining the list of attendees at the First International Congress of Physics, we can quickly build up a much more diverse picture of late nineteenth- and early twentieth-century science. Alongside scientists from Britain, France, and Germany, there were also those from Russia, Turkey, Japan, India, and Mexico. And they weren't just there to sit and listen. Rather, these international scientists presented their own research, challenging the idea that breakthroughs in physics could only be achieved in European laboratories.[5]

Peter Lebedev is a good example. At the meeting in Paris, he presented a paper on a recent experiment he had conducted back at the University of Moscow. Whilst by the late nineteenth century, most physicists accepted the existence of electromagnetic waves, many questions still remained unanswered. One of the most intriguing implications of Maxwell's original theory concerned the properties of light itself. According to Maxwell, if light was a wave, then it should also exert a force, as it would carry momentum. This seemed counterintuitive at first. How could light, which was surely completely immaterial, exert a physical force? But it all followed from the equations. However, before 1900, no one had actually been able to measure this force directly, as it was tiny. The audience in Paris therefore listened with much excitement as Lebedev described his experiments. By suspending a set of metal vanes in a vacuum, and exposing them to an electric lamp, Lebedev had been able to confirm that light really did exert a force. When the lamp was turned on, the vanes began to spin, much like when a breeze blows past a windmill.[6]

Following Lebedev, a number of other scientists spoke at the meeting. Hantaro Nagaoka, a Japanese physicist who we will meet again later in this chapter, gave a talk about his research into a phenomenon known as 'magnetostriction', whereby metals expand or contract in a magnetic field. There was also a group of Indian scientists at the meeting. Amongst them was a Bengali physicist named Jagadish Chandra Bose, another figure whose life we will explore in more detail later in this chapter. A pioneer of radio physics, Bose described for his audience in Paris some experiments he had conducted back in Calcutta. After trying to electrocute everything from a lump of metal through to a living plant, Bose concluded that there was no fundamental difference between organic and inorganic matter. After all, it seemed that everything responded in some way to electricity. For Bose, as for many scientists around 1900, the theory of electromagnetism really was a theory of everything. The fact that Maxwell's equations could be used to describe both the action of a nerve and the operation of a radio suggested that there was 'a fundamental unity' in nature, argued Bose.[7]

The presence of this diverse group of individuals in Paris in 1900 is an important reminder of a forgotten side to the history of the modern physical sciences. Over the course of the nineteenth century, scientists working in laboratories outside of Europe, including those in Russia, Turkey, India, and Japan, made a number of important contributions to the development of modern physics and chemistry. They came together in cities around the world to discuss their work and share ideas. It is in the nineteenth century that we start to see the first modern scientific conferences, many of which were organized to coincide with industrial exhibitions. The First International Congress of Physics was typical in this respect.

In the previous chapter, we saw how the world of capitalism and conflict shaped the development of the modern biological sciences. In this chapter, we explore that same theme, but from the perspective of the modern physical sciences. The growth of new industrial communication technologies in the nineteenth century helps explain why scientists became so interested in the properties of electricity and magnetism. Experimental telegraph lines had been set up in Britain and Germany in the first few decades of the nineteenth century. These worked by sending short bursts of electric current along a metal wire. The bursts of electricity corresponded to a code, usually Morse code, which could then be translated into a message by a human operator. The great advantage of this system was that information could be transmitted over long distances almost instantaneously. Following these early examples, telegraph lines began to expand internationally throughout the 1850s and 1860s, just at the time when James Clerk Maxwell was developing his theory of electromagnetism. The first transatlantic telegraph line, connecting Ireland to Newfoundland, was completed in 1858. This was followed in 1865 by a telegraph line connecting Britain to its colonies in India. Governments around the world quickly recognized the value of modern science for international communication, in times of both peace and war. Physicists and engineers suddenly found themselves in high demand, recruited to help advise on the construction of new telegraph lines as well as the introduction of radio receivers into the military.[8]

Alongside physics, chemistry was the other major industrial science of the era. Over the course of the nineteenth century, over fifty new chemical elements were discovered, many following the opening of new mines, or during the process of refining mineral ores. Recent breakthroughs in physics also helped, as scientists realized that an electric current could be used to separate out different chemical elements. But perhaps the most important breakthrough was the invention of the periodic table, in which all the chemical elements were ordered by atomic weight, beginning with the lightest element, hydrogen. First proposed by the Russian chemist Dmitri Mendeleev in 1869, the periodic table predicted the existence of many as-yet-unknown elements, as there were gaps waiting to be filled in, thus kickstarting a race to find them. There was a certain amount of national rivalry here. Scientists often chose to name new elements after the country of their birth. When the Russian chemist Karl Klaus discovered a new element in the middle of the nineteenth century, he called it 'ruthenium', from the Latin word for Russia. 'I named the new body in honour of my Motherland,' explained Klaus.[9]

There are plenty of similar examples of this kind of 'chemical nationalism'. Germanium, gallium, and polonium were all named after nations. In some cases, these nations were relatively new. (Germanium was discovered in 1886, a little over a decade after the Unification of Germany in 1871.) In other cases, the naming of an element predated the nation itself. Marie Skłodowska Curie chose to name polonium after her native Poland precisely because she hoped that the country would one day become an independent nation state. (At the time polonium was discovered, in 1898, Poland was still partitioned between Germany, Russia, and Austria–Hungary.)[10]

Nationalism and internationalism seemed to go together. After all, the nineteenth century was an era in which scientists travelled the world, trained at foreign universities, published in multiple languages, and met at international conferences. But it was also an era in which science was seen as a means to promote national strength, particularly when it came to industry and the military. In 1900, at the First International Congress of Physics in Paris, many were still optimistic about the future. 'So many new thoughts have been born, and so many friendships made or consolidated,' wrote one physicist after returning from the meeting in Paris. However, by 1914, with the outbreak of the First World War, the international order seemed to have broken down. In this chapter, we explore that tension between nationalism and internationalism as it played out between 1790 and 1914. The history of nineteenth-century physics and chemistry is ultimately best explained, not through a history of isolated European pioneers, but rather through a global history of nationalism, war, and industry. We begin with a storm brewing in northern Russia.[11]

I. War and the Weather in Tsarist Russia

Alexander Popov could see the storm approaching. The time had come to test his new invention. For many years, Popov had taught electrical science at the Russian Navy's Torpedo School, located at Kronstadt in the easternmost part of the Gulf of Finland. Now, in the spring of 1895, he planned to put his teaching into practice. Climbing a nearby tower, he launched a small balloon connected to a copper wire high into the sky. With lightning crackling in the distance, Popov connected the wire to a machine he called a 'storm indicator'. Just as he hoped, the machine sprung into life. Even though the storm was still over fifteen miles away, a small bell rang each time a bolt of lightning struck. Working for the navy, Popov immediately understood the potential for such an invention. It would allow ships at sea, as well as forecasters on land, to detect storms before they hit. But how did it work? The machine itself relied on the fact that lightning emits electromagnetic waves. What Popov had invented was a means to detect those waves at a distance. In doing so, he had constructed one of the first radio receivers in the world. In Tsarist Russia, the radio had its origins in the science of storms.[12]

Popov's machine built on the earlier work of the French physicist Édouard Branly. In 1890, Branly had reported his discovery that electromagnetic waves seemed to have some effect on metal filings. This led to the invention of what became known as the 'coherer'. This device, which formed the basis of all early radio receivers, consisted of a small glass tube filled with metal filings. On their own, the metal filings acted as a poor conductor of electricity. However, when an electromagnetic wave passed through the tube, the metal filings all aligned – they 'cohered' – and suddenly conducted electricity. In this way, early radio pioneers were able to detect electromagnetic waves. The only problem was that, in order to reset the detector, the tube had to be shaken by hand to mix up the filings again. Popov's great innovation solved this problem. His storm indicator used the current generated by the electromagnetic waves to power a hammer to simultaneously strike the glass tube, mixing up the metal filings again. In this way, the storm indicator was able to detect each individual emission of an electromagnetic wave, turning on and off with each lighting strike.[13]

The fact that the pioneer of the radio in Russia worked at a naval school is important. Physics in the nineteenth century was practical as much as it was theoretical, a product of industry as much as pure science. Born in 1859, Popov grew up in the Urals, close to the great Bogoslov Smelting Works, where toxic smoke could be seen billowing over the landscape. As a boy, Popov was fascinated by the machines in

29. Alexander Popov's 'storm indicator'. Note the small glass cylinder with rubber tubing above the bell. This is the 'coherer', which detects the radio waves and then automatically resets.

the nearby factories and mines. He even built a small electric alarm clock, proudly displaying it in his bedroom at home. Popov's taste for industrial science spurred him on to win a place at Saint Petersburg University, where he studied physics and mathematics between 1877 and 1882. However, Popov did not come from an especially wealthy family – his father was a poorly paid priest who had in fact hoped his son would study at a theological college. So alongside his university studies, Popov made money working for the new Elektrotekhnik Company in Saint Petersburg. He helped set up the lighting for a local pleasure garden and, in 1880, acted as a guide for an enormous industrial exhibition held in the city, in which companies from around the world displayed all the latest electrical machines: telegraph signals, electric lighting, and even an electrotherapy device which promised to cure various medical conditions.[14]

On graduating, Popov was offered the opportunity to teach at Saint Petersburg University. But the position didn't come with a good enough salary. Hoping to marry his sweetheart, and therefore in need of a stable career, Popov turned instead to the navy. In 1883, he joined the Torpedo School in Kronstadt as an instructor. For a budding scientist in nineteenth-century Russia, working for the navy not only paid better, it also offered improved facilities. The laboratory at the Torpedo School was equipped with advanced equipment as well as a library stocked with foreign scientific publications. Popov gave lectures to the trainees, who would go on to man torpedo boats, on everything from electromagnetism to the chemistry of explosives. It was in the Torpedo School laboratory that Popov first generated electromagnetic waves, demonstrating to his students how his storm indicator might also be used to communicate at sea. 'We may expect the application of these phenomena to be essentially useful in the Navy, both as beacons and for signalization between ships,' explained Popov. Prior to this, all communication at sea was done with flags and lights, much as it had been for centuries.[15]

Popov was justifiably proud of his invention. He was therefore shocked to learn that a very similar device was being promoted by a competitor. In 1897, skimming through the latest issue of a Russian engineering journal, Popov discovered that the Italian engineer Guglielmo Marconi was attempting to patent his own design for a radio receiver in Britain. Today, Marconi is widely credited with having invented the radio, but in fact – as Popov was at pains to point out – a number of other scientists were developing almost identical devices around the same time. 'Marconi's receiver was, in all its component parts, the same as my instrument, made in 1895,' complained Popov. Research into the practical uses of electromagnetic waves was clearly advancing at a rapid pace. With this in mind, Popov pushed to transform his storm indicator into a commercial system for radio signalling. He teamed up with the French engineer Eugène Ducretet, who began manufacturing Popov's radio detector in Paris. In 1898, using a modified version of Popov's design, Ducretet succeeded in detecting a radio wave sent between the Eiffel Tower and the Panthéon over two miles away. This was the first time that the Eiffel Tower was used as a radio mast, a function it continues to serve to this day.[16]

★

As we saw in the previous chapter, the second half of the nineteenth century was a period of renewed investment in the sciences in Tsarist Russia. This was just as true of the physical sciences as it was of the biological. Following Russia's defeat in the Crimean War of 1853–6, Tsar Alexander II was determined to modernize both the economy and the military. This meant establishing new laboratories, both in military schools and in universities, as well as encouraging the use of scientific research to solve industrial and military problems. Alexander II ultimately believed that the survival of the Russian Empire depended upon the application of modern science and technology. To celebrate his own coronation, held in Moscow in September 1856, Alexander II even ordered a military engineer to cover the Kremlin with electric lights. One set of lights, according to an official report, was arranged in the shape of 'a colossal crown . . . with fiery sapphires, emeralds and rubies'. This was an industrial vision of Tsarist power. For Alexander II, the future was electric.[17]

The laboratory at the Torpedo School in Kronstadt was just one of a number of new scientific institutions established in Russia during the second half of the nineteenth century. In 1866, Alexander II authorized the creation of the Russian Technical Society. Based in Saint Petersburg, it organized scientific meetings dedicated to topics including railway engineering, photography, and the electric telegraph. Alongside these meetings, the Russian Technical Society published a series of scientific magazines, including one titled *Electricity*. It also organized major industrial exhibitions, including the electrical exhibition that Alexander Popov worked at whilst studying in Saint Petersburg.[18]

Universities too began to invest more in the physical sciences, although they tended to lag behind the industrial and military schools. In 1874, a Russian physicist named Alexander Stoletov visited the University of Cambridge. Whilst there, he met James Clerk Maxwell, and attended the opening of the Cavendish Laboratory, the University of Cambridge's new centre for experimental physics. Inspired by the British example, Stoletov returned to his position at Moscow University, where he helped to expand and modernize the physics laboratory. By the late 1880s, the Department of Physics at Moscow University housed all the latest scientific equipment, including machines for generating electromagnetic waves. It was here that Peter Lebedev would later do his work on the 'pressure of light' that we encountered at the start of this chapter.[19]

As well as research into electromagnetism, the Tsars championed the development of modern chemistry. After all, chemistry was the most obviously practical of the physical sciences. Throughout the second half of the nineteenth century, Russian chemists were employed by the government to advise on everything from the manufacture of gunpowder to the distillation of vodka. At this time, Germany was widely recognized as the leading nation when it came to industrial chemistry. With this in mind, the Russian government sponsored hundreds of young scientists to train at German universities. Amongst these was Dmitri Mendeleev, perhaps the most famous Russian chemist of the era, who was sent to study at Heidelberg University in 1859. When Mendeleev returned to Russia in 1861, he took up a position at Saint Petersburg University, where he helped to modernize the chemistry course, introducing much more practical teaching in an expanded laboratory modelled on what he had seen in Germany. Mendeleev also helped establish the Russian Chemical Society in 1868, which began publishing its own Russian-language scientific journal the following year.[20]

Today, Mendeleev is best remembered for having invented the periodic table, in which all the chemical elements are arranged into eighteen groups, ordered by atomic weight. By leaving blank spaces where there wasn't a known element, he was able to predict the existence of new chemical elements, as well as their properties. However, what is often forgotten is that Mendeleev was not simply a theoretician. Rather, he was a practical man who believed that chemistry was essential for the industrial and military development of the Russian Empire. Chemistry was 'an instrument in the service of practical ends', argued Mendeleev in his influential textbook, *Principles of Chemistry* (1868–70). 'It opens the way to the exploitation of natural resources and the creation of new substances.' Ultimately, in order to understand Mendeleev's contribution to the development of modern chemistry, we need to move beyond the periodic table. Instead, we need to return to the world of industry and war that characterized nineteenth-century science.[21]

Dmitri Mendeleev raised his arm, giving the order to prepare the artillery. As he did so, a Russian naval officer loaded a shell into a nearby cannon. Mendeleev then lowered his arm, calling out 'Fire!' A split second later, the naval officer pulled a cord, firing the shell across an open field.

Mendeleev watched as the shell exploded in the distance. His new invention seemed to work. On a crisp morning in April 1893, Mendeleev conducted the first field tests of what he called 'pyrocollodion powder'. This was a new kind of smokeless gunpowder that he had been working on for the past three years. He had in fact been asked to develop this new gunpowder by none other than Tsar Alexander III himself. Concerned about the recent military advances made by other European states, Alexander III turned to Mendeleev, by this time one of the most famous chemists in the world. In order to support this work, Alexander III authorized the creation of the Naval Scientific-Technical Laboratory, located on a small island in the middle of the Neva River in Saint Petersburg. It was here that Mendeleev spent much of his time between 1890 and 1893, applying his knowledge of chemistry to the design of new explosives.[22]

The invention of smokeless gunpowder was one of the major military innovations of the late nineteenth century. Traditionally, gunpowder had been manufactured from a mixture of saltpetre, sulphur, and charcoal. But with advances in chemistry, military scientists began to explore alternative compounds which might be more powerful. These were typically manufactured from a mixture of nitroglycerine, which was first isolated in the 1840s, and various other chemicals. Most famously, Alfred Nobel – after whom the Nobel Prizes are named – made his fortune from the development of new chemical explosives, including a kind of smokeless gunpowder called Ballistite.[23]

As its name suggests, smokeless gunpowder produces very little smoke. This has an obvious advantage during combat, and particularly during naval battles, increasing visibility and facilitating the coordination of troops and ships. This, however, is not the only benefit. Smokeless gunpowder also produces a much more powerful blast. With traditional gunpowder, much of the fuel is wasted as it burns off as smoke, but with smokeless gunpowder, almost all of the fuel is converted into an explosion. This bigger blast has the effect of increasing the range, accuracy, and velocity of artillery shells, all a major advantage in naval combat, particularly once ships started to be built out of iron during the second half of the nineteenth century. Only a high-powered artillery shell could penetrate the ironclad hull of a modern battleship. For all these reasons, Alexander III was particularly keen for the Russian Navy to develop its own smokeless gunpowder.[24]

Working at the Naval Scientific-Technical Laboratory in Saint Petersburg, Mendeleev began by examining samples of existing smokeless gunpowder manufactured in Britain and France. In fact, he had earlier had an opportunity to inspect the Woolwich Arsenal in London, where he learned about a British variety of smokeless gunpowder called cordite. Analysing these samples, Mendeleev realized that he needed to create a new compound based on a mixture of carbon, hydrogen, nitrogen, and oxygen. He also hoped to improve on the French and British varieties, and create a gunpowder that was even more powerful and yet still produced very little smoke. His knowledge of the atomic weight of the different elements came in particularly handy here, as Mendeleev was able to work out the exact ratio of chemicals which would produce the maximum blast when ignited. By the end of 1892, he had succeeded in manufacturing a small quantity of his new smokeless gunpowder. This was 'a new product in chemical terms, deeply differing from regular gunpowder, and demanding a fundamental familiarity with chemical reactions and products', wrote Mendeleev in his notebook.[25]

Throughout his life, Mendeleev showed a keen interest in the military and industrial development of the Russian Empire. Sometimes he worked for the government, at other times private firms. Alongside gunpowder, he was heavily involved in the Russian oil industry. In the 1860s, he was employed by the Baku Oil Company to advise on the construction of a petroleum distillation plant. This was just after the Russian Empire had seized much of the region around the Caucasus, including modern-day Azerbaijan, from the Persian and Ottoman empires. The Tsars immediately claimed ownership of the oil-producing territories, and then sold long-term leases to private firms such as the Baku Oil Company. Once again, Mendeleev's knowledge of chemistry was put to industrial use, as he advised on how to separate out the different chemical products from crude oil, which could then be sold on for a profit. Later, in the 1870s, Mendeleev was even sent to the United States in order to report on the American oil industry. At the time, Russia was still importing the majority of its oil from the United States. By the end of the century, that relationship was completely reversed. Russia was supplying nearly 90 per cent of the world's crude oil, thanks in part to the advances made by industrial chemists such as Mendeleev.[26]

★

Although Dmitri Mendeleev was undoubtedly the most famous Russian scientist of the nineteenth century, he was by no means unique. Mendeleev's industrial vision of science was in fact typical of his generation. Amongst those who took a very similar approach to the study of the physical sciences in this period was a Russian chemist named Julia Lermontova. As we saw in the previous chapter, the nineteenth century was a time in which more and more women entered the world of professional science. Lermontova was one of a new generation of Russian women who undertook formal training in the physical sciences, fighting against the prejudices of the time in order to do so.[27]

Born in Saint Petersburg in 1846, Lermontova was the daughter of an army general. As a young girl, she had already demonstrated a passion for science, setting up a small chemistry laboratory in her family kitchen. Aged twenty, she decided to pursue a career in agricultural chemistry, applying to study at the Petrovskaya Academy of Farming and Forestry in Moscow. This was one of the new agricultural and industrial schools set up by Tsar Alexander II in the 1860s. However, despite Alexander II's pretensions to modernization, women were still excluded from the Russian higher education system. Lermontova was rejected from the Petrovskaya Academy after being bluntly informed that there were no places for women on the course.[28]

Undeterred, Lermontova decided to do what many other Russian women did in this period – she went and studied abroad. In 1869, Lermontova travelled to Germany and began attending lectures at Heidelberg University. At Heidelberg, she studied with many of the leading German chemists and physicists of the time, including Robert Bunsen, after whom the 'Bunsen burner' is named. Lermontova also spent time at the Chemical Institute of the University of Berlin as well as the University of Göttingen, where she was awarded a PhD in 1874. For a young Russian woman studying abroad, life was tough. She later recalled her time in Berlin, living in 'a bad apartment, with awful food, breathing unhealthy air'. But she was determined to succeed and enter the male-dominated world of industrial chemistry.[29]

By coincidence, Lermontova was in Germany at exactly the same time as Mendeleev. The two met in Heidelberg, and got talking. Mendeleev told Lermontova about his recent article on the periodic table. He also explained that he was having trouble getting all the elements in the right order, particularly the group of elements known as the 'platinum metals'. These metals were all clearly very similar, often found in the same ore and with a distinctive silver colour, and so Mendeleev knew that they needed to be grouped together. There was, however, a problem. The accepted atomic weights of the platinum metals, particularly iridium and osmium, did not fit with the order in the periodic table proposed by Mendeleev. Working in the chemical laboratory at Heidelberg University, Lermontova applied herself to solving this problem. Through a series of complex experiments, which involved repeatedly dissolving lumps of platinum ore in different chemicals, she was able to extract pure samples of both iridium and osmium. Then, using a technique developed by Bunsen, Lermontova carefully measured the atomic weights of each of the platinum metals. Pleased with the result, she decided to write to Mendeleev, who by this time had returned to Saint Petersburg. Mendeleev was delighted. He quickly updated the values for atomic weights in his textbook, *Principles of Chemistry*, reordering the platinum metals based on Lermontova's experiments.[30]

Lermontova returned to Russia in 1874. She went on to have a brilliant career, although her contributions to Russian science and industry have now largely been forgotten. She was elected as a member of the Russian Chemical Society in 1875, principally on the basis of her work on the platinum metals, which helped prove that Mendeleev was right about the periodic table. By knowing the correct atomic weights of each of these metals, Russian industrialists were also able to devise more efficient methods for processing platinum ore, which was mined throughout the nineteenth century in the Ural Mountains. Lermontova then spent much of her career working at Moscow University, helping to develop new techniques for analysing crude oil. She even invested some of her own money in a Russian oil company, based in the Caucasus. In 1881, in recognition of her work for the oil industry, she was elected as the first female member of the Russian Technical Society. Ultimately, Lermontova's career is an important reminder of the forgotten contributions made by Russian women to the world of industrial science in the nineteenth century.[31]

The outbreak of the First World War in 1914 revealed both the strengths and weaknesses of science under the Tsars. Starting in the 1860s,

Alexander II had initiated a series of reforms, aimed at modernizing Russian science and technology. This led to the creation of new laboratories as well as new industrial and military colleges. Many of the most successful Russian scientists of the nineteenth century were in one way or another involved with the world of industry and war. During the same period, Russian scientists studied abroad and began attending international conferences and industrial exhibitions. Alexander Popov attended the First International Congress of Physics in Paris in 1900, whilst Dmitri Mendeleev travelled to the United States in 1876 to attend the Philadelphia World's Fair.[32]

Yet despite all these advances, it soon became clear that Tsarist Russia was no match for the German industrial and military machine. When the border closed with Germany in August 1914, Russian scientists suddenly found themselves cut off. It was no longer possible to import essential scientific equipment or chemicals, most of which were manufactured in Germany at the time. 'Until now our country has made no serious effort to produce its own scientific and educational instruments and to free itself from the stranglehold placed on it by Germany,' complained one Russian scientific journal in 1915. Some attempts were made to mobilize Russian scientists for the war effort. In 1916, the government formed the War Chemicals Committee, comprised of members of the Russian Physical and Chemical Society. The War Chemicals Committee was tasked with manufacturing important industrial and military chemicals that had previously been imported from Germany. These included chemical weapons such as cyanide, arsenic, and chlorine gas.[33]

It was all too little too late. Things only got worse when, in November 1917, the Bolsheviks – a group of revolutionary socialists – marched on the Winter Palace in Saint Petersburg. This was the beginning of the Russian Revolution, which culminated in the execution of the last Tsar, Nicholas II, and his family in July 1918. As the chaos of the Russian Revolution unfolded, with fighting on the streets of Moscow and Saint Petersburg, scientists in Russia found themselves more and more isolated. Science in Tsarist Russia was born out of a world of nationalism, industry, and war. It was ultimately destroyed by that same world as well. In the following section, we explore the history of the physical sciences in another empire which, despite efforts to reform throughout the nineteenth century, met a very similar fate.

II. Ottoman Engineering

Sultan Abdulmejid I watched as an American engineer began setting up an experimental telegraph line. In August 1847, John Lawrence Smith, a graduate of Yale University, placed a small electrical machine at the entrance of the Beylerbeyi Palace, located on the outskirts of Istanbul. He then ran a long copper wire from the machine, through the gilded doorway, into the main reception room of the palace. Connecting the wire to another machine, Smith announced that the demonstration was ready to begin. The equipment, based on the design of the American telegraph pioneer Samuel Morse, had been sent all the way from the United States. Smith, who was working for the Ottoman sultan as a mining engineer at the time, promised that the electric telegraph would allow information to be 'transmitted at any distance instantaneously'.[34]

Once the equipment was all set up, Smith began to explain the workings of the telegraph to the Ottoman sultan. 'His Majesty understood very well the properties of the electric fluid,' noted an American diplomat who was present at the time, referring to the movement of the electric current along the telegraph wires. Smith then asked Abdulmejid for a message that he would like to be sent between the two machines. Once the message was confirmed, Smith began to type it out in Morse code. ('Has the French steamer arrived? And what news from Europe?') As promised, the message was relayed along the telegraph line into the reception room, before being printed out on a strip of paper as a series of dots and dashes. Smith then translated the message into Ottoman Turkish. The sultan was impressed, and immediately saw the potential of this 'wonderful invention' to transform communication across the Ottoman Empire. In fact, Abdulmejid was so impressed that he wrote personally to Morse in the United States, enclosing a diamond-encrusted decoration and praising the invention of the electric telegraph, 'a specimen of which has been exhibited in my imperial presence', wrote the sultan.[35]

Over the following years, Abdulmejid ordered the construction of thousands of miles of telegraph lines across the Ottoman Empire. The first lines were installed during the Crimean War of 1853–6, in which the Ottoman Empire fought against and ultimately defeated the Russian

Empire. The British, who supported the Ottomans, helped install a telegraph line between Sevastopol and Istanbul. These lines were then used to coordinate military action, ultimately contributing to the Ottoman victory. The electric telegraph clearly presented a great advantage in terms of military power, along with administration, something not lost on Abdulmejid. Shortly after the end of the Crimean War, the Ottomans established a dedicated School of Telegraphic Science in Istanbul, as well as a factory for manufacturing telegraph equipment. By 1900, Ottoman engineers had built over 20,000 miles of telegraph lines, connecting disparate provinces to the centre of imperial rule in Istanbul. Prior to this, almost all communication was done by post. A message sent from Cairo to Istanbul could take days, even weeks. Now it took a matter of seconds.[36]

Much as we saw in Tsarist Russia, the nineteenth century was a period of reform in the Ottoman Empire. Although Istanbul had once been a centre of scientific advance, particularly during the sixteenth and seventeenth centuries, this was no longer the case by the late eighteenth century. A series of military defeats in the second half of the eighteenth century, particularly the Russo-Turkish War of 1768–74, had revealed the limits of Ottoman power. This sense of weakness worsened during the early decades of the nineteenth century following the Greek War of Independence of 1821–9, in which yet another Ottoman territory broke away from Istanbul. Worried about the spread of European imperialism, as well as unrest in the provinces, the Ottomans established a number of new scientific institutions designed to modernize the military. These included the Naval Engineering School as well as the Military Engineering School, both established in 1775 as a direct response to the Ottoman defeat during the Russo-Turkish War. Ottoman military officers were now expected to learn modern mathematics, chemistry, and physics. This kind of scientific knowledge became increasingly important, particularly following the introduction of steamships into the Ottoman Navy in the 1820s.[37]

This early period of reform was then followed by a much more extensive programme of modernization. When Abdulmejid I came to power in 1839, he initiated a series of reforms known as the *Tanzimat* (literally, 'reorganization'). Alongside the electric telegraph, the Ottomans began building railways, with the first line, connecting Cairo to Alexandria, opening in 1856. As part of the *Tanzimat*, a number of new scientific institutions were also established, whilst existing ones were expanded. The Imperial School of Medicine, originally founded in Istanbul in 1827, moved to a new building featuring a modern chemistry laboratory in 1839. There was also a new School of Industrial Arts, established in Istanbul in 1868, where many influential Ottoman engineers later trained. This period of reform, which lasted up until the late 1870s, mirrored what we saw in Tsarist Russia. Indeed, for much of the nineteenth century, the Russian and Ottoman empires struggled against one another for military, industrial, and scientific dominance in Central Asia.[38]

One of the most important scientific institutions established during the *Tanzimat* period was the Ottoman University of Istanbul. Originally founded in 1846, the Ottoman University went through various changes of name throughout the nineteenth and twentieth centuries. However, during the reign of Abdulmejid I it was widely referred to as the 'House of Sciences'. Located in a neoclassical building next to the Topkapı Palace, the Ottoman University featured a large lecture theatre, extensive library, and modern scientific laboratory. It even had a dedicated Department of Natural Sciences, where many leading Ottoman scientists trained, and later a School of Civil Engineering. The Ottoman University, according to one official report, would promote 'the dissemination and development of all the sciences'.[39]

Amongst the lecturers at the Ottoman University was a chemist named Derviş Mehmed Emin Pasha. His career was typical of a new generation of Ottoman scientists. Born in Istanbul in 1817, Mehmed Emin entered the world of science through the Military Engineering School, where he studied in the early 1830s. It was here that he learned the basics of chemistry and physics, such as how to distinguish between an acid and an alkali, as well as more practical matters, such as how to manufacture gunpowder. Following graduation, Mehmed Emin was selected to go and study abroad. Throughout the nineteenth century, many promising young Ottoman scientists and engineers were sent to study in Europe, typically in Britain, France, or Germany. Much like in Tsarist Russia, this was part of a broader strategy to build up scientific expertise by learning from imperial rivals. Mehmed Emin travelled to Paris with a group of Ottoman students in 1835. Some studied medicine, others engineering. Mehmed Emin enrolled at the prestigious School of Mines, where he took courses in chemistry and geology. Each morning he would stroll through the streets of Paris, across the Jardin du Luxembourg, to his lectures at the Hôtel de Vendôme. By the time he graduated, Mehmed Emin was described by contemporaries as 'well versed in mathematics and likewise in chemistry, physics and minerology'.[40]

After five years studying in Paris, Mehmed Emin returned to Istanbul, where he took up a position at the Military Engineering School. It was whilst teaching there that he published his first major scientific work, a textbook titled *Elements of Chemistry* (1848). This was the first modern chemistry textbook written in Ottoman Turkish. It described all the breakthroughs of the late eighteenth and early nineteenth centuries, including atomic theory as well as the use of modern chemical notation. As was typical of the time, Mehmed Emin emphasized the practical value of the physical sciences. Chemistry would aid in 'the acquisition of new industries and the attainment of numerous benefits', he explained. There was also an element of nationalism to Mehmed Emin's textbook. Despite having trained abroad, he was determined that Ottoman scientists should write and study in their own language, not English or French. In *Elements of Chemistry*, all the chemical formulae were written in Ottoman Turkish. There was apparently no place for 'European chemical terms in Turkish chemistry books'.[41]

Mehmed Emin spent the rest of his life in the service of the Ottoman Empire, enjoying a somewhat eclectic career. At one point he worked as a mining engineer, then as a military surveyor helping to determine the border between the Ottoman and Persian empires. He also began teaching at the new Ottoman University. One series of public lectures, delivered in the early 1860s to celebrate the opening of a new university building, was reported with great enthusiasm in local newspapers at the time. At 11 a.m. on 13 January 1863, 300 people packed into the new lecture hall at the Ottoman University to hear Mehmed Emin speak. The audience included a number of leading statesmen, many of whom wanted to learn about how modern science could aid in the development of the Ottoman Empire. He started the first lecture by generating a few electrical sparks using an induction coil. 'Sparks of fire emerged from special instruments,' reported one Ottoman newspaper shortly afterwards. The 'electrical force' was then 'transmitted into a man's body via a thin wire'. According to the same newspaper, 'whatever part of his body that the wire touched emitted blue sparks'. During the demonstration, Mehmed Emin explained the basic principles behind the flow of electricity, noting how this could be put to all kinds of practical uses, such as in the electric telegraph.[42]

Following the lecture, one of the Ottoman statesmen in the audience rose to speak. The Grand Vizier, Mehmed Fuad Pasha, applauded Mehmed Emin's knowledge of physics, and wholeheartedly endorsed the idea that the development of modern science was vital to the future of the Ottoman Empire. 'The difference between old physics and new physics is like the difference between a sailboat and steamboat,' explained the Grand Vizier, who was one of the leading advocates of the *Tanzimat*. It was the 'duty of the state', he argued, to support the growth of science and industry. Aware that there were a number of Islamic clerics in the audience, the Grand Vizier also emphasized the religious value of modern science. Recent breakthroughs in physics and chemistry built on a long Islamic tradition, he argued. After all, medieval Islamic thinkers had written a number of important early works of chemistry, whilst many modern chemical terms, such as 'alkali', were derived from Arabic. Here, we see how the idea of an Islamic 'golden age' was already starting to prove popular amongst Ottoman modernizers. According to the Grand Vizier, the theory of electric current was just another example of 'the philosophy of the divine'.[43]

This was all part of a broader strategy to portray the modernization of the Ottoman Empire as part of the modernization of Islam itself. Before long, Muslims living in Istanbul were travelling to Mecca by railway and steamship, whilst prayer times in Cairo started to be synchronized using an electric telegraph signal. And so, whilst today we often think of religion as in tension with modern science, this was certainly not the case in the Ottoman Empire. The Grand Vizier, like many others, saw the industrial sciences of the late nineteenth century as part of a new Muslim modernity.[44]

At the beginning of 1868, the Ottoman sultan ordered the construction of a new astronomical observatory, to be built on a hill overlooking the Bosporus. Known as the Imperial Observatory, this was the first dedicated astronomical observatory built in Istanbul since the sixteenth century. As with earlier observatories, such as the one we learned about in chapter 2, one of the functions of the Imperial Observatory was to help compile the Islamic calendar. However, alongside astronomy, the Imperial Observatory also functioned as a station for monitoring both the weather and earthquakes. Thanks to advances in physics and chemistry, scientists understood the workings of the atmosphere in much greater detail. It was now possible to track and even predict the weather with a degree of accuracy previously unheard of. This was clearly of great value to the Ottoman state, particularly when it came to planning military or naval campaigns. By the late 1870s, the Imperial Observatory sat at the heart of a network of meteorological stations stretching right across the Ottoman Empire, all connected together by electric telegraph lines.[45]

Alongside meteorology, the science of seismology underwent a revolution during the nineteenth century, as new ideas from chemistry and physics helped scientists better understand the causes of earthquakes. Istanbul was one of the best places in the world to do this research. Sitting along the fault line between Europe and Asia, the region occupied by the Ottoman Empire regularly experienced serious earthquakes. The director of the Imperial Observatory, an Ottoman scientist named Aristide Kumbari, lived through the most damaging earthquake to hit Istanbul during the nineteenth century. At 12.24 on 10 July 1894, just as many Muslims were returning from their midday prayers, the ground began to tremble. A series of shocks, each lasting around fifteen seconds, reduced much of Istanbul to rubble. Hundreds were killed, thousands injured, and many buildings were destroyed, including a number of mosques. 'There is scarcely a street in the city which does not show signs of the destructive effects of the earthquake,' reported the international news agency Reuters.[46]

The Ottoman sultan, Abdulhamid II, immediately called for Kumbari. The earthquake presented both a danger and an opportunity for the sultan. Much like in the early modern period, an earthquake, like an unforeseen astronomical event, had the potential to prompt a political crisis. The public might lose confidence in the sultan as protector and ruler. However, the earthquake also provided an opportunity for Abdulhamid: now was the perfect time to demonstrate the strength of Ottoman science to the international community, as well as the virtues of modernization to the residents of Istanbul. With this in mind, Abdulhamid ordered Kumbari to prepare a report on the cause of the earthquake.[47]

Born to a Greek family in Istanbul in 1827, Kumbari was typical of the new generation of modernizing Ottoman scientists. He had studied mathematics at the University of Athens, before travelling to Paris to undertake further scientific training. By the time he returned to Istanbul in 1868, Kumbari was familiar with all the latest breakthroughs in seismology. In preparing his report, Kumbari was also assisted by a Greek physicist named Demetrios Eginitis. For the next four weeks, the pair of scientists travelled around the Ottoman Empire aboard a steamship provided by the sultan. They assessed the damage in different regions, sifting through the wreckage and taking photographs, trying to determine the direction of the shockwaves and therefore the location of the epicentre of the earthquake. Kumbari also gathered reports from local meteorological stations, as well as survivors of the earthquake. Further reports were collected internationally, sent via electric telegraph from scientists as far away as Paris and Saint Petersburg.[48]

The final report, presented to the Ottoman sultan and simultaneously published in a prestigious French scientific journal, represented one of the most detailed studies of an earthquake to date. At this time, scientists did not have a fully worked out theory of plate tectonics, but thanks to recent advances in physics and chemistry, there was a broad understanding that earthquakes were caused by movements in the core of the Earth. Drawing on all the different seismic data, Kumbari and his team presented a map of the region around Istanbul, indicating the direction of the shockwaves, which moved from north to south. They were also able to determine that the earthquake had been caused by a crack in the Earth's crust beneath the Sea of Marmara, just offshore from Istanbul. Kumbari and Eginitis then concluded with a warning. This would not be last earthquake to hit Istanbul, they noted. The 'geological evolution' of the region was far from finished.[49]

Much like in Tsarist Russia, science in the Ottoman Empire was made in a world of capitalism and conflict, and it was destroyed by that same world as well. When the First World War broke out, the Ottoman sultan decided to forge an alliance with Germany and the other Central

Powers. This seemed sensible enough, given the clear military and industrial strength of Germany at the time. The hope was that the Germans might help accelerate Ottoman scientific and industrial development. As one Turkish newspaper declared enthusiastically, 'we need a battalion of teachers . . . we need to adopt the German education system, German economic ideas, discipline and order'. As it happens, Germany did send a 'battalion of teachers'. In 1915, a group of German scientists arrived in Istanbul to help teach at the Ottoman University. They were led by a chemist named Fritz Arndt, who had previously taught at the University of Breslau. As in Tsarist Russia, the Ottoman sultan hoped that modern science and technology would help bring about victory on the battlefield.[50]

In the end, no amount of science could change the basic military challenges faced by the Ottoman Empire, squeezed as it was between the Allied Powers of Britain, France, and Russia. In November 1918, British troops entered Istanbul. Arndt and the other German scientists had long since fled. This marked the beginning of the end of the Ottoman Empire, which was partitioned following the First World War. The last Ottoman sultan was eventually removed in 1922. The demise of the Ottoman Empire was then followed by a new era of conflict in the Middle East, a story that we pick up again in later chapters. But for now, we move on with the history of chemistry and physics in the nineteenth century, exploring the relationship between science, nationalism, and war in British India.

III. Making Waves in Colonial India

Jagadish Chandra Bose had spent all day carefully setting up his equipment. That evening, he was due to deliver a lecture at the Royal Institution in London on the topic of 'Electro-Magnetic Radiation'. Founded in 1799, the Royal Institution was one of the premier scientific venues in nineteenth-century Britain. Many of the most famous scientists of the Victorian era had made their name speaking there. Now, in January 1897, an audience of over 500 of Britain's leading thinkers sat in awe as Bose – the first Indian scientist to be invited to speak at the Royal Institution – demonstrated the power of what he called 'electric rays'.[51]

Bose began the lecture by generating a few short bursts of electricity, turning on and off a battery connected to a spark coil. 'A flash of radiation for an experiment is obtained from a single spark,' noted Bose. He reminded his audience that 'electric rays', or radio waves as we would now call them, are 'invisible . . . we cannot see the waves that are produced'. How then could the audience be sure that radio waves existed? Bose pointed to a small device sitting on the wooden table in front of him – a radio receiver he had built back in India. This 'extremely sensitive' machine would allow scientists to detect 'electric radiation', explained Bose. He then proceeded to show the device in action. The experiment was simple but, for a Victorian audience, utterly incredible. When Bose turned on the radio transmitter, the receiver on the other side of the lecture hall rang a bell.[52]

Bose clearly had a way with words, describing for his audience an 'ethereal sea in which we are all immersed . . . agitated by these multitudinous waves'. He went on, painting an intoxicating image of the physical universe beyond our senses, describing the different parts of the electromagnetic spectrum, from infrared through to light and radio waves:

> As the ether note rises still higher in pitch, we shall for a brief moment perceive a sensation of warmth. As the note still rises higher, our eye will begin to be affected, a red glimmer of light will be the first to mark its appearance . . . As the frequency rises still higher our organs of perception fail us completely; a great gap in our consciousness obliterates the rest. The brief flash of light is succeeded by unbroken darkness.

Well aware of the significance of the occasion, Bose concluded his lecture with an appeal to bridge the gap between European and Indian science. He expressed his sincere hope that 'at no distant time it shall neither be the West nor East, but both the East and the West, that will work together, each taking her share in extending the boundaries of knowledge, and bringing out the manifold blessings that follow in its train'. And with that, the audience rose to its feet in applause, eager to follow this enigmatic Indian physicist into the hidden world of electromagnetism.[53]

Bose had travelled a long way to deliver his lecture at the Royal Institution. He was born in 1858 in British India, in a small town north of

Dacca, in what is today Bangladesh. Bose's father worked as a magistrate for the colonial government, and sent his son to a local Bengali school until the age of eleven. There was little opportunity to study science, let alone make breakthroughs in physics. Throughout this period, the British typically discouraged Indian involvement in scientific research. Sir Alfred Croft, the Director of Public Instruction in Bengal during Bose's formative years, unashamedly declared that Indians were 'temperamentally unfit to teach the exact method of modern science'. This was typical of the kind of racism that Indian scientists faced under British rule.[54]

With little support from the colonial government, a group of Indian intellectuals decided to take things into their own hands. The campaign was led by Mahendralal Sircar, a wealthy Bengali doctor and champion of Indian scientific education. In 1876, after nearly a decade of drumming up financial and political support, Sircar founded the Indian Association for the Cultivation of Science in Calcutta. This new institution, complete with lecture hall, library, and small laboratory, offered courses in physics and chemistry. The Indian Association, according to Sircar, would 'enable the Natives of India to cultivate Science in all its departments'. The timing couldn't have been better. Bose had recently arrived in the city, having just passed the entrance exam to study at the University of Calcutta. Like the Royal Institution in London, the Indian Association ran regular evening lectures on topics ranging from thermodynamics to electricity. And so, whilst studying for his formal university degree during the day, Bose spent his evenings in the lecture hall and laboratory at the Indian Association. It was here that he got his first exposure to the world of physics.[55]

After completing his degree, Bose was hooked on physics, but his father wanted him to train as a doctor, certainly a more secure career path for a young Bengali graduate at this time. A compromise was reached. Bose would travel to Britain to study Natural Sciences at the University of Cambridge. This would allow him to pursue his passion for science, but at the same time provide the necessary preparation for taking a medical degree. Thankfully, Bose had the right contacts. His brother-in-law had studied at Cambridge a few years earlier, and secured a place for him at Christ's College. As in Calcutta, Bose's arrival was well timed. He entered Cambridge in 1882, just as the university was transforming the teaching of the sciences, introducing more practical classes on experimentation. Only a few years earlier, the British pioneer of electromagnetic theory, James Clerk Maxwell, had founded the Cavendish Laboratory. Bose now had an opportunity to study in one of the most advanced physics laboratories in the world. It was here that he really got to grips with the science of radio waves.[56]

When Bose returned to Calcutta in 1885, he quickly abandoned any plans of training as a doctor. With letters of reference from the Cavendish Laboratory, he was immediately appointed as the first Indian Professor of Physics at Presidency College, part of the University of Calcutta. He was, however, paid only a third of the salary of his European counterparts at the same institution – another reminder of the injustices of colonial rule. Determined to fight against these prejudices, Bose returned to the Indian Association, but this time as a lecturer rather than a student. At the Indian Association, Bose was able to perfect his lecturing style, inspiring a new generation of Indian scientists. He also began serious research into the properties of electromagnetic waves, making use of the Indian Association's laboratory which had recently been expanded. By this time, it was well known that electromagnetic waves existed. The challenge, however, was to prove experimentally that different kinds of electromagnetic waves – whether that was light or a radio wave – all exhibited the same physical properties. To do this, scientists needed to show that radio waves were capable of polarization (splitting into vertical and horizontal components) and refraction (changing speed and direction when travelling through different substances). If they could do this, they could be sure that radio waves and light were essentially the same thing.[57]

Bose's great originality lay in developing the instruments needed to perform these experiments. This wasn't always so easy in Calcutta, what with the heat and humidity alongside limited access to specialist resources. Bose had to train a local Bengali tinsmith to build his scientific instruments from scratch, as he couldn't afford to import expensive machines from Europe. But these difficult conditions ultimately spurred Bose on to new discoveries. He used whatever he could get his hands on. When searching for a material that might polarize electromagnetic waves, Bose turned to the industrializing world of late nineteenth-century India. Bengal was at the centre of the international jute trade, with thousands of mills processing the plant for export. By placing a section of 'twisted jute' between the radio transmitter and receiver, Bose discovered that the overlapping threads of this everyday vegetable fibre could be used to polarize radio waves.[58]

Similarly, when Bose was experimenting with a new radio receiver, he found that the iron filings used to detect the waves simply turned to rust in the Indian climate. Bose replaced the filings with steel wire, and had this coated in cobalt to protect it from the humidity. This solved the problem. However, it also led Bose to the discovery that it was only the surface coating which affected the sensitivity of the receiver, not the underlying metal. This breakthrough, born out of the challenges of doing science in a tropical environment, resulted in an article published in the prestigious *Proceedings of the Royal Society of London*. It also helped advance the field of radio communication. With engineers taking increasing interest in the prospect of using electromagnetic waves to transmit messages, developing a sensitive and reliable radio receiver was high on the agenda, the first step in the manufacture of a commercial system of wireless telegraphy. Bose's new design could be adapted 'for practical and possibly money-making purposes', noted one engineering journal in London.[59]

Ever the performer, Bose arranged for a public demonstration of his instruments at Calcutta Town Hall in 1895. This was no ordinary physics lecture. Instead, Bose set up an elaborate series of contraptions designed to prove, not only the existence of radio waves, but that they could be used to transmit signals. In many ways it was a more dramatic forerunner to his Royal Institution lecture two years later. In one room, Bose set up his transmitter and then, in another, over seventy-five metres away, he connected his receiver to a bell and a small pot of gunpowder. He then asked Sir Alexander Mackenzie, the Lieutenant-Governor of Bengal, to sit in a chair between the two rooms, right in the path of the radio waves. The experiment was simple but spectacular. Bose fired up his electromagnetic transmitter. Instantly, the receiver jumped into life. The bell started ringing and the gunpowder sparked with an almighty bang. The radio waves had passed right through the two walls, as well as the body of Sir Alexander Mackenzie, who was suitably impressed with what was the first public demonstration of radio telegraphy in India. The news of the experiment in Calcutta soon reached Europe. And it was this success that brought Bose an opportunity to return to Britain and deliver his famous lecture at the Royal Institution in 1897.[60]

Jagadish Chandra Bose went on to become one of the most famous physicists of the late nineteenth century. Following his lecture at the Royal Institution, he was invited to speak across the world, including at the Prussian Academy of Sciences in Germany and Harvard University in the United States. As we saw at the start of this chapter, Bose also attended the First International Congress of Physics in Paris in 1900. He published in leading scientific journals, received a number of patents for his radio designs, and in 1920 was made a Fellow of the Royal Society. Yet despite all these achievements, Bose has today largely been forgotten outside of India. This in part is a consequence of the legacies of colonialism and racism, something Bose spent much of his own life fighting against. But it is also a consequence of the failure to consider the history of science in India as part of a broader global story. Much as we've seen elsewhere, the development of the modern physical sciences in India was fundamentally shaped by the growth of industry, nationalism, and war.[61]

In order to understand the history of science in late nineteenth-century India, we need to begin by looking at the changing nature of colonial rule. In 1858, the year Bose was born, the British Crown took formal control of the Indian colonies, which had previously been governed by the East India Company. This marked the beginning of the British Raj, which lasted until Indian independence in 1947. With formal colonial rule came a number of new scientific institutions. Just prior to the formation of the British Raj, the East India Company had established the first three universities in India, in the cities of Calcutta, Madras, and Bombay. These were then followed by a number of new universities created under the British Raj, including the University of the Punjab in 1882 and the University of Allahabad in 1887. The expansion of higher education in India was part of a colonial programme to supply graduates to work in the Indian Civil Service. Many of the posts required some kind of scientific training, whether that was working for the Geological Survey of India analysing mineral ores or in the Meteorological Department monitoring the weather.[62]

At the same time, the expansion of colonialism in India coincided with a period of industrialization. This was again partly a consequence of the transition from East India Company to Crown rule. Whereas the East India Company held a monopoly on trade with India, the British Raj opened up the region to much greater capital investment. British and Indian investors put their money into railways and factories, and by the early twentieth century the colonial government was explicitly promoting 'the strength which an industrialised India will bring to the power of Empire'. By 1900, Calcutta had been transformed into an industrial metropolis, with steamships going up and down the Hooghly River and jute mills supplying the world market for cloth and cordage. As elsewhere, electricity was widely considered a marker of industrial modernity. Telegraph lines criss-crossed the country, connecting India to the wider British Empire, whilst various private companies began manufacturing and installing electric lights in Indian cities. In fact, Bose himself advised on the introduction of the first electric streetlights in Calcutta in 1891.[63]

Despite the growth of science and industry under colonial rule, opportunities for Indians to conduct original research were still relatively rare. Leadership positions within colonial scientific departments were reserved for British scientists. The same was true of the majority of teaching positions at Indian universities. When Bose became Professor of Physics in 1885, he became the first – and for a few years, the only – Indian teaching science at the University of Calcutta. Bose himself later recalled the racism he faced, complaining that there was 'a strong doubt, not to say prejudice, against the capacity of an Indian to take any important position in science'. The racism that underpinned colonial rule was structural. Indian scientists were paid between a third and two thirds of the salaries of their British counterparts, whilst the Director of Public Instruction in Bengal openly referred to 'the degradation of the national intellect among the Hindus'. Even by the 1920s, Indians comprised less than 10 per cent of the scientific personnel employed by the colonial government, despite obviously making up the vast majority of the population.[64]

Over time, the injustices of British rule sparked the growth of anticolonial nationalism in India. This is a theme that we explore in more detail in later chapters, particularly with the rise of the anticolonial movement in the early twentieth century. However, even in the nineteenth century, there was growing unease with the British Raj. Indians began to come together, forming political associations and campaigning for better treatment and representation. As elsewhere, the development of nationalism in India had a profound effect on the sciences. The Indian Association for the Cultivation of Science, where Bose studied and later taught, is a good example. It was established in the very same year as the Indian National Association, one of the first political societies dedicated to the cause of Indian nationalism. The Indian Association for the Cultivation of Science was intended to serve a very similar goal. 'I want it to be solely native and purely national,' wrote its founder, Mahendralal Sircar. The problem of Indian scientific achievement, argued Sircar, lay in 'the want of opportunity, want of means, and want of encouragement, not in a defective moral nature'. The new laboratory, paid for by Sircar and his supporters, would provide the space and equipment Indians needed to conduct original scientific research. There was now 'nothing to disqualify the natives of India from taking their share in the advancement of the natural sciences'.[65]

The Indian Association for the Cultivation of Science provided a home for a new generation of Indian scientists. Amongst these was a Bengali chemist named Prafulla Chandra Ray. His career neatly illustrates the way in which industry, nationalism, and war all came together to shape the development of modern science in colonial India. Born in 1861, Ray studied at a small village school in eastern Bengal, before moving with his parents to Calcutta in 1870. After a few years at various English schools in the city, Ray entered Presidency College in the University of Calcutta. Like Jagadish Chandra Bose, Ray also attended lectures at the Indian Association for the Cultivation of Science. He was wowed by the chemistry demonstrations given by Tara Prasanna Rai, who – when not teaching at the Indian Association – worked as an assistant chemical examiner for the colonial government. Inspired by Rai's lectures, Ray went back to his student hostel and set up what he later described as 'a miniature laboratory'. On one occasion, Ray even admitted to causing 'a terrible explosion' after igniting a mixture of hydrogen and oxygen.[66]

In the summer of 1882, Ray graduated from the University of

Calcutta. Like many Indian scientists in this period, he then travelled to Britain to undertake further training. Arriving in London in August that year, Ray stayed with Bose, who had also recently arrived in the imperial metropolis. After a few weeks, Bose headed to Cambridge, whilst Ray caught the train up to Scotland, where he was due to begin his studies at the University of Edinburgh. Although he found the cold climate somewhat of a shock, Ray thoroughly enjoyed his time in Scotland. Every morning he would eat a bowl of porridge, cooked by his Scottish landlady. Wrapped up in a woollen coat and scarf, Ray then 'trudged along the pavement covered in snow' to his lectures.[67]

After graduating from the University of Edinburgh with a DSc in 1888, Ray returned to India. A year later, he was appointed as Assistant Professor in Chemistry at Presidency College, where he joined Bose as one of the small number of Indian teaching staff. At this time, the colonial government didn't really imagine that Indian scientists would conduct any meaningful research. Rather, Ray and Bose were expected to teach the basics of modern science, mainly to support students who would then go on to work for the Indian Civil Service. The University of Calcutta didn't even award doctorates at this time. As such, the chemistry laboratory at Presidency College was poorly equipped and frankly dangerous. Ray later recalled how 'there were no flues for drawing off the noxious gases and the ventilating arrangements were most rudimentary'. At times it could be difficult to breathe. 'While the practical classes were in full swing, the atmosphere . . . thickly laden with fumes, became suffocating and highly injurious to health,' complained Ray. After much hassling, Ray did manage to convince the colonial government to provide some extra funds. In 1894, Presidency College opened its new chemistry laboratory, modelled on that of the University of Edinburgh, which now featured proper ventilation, work benches, and storage for all the chemicals.[68]

It was in this new laboratory in Calcutta that Ray began his most important scientific work. He had recently read an English translation of Dmitri Mendeleev's *Principles of Chemistry*, 'a classic in the domain of chemical literature', according to Ray. Inspired by Mendeleev's writings, Ray began searching for new chemical elements. 'I had taken up the analysis of certain rare Indian minerals, in the expectation that one or two new elements might turn up and thus fill in the gap in

Mendeleff's Periodic System,' he wrote. Through a friend working at the Geological Survey of India, Ray was able to get hold of various metal ores, all with the hope of discovering a new chemical element.[69]

In the end, Ray didn't find a new element, but he did discover an entirely new kind of compound, one that proved exceptionally important for the development of industrial chemistry. In 1894, whilst working in the chemical laboratory at Presidency College, he mixed together a flask of water, nitric acid, and mercury. After 'about an hour', he noticed the formation of some 'yellow crystals' on the surface of the mixture. These crystals turned out to be a previously unknown chemical compound called 'mercurous nitrite'. Ray quickly realized that there were many other kinds of 'nitrites' that could be formed – usually based on a reaction involving nitric acid. This discovery, which was reported in leading scientific journals in Europe, including *Nature*, ultimately opened up a whole new field of chemical research, now known as 'nitrite chemistry'. Scientists around the world began searching for other similar compounds, many of which turned out to have practical uses. Today, nitrites are used in everything from food preservation to pharmaceuticals.[70]

Just around the time of his discovery of mercurous nitrite, Ray himself entered the world of industrial chemistry. In 1893, Ray invested 3,000 rupees of his own money in a chemical factory on the outskirts of Calcutta. The factory, in which Ray installed a dedicated laboratory, soon came to be known as the Bengal Chemical and Pharmaceutical Works. At this time, the majority of chemicals and medicines used in India were still imported from Britain. Ray's idea was to begin manufacturing these locally, and so save on the import cost. Ray also hoped to make India – and by implication, Indians – less dependent on Britain. The Bengal Chemical and Pharmaceutical Works was therefore an industrial equivalent to the Indian Association for the Cultivation of Science. It was to be a 'model institution', argued Ray, one that would demonstrate the autonomy of Indian science and industry. As we'll see in the following chapter, these early experiments in 'self-sufficiency' served as a precursor to the anticolonial campaigns of the twentieth century, something Ray himself later became involved in.[71]

Ray's career was undoubtedly shaped by the wider world of international and industrial science. He studied in Scotland, attended scientific meetings in France and Germany, and read English translations of Russian textbooks. Yet Ray never forgot where he had come from. Throughout his life, he championed the value of Indian culture as a source for the development of modern science. He was also deeply involved in a religious movement, known as Brahmoism, to reform and revitalize Hinduism. There are many parallels here with the story of the Ottoman Empire, where the adoption of modern science was promoted as part of the reform and modernization of Islam. 'I am as proud of the glories of the Hindus of old as anybody,' wrote Ray in 1910, before going on to describe 'the contributions of the ancient Indians to the science of chemistry'.[72]

In fact, at exactly the same time that Ray was experimenting with mercury and nitric acid, he was also writing a two-volume work titled *A History of Hindu Chemistry* (1902–4). Based on his reading of ancient Sanskrit texts held at the Asiatic Society of Bengal, as well as those he had collected in the holy city of Benares, Ray made the case that the

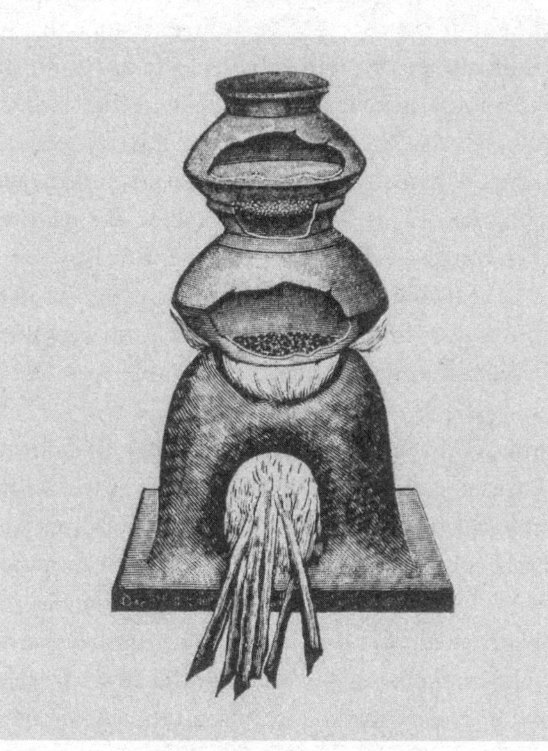

30. A traditional Indian technique for the 'extraction of mercury', from Prafulla Chandra Ray, *A History of Hindu Chemistry* (1902–4). Ray discovered mercurous nitrite in 1894 by following a similar method.

people of ancient and medieval India had a sophisticated understanding of chemistry. He even began manufacturing traditional Indian medicines according to the ancient system known as Ayurveda at the Bengal Chemical and Pharmaceutical Works. As Ray explained, 'all that was needed was that their active principles should be extracted according to scientific up-to-date methods'.[73]

One medieval text in particular fascinated Ray. Written in Sanskrit in the twelfth century, the *Treatise on Metallic Preparations* described the manufacture of various medicinal compounds. This text, according to Ray, contained 'a wealth of information and chemical knowledge'. Strikingly, much of it concerned the chemistry of mercury, which was widely used in traditional Indian medicine. It therefore seems likely that Ray's own interest in the chemistry of mercury, which led him to the discovery of mercurous nitrite, was directly inspired by his reading of this medieval Sanskrit text. Ultimately, Ray provides another reminder of the complex process of cultural exchange that underpinned the emergence of modern science. He was a factory-owner, but also a devout Hindu; a modern scientist, but also a scholar of ancient Sanskrit; an Indian nationalist, but also someone who later accepted a knighthood from the British government. Today, it is easy to see Ray as a set of contradictions. This is probably one of the reasons why he so rarely features in histories of modern science. But Ray was in fact typical of the world of late nineteenth-century science – a world in which the growth of industry and nationalism led to the entanglement of a variety of different scientific cultures.[74]

When the First World War broke out in 1914, Britain mobilized its empire in an effort to defeat the Central Powers. Over a million Indian soldiers fought in the war, many of whom lost their lives far from home on the Western Front. Alongside the military, Britain also turned to Indian science and technology. As we've seen, a number of Indian scientists had already made important contributions to the development of modern physics and chemistry, but they did so with little direct support from the colonial government. The outbreak of the First World War changed everything. After years of relative indifference, the colonial government began to invest in new scientific and industrial institutions. Four new universities were established in India during the war, at Benares, Mysore, Patna, and Hyderabad. These new universities, which included modern physics and chemistry laboratories, were designed to support the efforts of the Indian Industrial Commission, established in 1916, and the Indian Munitions Board, established in 1917. India would now contribute directly to the war effort, not just by supplying soldiers, but also by supplying explosives and chemicals.[75]

Indian scientists were asked to play their part as well. Prafulla Chandra Ray himself served on the Indian Industrial Commission, whilst the Bengal Chemical and Pharmaceutical Works was repurposed to manufacture gunpowder and military medicines. In fact, it was for his service during the First World War that Ray was awarded a knighthood in 1919. 'The late war called for every ounce of scientific knowledge,' noted Ray shortly afterwards. 'The scientific battle has been fought by laboratory men,' he concluded. In the following section, we explore a very similar history – a history of industry, nationalism, and war – but from the perspective of a very different imperial state.[76]

IV. Earthquakes and Atoms in Meiji Japan

At 6.38 in the morning, the clocks stopped and the ground began to shake. In Tokyo, buildings started to crumble, whilst outside of Osaka, a large steel bridge collapsed into a nearby river. On 28 October 1891, Japan experienced the strongest earthquake in its history. Over 7,000 people were killed, more than 100,000 were made homeless, and much of the region around the south coast of Honshu was left in ruins. As we saw in the previous chapter, the Meiji Restoration of 1868 brought about a transformation in Japanese society. Ironically, this was one of the reasons that the 1891 Nobi earthquake proved so damaging. Industrialization and urbanization had led to more and more Japanese citizens living in densely populated cities, all connected by railways and by electric telegraph lines, many of which were destroyed during the earthquake.[77]

Much as we saw in the Ottoman Empire, the 1891 Nobi earthquake presented a moment of potential crisis for the state. Despite decades of investment in modern science and technology, the Meiji government had been unable to protect its citizens from the devastating impact of this natural disaster. If the Japanese people were going to be convinced of the power of modern science to transform the nation for the better, now was the time to demonstrate it. The government immediately ordered the formation of the Earthquake Investigation Committee. Led by a Japanese scientist named Aikitsu Tanakadate, the Earthquake Investigation Committee spent the following year travelling across the country, surveying the damage. Tellingly, Tanakadate himself was trained, not as a geologist, but as a physicist. He believed that recent breakthroughs in physics could help scientists, not only understand the cause of earthquakes, but possibly even predict them too.[78]

Born in 1856 in northern Honshu, Tanakadate was typical of a new generation of modernizing Japanese scientists who grew up in the years following the Meiji Restoration. His father was a samurai, and initially Tanakadate had been schooled in the traditional arts of calligraphy and swordsmanship. However, the Meiji Restoration severely weakened the political power of the samurai. It soon became clear that they would need to reinvent themselves in order to survive the nineteenth century. And so, rather than pursuing a traditional samurai education, Tanakadate studied physics at the University of Tokyo, graduating with a BSc in 1882. Like many former samurai, he saw in modern science a means to bring the art of war into the industrial age. Indeed, when Tanakadate was taught science at the University of Tokyo, it was usually through some kind of military or industrial case study. The basics of physics and chemistry were explained through the operation of artillery, whilst students were regularly taken on trips to local factories.[79]

Like many Japanese scientists, Tanakadate spent time studying abroad. In 1888, the Meiji government sent him to the University of Glasgow in Scotland, where he worked for two years in the laboratory of the renowned British scientist Lord Kelvin. A pioneer of physics, as well as an accomplished engineer who contributed to the development of the electric telegraph, Lord Kelvin was an ideal mentor. Tanakadate quickly learned about the latest scientific breakthroughs, particularly in the field of electromagnetism. He also got to tour the nearby factories and shipyards, witnessing first hand the industrial world of Victorian Britain. And it was during his time in Glasgow that Tanakadate published his first scientific articles, all of which were about magnetism. This, as we'll see, proved crucial for his later research into the causes of earthquakes.[80]

In the summer of 1891, Tanakadate returned to Japan to take up a position as Professor of Physics at the University of Tokyo. Just a few months later, the Nobi earthquake struck. The Meiji government immediately recruited Tanakadate to lead the Earthquake Investigation Committee. It was finally time to put his scientific knowledge into practice. Tanakadate's idea was to conduct both a geological and a geomagnetic survey of Japan. The geological part, identifying fault lines and areas of varying seismic activity, was relatively straightforward, and not so different from the kind of work we saw in the Ottoman Empire following the Istanbul earthquake. However, the geomagnetic part was much more original. Since the early nineteenth century, scientists had known that the magnetic field of the Earth varies across the planet. This means that the direction of the north pole ('true north') and the direction that a compass needle points ('magnetic north') are not necessarily identical, depending on where you are. The cause of this variation in the magnetic field was widely debated throughout the nineteenth century. However, most scientists agreed that it must have something to do with the presence of metallic elements in the Earth's crust.[81]

From the 1830s onwards, there were various attempts to map the variation of geomagnetism across the planet. This was usually done with the practical aim of helping to better calibrate scientific and navigational equipment. Tanakadate himself had been involved in an earlier geomagnetic survey. In 1887, just before he left for Glasgow, he had been recruited by one of his tutors at the University of Tokyo to help map 'the general magnetic characteristics of Japan'. Over the course of six months, Tanakadate travelled more than 3,000 miles, sometimes by steamship, at other times by railway, taking hundreds of measurements. He reached as far north as the Korean Peninsula, which had recently been colonized by Japan, and as far south as the Bonin Islands, a Japanese colonial territory in the Pacific Ocean. At each location, he needed to determine the exact latitude and longitude by astronomical observation. This gave the direction of 'true north'. He also needed to take a reading for 'magnetic north' using a special compass powered by an electromagnet. Tanakadate then calculated the difference between the two values, giving what is known as the 'magnetic declination' or 'magnetic variation' of the particular location. This was what scientists, engineers, and navigators needed to calibrate their equipment in Japan.[82]

Tanakadate's training as a physicist gave him a unique perspective on

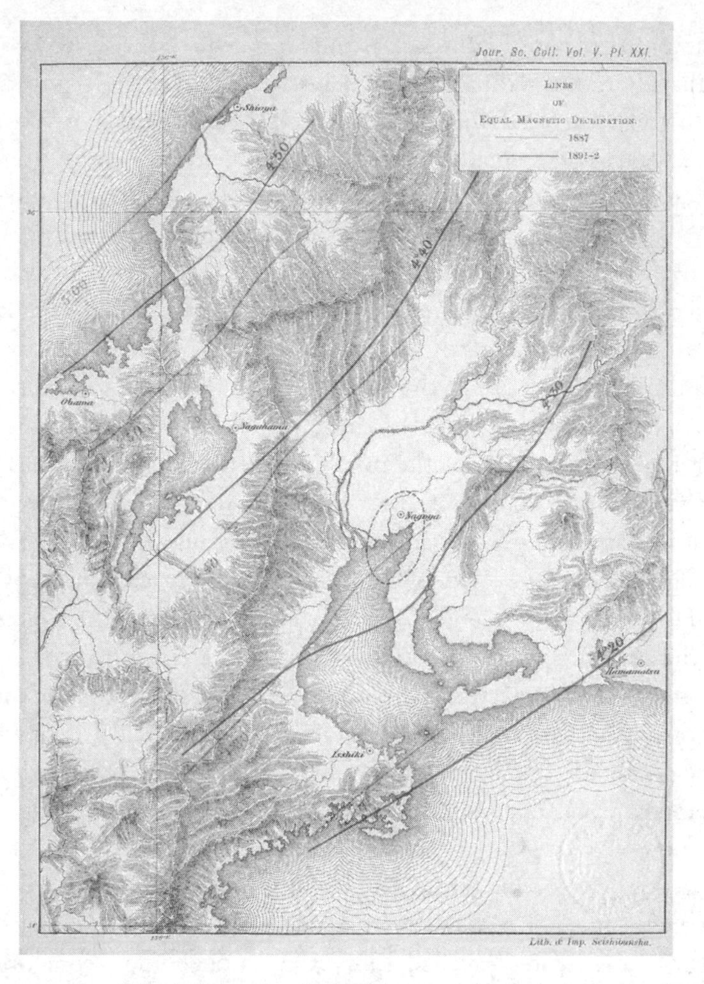

31. A map produced by Aikitsu Tanakadate showing the disturbance in the Earth's magnetic field around the site of the 1891 Nobi earthquake. Note the deflection of the line of 'equal magnetic declination' close to the circled city of Nagoya.

the science of earthquakes. Whereas most seismologists thought of earthquakes in terms of geology, Tanakadate thought of them in terms of electromagnetism. His hypothesis, which turned out to be correct, was that earthquakes might cause a localized disturbance in the Earth's magnetic field. Tanakadate also suggested that it might be possible to predict earthquakes by carefully monitoring changes in geomagnetism. The 1891 Nobi earthquake presented the perfect opportunity to test this hypothesis. Tanakadate had already completed a recent geomagnetic survey of Japan. All he needed to do was repeat the survey around the site of the earthquake, and see if there was a change in the geomagnetic field. Sure enough, he was proved right. His final report, published by the University of Tokyo in 1893, clearly showed that earthquakes did cause 'changes of the magnetic state'. Tanakadate even included a series of maps, comparing the lines of magnetic declination before and after the earthquake of 1891. Around the city of Nagoya, which was at the epicentre of the earthquake, there is an obvious shift in the line. This was clear evidence of 'the effect of seismic events upon the magnetic elements of a country', he concluded.[83]

Much like in the Russian and Ottoman empires, the nineteenth century was a period of reform in Japan. This was largely in response to the threat of European and American imperialism. In 1853, the United States Navy blockaded Edo Bay, forcing the Tokugawa shogunate to open up to American trade, whilst in 1863 the British Royal Navy bombarded the coast of Satsuma following the killing of a local English merchant. Worried about this kind of gunboat diplomacy, the Japanese government established a number of new scientific and military institutions, including the Nagasaki Naval Academy in 1855, where naval officers were taught engineering and physics. Shortly after this, steamships were introduced into the Japanese Navy for the first time. Government investment in modern science and technology then accelerated in the years following the Meiji Restoration of 1868. There was, as elsewhere, an element of nationalism to all this. 'The only way to maintain the nation's strength and to guarantee the welfare of our people in perpetuity is through the results of science,' argued the Japanese Prime Minister in 1886. As we've seen, the idea of a struggle between nations was at the heart of how many statesmen and scientists understood the world at this time.[84]

The Meiji Restoration coincided with a period of rapid industrialization in Japan, something promoted directly by the government. 'Today, it is our urgent business to develop industries and lay the foundation of our wealth. We must investigate industrial applied science and train industrial scholars,' argued the Ministry of Education in 1885. Many Japanese scientists contributed to this industrial vision. Alongside his work on earthquakes, Aikitsu Tanakadate advised on the manufacture of magnets as well as on the introduction of balloons into the Japanese Navy. Tellingly, the vast majority of PhDs awarded by the University of Tokyo between 1888 and 1920 were in physics and engineering. Tanakadate himself saw the physical sciences as essential to the strength of the nation, writing that he wanted to 'master physics, which is the basis of all science, so as to make up on full measure for our country's deficiencies'.[85]

Chemistry too played an important role in the growth of industry in Japan. A number of European chemistry books had been translated into Japanese during the early nineteenth century. In Edo, a doctor named Yoan Udagawa published a book titled *An Introduction to Chemistry* (1837), which was based on the work of the famous French chemist Antoine Lavoisier. Alongside modern chemical terminology, Udagawa's book introduced Japanese readers to the more practical side of modern chemistry, such as how to build an electric battery. This early interest in industrial chemistry was then reinforced following the Meiji Restoration. The University of Tokyo featured a dedicated chemistry laboratory, which was intended 'to prepare students to improve the industries prospering in Japan'. This view of chemistry as a practical subject was shared by many Japanese scientists. Toyokichi Takamatsu, Professor of Applied Chemistry at the University of Tokyo, described the 'main object' of his research as 'producing useful goods from raw materials'.[86]

By far the most successful industrial chemist in Meiji Japan was a man named Jokichi Takamine. Born in 1854, Takamine, like many Japanese scientists in this period, was the son of a samurai. However, following the Meiji Restoration, he too decided to follow a different path, enrolling at the Imperial College of Engineering in Tokyo to study a degree in chemistry. Following graduation, he was selected to undertake further training abroad. Between 1880 and 1882, he studied at Anderson's College (now the University of Strathclyde) in Glasgow. This was the beginning of a long and successful career as an industrial chemist, one that made Takamine incredibly rich. On returning to Japan in 1883, he worked for a few years at the Ministry of Agriculture and Commerce, helping to modernize traditional Japanese industries such as sake brewing.[87]

It was in fact during his time working at a sake brewery that Takamine made an important discovery. Sake, a kind of Japanese rice wine, is traditionally fermented using a particular fungus called *koji*. Employing a technique he had learned in Glasgow, Takamine managed to extract the chemical produced by this fungus, and realized that it could be put to all kinds of industrial uses. First off, he experimented with making whiskey, replacing malt with *koji* to reduce the fermentation period from six months down to just a couple of days. He then began selling the same chemical extract as a medicine for indigestion, under the tradename 'Taka-Diastase'. By the early 1900s, Takamine was one of the most famous industrialists in the world. He owned factories in Japan and the United States, and was reportedly worth over $30 million – close to $1 billion in today's money. Like many of the figures in this book, his success was due in large part to the way in which he combined knowledge from different cultures. And it all started with a bottle of sake.[88]

Following the Meiji Restoration, a number of Japanese scientists made important contributions to the development of modern physics and chemistry. But one individual went a step further, transforming our understanding of the very nature of matter itself. Hantaro Nagaoka, like the other Japanese scientists we've encountered in this chapter, was the son of a samurai. Born in 1865, he was exposed from an early age to European science. His father supported the Meiji Restoration and, at the request of the Emperor, had travelled to Europe as part of the Iwakura Mission of 1871. The purpose of the mission was twofold. First, to develop diplomatic relations with other nations, and second, to collect information on European science and industry in order to further the programme of reform back in Japan. Nagaoka's father was impressed by what he saw, returning to Japan with scientific books purchased in England for his children. Encouraged by his father, Nagaoka entered the University of Tokyo in 1882 to study physics.[89]

Nagaoka then followed a well-established path. Between 1893 and 1896, he studied in Germany and Austria, where he met many leading European physicists. During this time, Nagaoka really flourished as a researcher. As was typical of the international nature of physics in this period, Nagaoka published scientific articles in English, French, German, and Japanese. However, he was not content to simply replicate the science being done in Europe. Rather, Nagaoka wanted to show that Japan could lead the world in scientific research, much as it had in the early modern period. 'I did not plan to follow the work of others or to devote my life to importing learning from abroad,' he explained. In private, Nagaoka was even more candid about the competitive nationalism that underpinned his desire to study physics. 'There is no reason why the Europeans will be so supreme in everything,' he wrote in a letter to his friend, the physicist Aikitsu Tanakadate.[90]

When Nagaoka returned to the University of Tokyo in 1896 he was immediately promoted to a professorship. And it was in Japan that he made his most important breakthrough. On 5 December 1903, Nagaoka presented a paper at the Tokyo Mathematico-Physical Society in which he described 'the actual arrangement in a chemical atom'. For centuries, scientists had puzzled over the nature of matter. And throughout the nineteenth century, there had been vigorous debate concerning its basic structure. Nagaoka finally put this debate to rest, kickstarting the field of atomic physics in the process. Based on a series of complex mathematical calculations, he proved that the atom must consist of a group of negatively charged electrons orbiting a large 'positively charged particle'. The way to think about this, explained Nagaoka, was to imagine the planet Saturn. The positively charged particle in the centre was like the planet itself, whilst the negatively charged electrons were like the rings. Crucially, Nagaoka was able to show that this 'Saturnian system', as he called it, was physically stable.[91]

What inspired this fundamental breakthrough? On the one hand, there was clearly the influence of the time Nagaoka spent in Europe. As we saw at the start of this chapter, he actually attended the First International Congress of Physics in Paris in 1900, where he got to meet J. J. Thomson, the British scientist who discovered the electron. However, there was also something particular about life in Japan that shaped Nagaoka's thinking. Just before he left for Germany, he had been recruited by the Earthquake Investigation Committee to help Tanakadate with his work on the 1891 Nobi earthquake. For six months, Nagaoka trekked across Japan with Tanakadate, up and down mountains, measuring the precise geomagnetic impact of the earthquake. Nagaoka was even listed as a co-author on Tanakadate's final report. It was ultimately this experience of earthquakes in Japan which

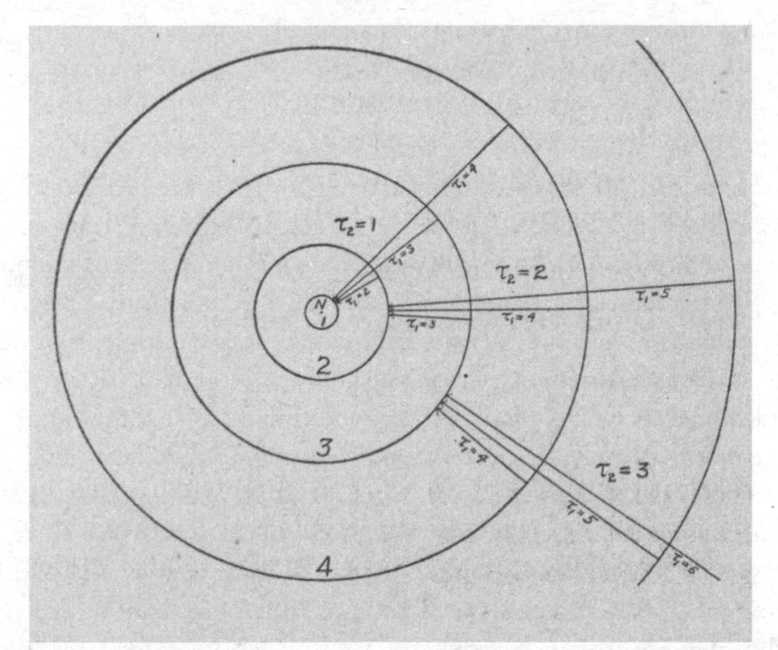

32. An illustration of the 'Saturnian atom' from Ernest Wilson, *The Structure of the Atom* (1916), which also referenced Hantaro Nagaoka. Note the central positively charged nucleus surrounded by rings of orbiting electrons.

fundamentally shaped how Nagaoka thought about the physics of the atom.[92]

In early 1905, Nagaoka published another article identifying what would happen when an electromagnetic wave interacted with the centre of an atom. Strikingly, he returned to the science of seismology to describe the effect. Nagaoka argued that the large positively charged particle at the centre of an atom was rather like a 'mountain or mountain range'. An electromagnetic wave passing through the centre of an atom would therefore be dispersed, suggested Nagaoka, much like when a seismic wave passes through a mountain during an earthquake. Between 1905 and 1906, Nagaoka even published a pair of articles directly comparing 'the dispersion of seismic waves' to 'the dispersion of light'. Ultimately, Nagaoka is another good example of how different cultures, and different scientific disciplines, were being brought together in the years around 1900, and therefore how science is a product of global cultural exchange. Nagaoka combined ideas from physics and chemistry, whilst drawing on his experience in both Europe and Japan. In doing so, he made one of the most important scientific breakthroughs in modern physics.[93]

Today, the discovery of the structure of the atom is usually attributed to the British physicist Ernest Rutherford. This is a prime example of the way in which non-European scientists have been written out of the history of modern science. Rutherford's influential article, announcing the structure of the atom, was published in 1911, well after Nagaoka had published a series of articles on exactly the same topic. Not only that, but Rutherford himself was well aware of Nagaoka's research, and made no secret of it. Indeed, the two met and discussed their ideas. In September 1910, Rutherford happily showed Nagaoka around his laboratory at the University of Manchester, explaining that he was doing the experimental work to confirm the structure of the atom. Then, in February 1911, Rutherford wrote to Nagaoka, letting him know about his forthcoming article. 'You will notice that the structure assumed in my atom is somewhat similar to that suggested by you in a paper some years ago,' wrote Rutherford. Sure enough, when Rutherford published his results in 1911, he included a reference to Nagaoka's original 1904 article. Here, in Rutherford's footnotes, we can once again uncover a hidden history of modern science. A history which is neither simply British, nor Japanese, but a combination of the two.[94]

Japan was very much part of the wider world of late nineteenth-century science. Almost all Japanese scientists in this period spent some time studying in Europe, and many continued to attend international scientific conferences, right up until the outbreak of the First World War. However, much as we've seen throughout this chapter, internationalism and nationalism typically went hand in hand. The connection between science, nationalism, and war was particularly strong in Japan, as the majority of scientists who trained in the years immediately following the Meiji Restoration of 1868 came from samurai families. These samurai-scientists combined a traditional belief in the importance of military strength with a renewed sense of the value of modern science and technology. 'To enrich the nation and strengthen the military, one must bring to perfection the sciences of physics and chemistry,' wrote another former samurai working at the University of Tokyo.[95]

As elsewhere, this delicate balance between internationalism and nationalism did not last. In August 1914, Japan entered the First World War on the side of the Allied Powers, quickly seizing many German colonial territories across East Asia and the Pacific. Some Japanese scientists, particularly those who had trained in Germany, felt uneasy about the war. Still, they played their part. Aikitsu Tanakadate advised the Japanese military on the design of aeroplanes, whilst Jokichi Takamine helped establish a dedicated industrial research institute in order to produce war chemicals. Unlike in the case of the Ottoman and Russian empires, the First World War did not bring about a political crisis in Japan. Rather, as we'll see in the following chapters, Japan emerged from the First World War as a major scientific, military, and industrial power in East Asia.[96]

V. Conclusion

In order to understand the history of modern science, we need to think in terms of global history. This is especially true of the physical sciences during the nineteenth century. Whether you worked in Russia, Turkey, India, or Japan, to be a scientist meant travelling the world, publishing in different languages, and reaching audiences in different countries. As a consequence, the world of scientific publishing was much more linguistically diverse than today. Japanese scientists published in German, whilst Russian scientists read French. On meeting Aikitsu Tanakadate in 1890, the German physicist Heinrich Hertz was only half-joking when he remarked that 'it looks as if from now on we are going to have to learn Japanese. What a problem that is going to be.'[97]

Wherever they studied, the scientists uncovered in this chapter all made major contributions to the development of the modern physical sciences. That was more often acknowledged at the time than in the present. Today, few outside of their home countries have heard of Jagadish Chandra Bose, Hantaro Nagaoka, or Peter Lebedev. But in the nineteenth century, leading European scientists took them very seriously. Ernest Rutherford cited Nagaoka in his famous paper on the structure of the atom, whilst Lord Kelvin admitted that it was Lebedev's experiments on the pressure of light which had finally convinced him of the truth of James Clerk Maxwell's theory of electromagnetism.[98]

Science in the nineteenth century was an industrial enterprise. Many physicists and chemists spent time working for companies and governments, helping to design factories and build telegraph lines. Even the famous 'Maxwell's equations', still used by physicists today, were first developed, not by Maxwell, but by a telegraph engineer looking for a quicker way to do his calculations. Still, the industrial age did not sweep away all earlier ideas, and many nineteenth-century scientists continued to draw on existing cultural traditions in their work. Prafulla Chandra Ray's discovery of mercurous nitrite was inspired by his reading of ancient Sanskrit texts, whilst Jokichi Takamine revolutionized the world of industrial chemistry through his knowledge of sake brewing.[99]

During the nineteenth century, governments around the world equated scientific prowess with military and industrial strength. This was certainly what motivated renewed investment in modern science in both Meiji Japan and the Ottoman Empire. Nationalism and internationalism always seemed to go together. Indeed, at the same time as many scientists were collaborating, others were going to war. In the 1860s, the Meiji government sent some Japanese students to study in Tsarist Russia. By 1904, the two nations were fighting over Manchuria. This pattern of conflict continued, culminating in the outbreak of the First World War in 1914. In the following chapters, we push forward, exploring how modern science developed in the aftermath of this global conflict. The ideological battles of the twentieth century would go on to transform, not only world politics, but also our understanding of the universe, and even life itself.[100]

PART FOUR

Ideology and Aftermath, *c.* 1914–2000

7. Faster Than Light

Steaming up the Yangtze River, the SS *Kitano Maru* approached Shanghai. On board, was one very important passenger. On the morning of 13 November 1922, Albert Einstein arrived in China. As he stepped down the gangway onto the embankment in Shanghai, Einstein was met by a crowd of journalists and photographers. The newspapers had been tipped off. Einstein was about to receive some exciting news. There, on the banks of the Yangtze, he was handed a telegram informing him that he had been awarded the Nobel Prize in Physics. For Einstein, this was a momentous day, one that confirmed his status as one of the most important scientific thinkers of the twentieth century. However, having just arrived in Shanghai, Einstein had little time to reflect on the significance of his achievement. He was quickly whisked away on a tour of the city. Writing in his diary, Einstein recalled the 'terrific bustle' of early twentieth-century Shanghai, 'swarming with pedestrians, rickshaws, caked with dirt of every kind'. For lunch, he was taken to a local restaurant in which he attempted, somewhat clumsily, to eat with chopsticks. That evening, he was guest of honour at the house of the wealthy businessman and modernist artist Wang Yiting. After dinner, Einstein gave a short speech, remarking that 'as to Chinese youths, I believe they are bound to make a great contribution to science in the future'.[1]

The next morning, Einstein boarded the SS *Kitano Maru* again. His brief stop in Shanghai was part of a five-month tour of Asia. By this point, Einstein had already visited Ceylon, Singapore, and Hong Kong, and next on his itinerary was Japan. On 17 November 1922, Einstein arrived in Kobe. By the early 1920s, Japan had transformed into a modern industrial economy. Einstein travelled the length of the country by railway, followed by 'throngs of people and photographers with flashbulbs'. In Kyoto, Einstein delivered a lecture to a packed audience on his most important scientific breakthrough – the theories of special and general relativity. In the lecture, he explained his radical idea that the passage of time was not constant, but rather varied according to the relative velocities of different observers. This was the consequence of a simple but profound observation: nothing in the universe could travel faster than the speed of light. Einstein then explained that gravity had a similar effect on time. Observers in strong gravitational fields would experience the passage of time at a slower rate than those in weak gravitational fields. All this amounted to a complete rejection of the earlier world of Newtonian physics. Rather than thinking of space and time as separate and constant, Einstein showed that space and time could bend and distort. This was a revolutionary theory, one that had profound implications for the entire field of physics. After all, almost every scientific experiment conducted up to that point relied on the idea that space and time remained constant. A scientist measuring the speed of an object, for example, would simply divide the distance travelled by the time taken to travel that distance. But how could scientists measure speed accurately if space started to contract and time began to slow down?[2]

The lecture, which Einstein gave in German, was quickly translated and published by the Japanese physicist Jun Ishiwara. Having studied physics in Berlin, Ishiwara was one of the few people outside of Europe who really understood relativity at the time. Einstein clearly respected Ishiwara, even agreeing to work with him on a joint article for publication in the *Proceedings of the Japan Academy*. By all accounts, Einstein thoroughly enjoyed his time in Japan. When not delivering scientific lectures, he went hiking in the forests of Nikko and even attended the annual chrysanthemum festival in the gardens of the Imperial Palace in Tokyo. 'One cannot help loving and revering this country,' Einstein noted in his diary.[3]

On leaving Japan, Einstein began the return leg of his tour. He stopped off briefly in Malacca and Penang, before crossing the Indian Ocean to reach the Suez Canal. At Port Said, he disembarked once more. From there Einstein boarded a train to Jerusalem. Only a few months earlier, in July 1922, the League of Nations had approved the establishment of the Mandate for Palestine. This new territory would provide a 'national home for the Jewish people', a precursor to the modern State of Israel. As a Jew, Einstein had experienced antisemitism in Germany. He was accused of promoting 'Jewish physics' in the press, whilst his lectures were often picketed and disrupted by the antisemitic Anti-Einstein League. A number of prominent German Jews had also been attacked in the months before Einstein left for Asia. By the summer of 1922, he no longer felt safe. 'It seems I belong to that group of people whom the radical right plan to assassinate,' Einstein noted in a letter to the physicist Max Planck. This was in fact one of the reasons he had chosen to embark on a voyage to Asia. The hope was that things might calm down after a few months abroad.[4]

Einstein had long campaigned for the creation of a Jewish homeland. As early as 1919, he wrote of his 'great confidence in a positive development for a Jewish state . . . a little patch of earth on which our brethren are not considered aliens'. He was therefore delighted to finally arrive in Palestine. On 3 February 1923, Einstein toured the Old City of Jerusalem, visiting the Dome of the Rock and the Wailing Wall. He then gave a public lecture at the recently established Hebrew University of Jerusalem. Einstein began the lecture in Hebrew, before switching to German for the scientific content. The presence of Einstein, by that time one of the most famous scientists in the world, meant a great deal to the Jewish leaders in Palestine. Two years later, in 1925, the Hebrew University established the Einstein Institute of Mathematics in his honour. Einstein was even invited to move to Jerusalem and take up a position at the Hebrew University. Given the rise of antisemitism in Germany, he considered the offer seriously. But in the end, he decided he could not accept. ('The heart says yes, but the mind says no,' Einstein wrote in his diary.) He was not ready to leave Europe for good, at least not yet.[5]

When Einstein arrived back in Berlin, the political climate was no better. Hyperinflation was undermining the German economy, whilst the Nazi Party was growing in terms of both membership and influence. Over the following years, the political situation continued to worsen. On 30 January 1933, Adolf Hitler was appointed Chancellor of Germany. He quickly passed a series of antisemitic laws designed to discriminate against the Jewish population. German Jews were stripped of their citizenship, excluded from state schools, and targeted with forced sterilization. Einstein had seen it coming. A month earlier, in December 1932, he had left Berlin for the last time. He travelled to the United States, renounced his German citizenship, and took up a position at Princeton University. Einstein never returned to Germany. 'I did not wish to live in a country where the individual does not enjoy equality before the law and the freedom to say and teach what he likes,' he explained in his resignation letter to the Prussian Academy of Sciences.[6]

Albert Einstein is often thought of as an isolated genius, a man largely divorced from the wider intellectual and political world. True, Einstein's theories of special and general relativity transformed how scientists understood the physical universe. But he was by no means isolated. In fact, Einstein travelled the globe, promoting his ideas in cities stretching from Shanghai to Buenos Aires. Not only that, but Einstein also worked with scientists from many different countries around the world. All this reflected Einstein's deep political belief in the value of international collaboration. Following the First World War, he thought it was more important than ever that scientists come together to promote what he called 'mutual cooperation, mutual advancement'. 'I believe no one should shirk the political task . . . of restoring the unity between nations that has been completely destroyed by the world war,' explained Einstein. With this in mind, he joined the International Committee on Intellectual Cooperation. Set up by the League of Nations in 1922, the International Committee was charged with promoting closer ties 'between the scientific and intellectual communities of various countries'. Other members of the International Committee included the Indian physicist Jagadish Chandra Bose as well as the Japanese physicist Aikitsu Tanakadate, both of whom we met in the previous chapter.[7]

Einstein's interest in the international world of science and politics was typical of his generation. Many other physicists in this period spent a great deal of time travelling the world. In late 1929, the German physicist Werner Heisenberg visited India. He had been invited by the Bengali physicist Debendra Mohan Bose, nephew of Jagadish Chandra Bose. Heisenberg was a pioneer of another major field in early twentieth-century physics: quantum mechanics. To an audience of Indian scientists, he explained how there seemed to be a minimum discrete amount of energy that could be exchanged between atoms, or in any physical interaction. This was known as the 'quantum' of energy. That might sound unimportant, but a number of exceptional consequences followed. As we saw in chapter 6, physicists in the nineteenth century had understood light as a kind of electromagnetic wave. However, quantum mechanics showed that this wasn't strictly true. Rather, light needed to be thought of as both a wave *and* a particle. Things only got more strange in the following years, as physicists started to question the relationship between cause and effect and even the nature of scientific observation. Heisenberg himself was famous for having formulated the 'uncertainty principle', in which he showed that there was an ultimate limit to the precision of any physical measurement, hence why cause and effect sometimes got mixed up. This wasn't just a question of the accuracy of available scientific instruments, but rather a fundamental property of the universe.[8]

Other influential scientists made similar trips, including the British physicist Paul Dirac and the Danish physicist Niels Bohr. Dirac, who developed the first relativistic theory of the electron, gave a series of lectures in Japan in 1929, before taking the Trans-Siberian Railway all the way from Vladivostok to Moscow in order to attend a meeting of Soviet scientists. Similarly, Bohr, who proposed the first quantum model of the atom, spent two weeks in China in the spring of 1937. Bohr's lectures, which he gave at Shanghai Jiao Tong University, were broadcast live on the radio across China. In June 1937, he returned to Europe. A month later, the Japanese Army launched an assault on Beijing, marking the beginning of the Second Sino-Japanese War, which soon became part of the wider conflict of the Second World War.[9]

The first half of the twentieth century was a period of major social and political upheaval. In China, the Revolution of 1911 brought an end to the Qing dynasty, whilst in Russia the Bolsheviks seized power in the October Revolution of 1917. Even those countries which did not undergo a revolution witnessed equally significant political shifts. Following the First World War, the Ottoman Empire collapsed, prompting fierce political and religious conflict in the lands around Palestine. In Japan, the death of the emperor in 1912 marked a period of increasingly liberal politics, whilst in India the anticolonial movement became much more assertive, particularly after the Partition of Bengal in 1905.

This, then, is our next key moment in global history. Fascists, socialists, nationalists, suffragists, and anticolonial fighters all played a part in transforming the political world in the decades following 1900. This world of politics had a profound effect on the world of science, not just in Europe, but right across the globe. In this chapter, we explore the connection between physics and international politics in the early twentieth century. Along the way we uncover the important contributions made by scientists from countries which do not normally feature in the history of modern physics. In the following chapter, we move forward, charting the impact of the Cold War and decolonization on the development of modern genetics. Ultimately, to understand the history of twentieth-century science, we need to look to the global political debates which defined the age.

I. Physics in Revolutionary Russia

Every summer, Peter Kapitza travelled to Leningrad to visit his mother. August 1934 began no differently. However, as he went about his business, collecting groceries and visiting old friends, Kapitza started to notice that something was wrong. Everywhere he went, he was being followed. The secret police were onto him. For the last ten years, Kapitza had been working at the Cavendish Laboratory at the University of Cambridge, where he had done his PhD. He had already made some impressive discoveries. In early 1934, just before he left for Russia, Kapitza had become one of the first scientists in the world to successfully produce liquid helium in large quantities. This was exceptionally hard to do, as the helium needed to be repeatedly compressed and then cooled to very low temperatures in order to enter a liquid state. Kapitza had also recently been appointed as the director of a new research centre dedicated to cutting-edge physics, the Mond Laboratory in Cambridge. All this brought Kapitza international recognition. However, it also brought him to the attention of the Soviet authorities.[10]

In September 1934, Joseph Stalin himself signed an order which required Kapitza to remain in Russia. 'Kapitza may not be arrested officially, but he must be retained in the Soviet Union and not let return to England,' wrote Stalin. Sure enough, when Kapitza tried to leave for Cambridge, he was quickly detained. His passport was confiscated and he was informed that he was not to undertake any foreign travel. Stalin's decision to detain Kapitza was partly a response to the actions of another prominent Soviet scientist. George Gamow, an expert in quantum mechanics, had recently fled to the United States under the cover of attending a conference in Europe. At the same time, the Soviet government was increasingly concerned that Russian scientists working abroad might be acting as spies, or contributing to the military development of foreign powers.[11]

At first, Kapitza was completely despondent. In a letter to his wife, who was back in Cambridge, he complained that 'life is so empty now . . . I rage and want to tear my hair and scream'. How could he do any proper scientific work now that he was separated from his laboratory? 'I think I am beginning to go mad,' he wrote, 'I sit here all alone. And what for? I don't understand.' For a few months, Kapitza's colleagues in Cambridge did what they could to help. The director of the Cavendish Laboratory, Ernest Rutherford, wrote to the Russian ambassador in London, whilst Paul Dirac travelled to Moscow in the hope of securing Kapitza's release. In the end, it all came to nothing. Stalin had made up his mind. Russian scientists were to remain in Russia, where they could best serve the Soviet Union.[12]

After a while, Kapitza began to accept his new fate. 'The injustice done to me must not blind me,' Kapitza wrote in a letter to Niels Bohr. If he couldn't return to Cambridge, then he would have to make the most of what he could do in Russia. Towards the end of 1934, Kapitza agreed a compromise. He would remain in Russia, and support the scientific development of the Soviet Union. In return, the Soviet government would provide Kapitza with the equipment and the space he needed to conduct serious scientific research. With this in mind, Kapitza was appointed as director of a brand new research centre: the Institute for Physical Problems, based in Moscow. The Soviet government also agreed to pay £30,000 to purchase the experimental equipment Kapitza needed from the Mond Laboratory in Cambridge. This included a pair of exceptionally strong electromagnets as well as a helium liquefier.[13]

The investment paid off. On 1 January 1938, Kapitza published a short article in *Nature*, announcing the discovery of what became known as 'superfluidity'. In the article, he described an experiment he had conducted in Moscow to measure the viscosity of liquid helium. Essentially, this involved measuring how easily liquid helium would flow through a very small gap from one container to another. As Kapitza remarked,

33. Liquid helium entering the 'superfluid' phase. In this phase, the liquid helium begins to flow over the sides of the glass cup, hence the small droplet at the base.

'the results of the measurements were rather striking'. When helium was cooled to just below its boiling point, at around –269°C, it acted like a normal liquid, and flowed at a constant but relatively slow rate between the two containers. However, when helium was cooled further, and approached absolute zero (equivalent to –273°C), it suddenly began to flow at an incredible rate, acting like a 'superfluid'. This was the beginning of a whole new field of study, known as low-temperature physics. Kapitza had discovered that, when certain substances are cooled to very low temperatures, they exhibit strange new properties. Some of Kapitza's experiments seemed to defy the known laws of physics. When sufficiently cooled, liquid helium would creep up the sides of a glass container, even moving between spaces that seemed completely sealed. All this required an entirely new explanation of how molecules interacted.[14]

*

Peter Kapitza went on to win the Nobel Prize in Physics for his discovery of superfluidity. He was the first of a new generation of Russian scientists who made a series of major breakthroughs over the course of the twentieth century. Much of this work was in the new fields of quantum mechanics and relativity. However, Kapitza's life also points to the two different sides of Soviet science. On the one hand, the Russian Revolution of 1917 brought about increased investment in the sciences. Yet on the other hand, science in the Soviet Union was often subject to political and ideological interference.

As we saw in previous chapters, scientists working in Tsarist Russia made a number of important contributions to the development of modern science. From Peter the Great in the seventeenth century through to Alexander II in the nineteenth century, the Tsars championed science as a means to modernize and strengthen the Russian Empire. The Russian Revolution was not therefore a complete break from the past. However, the scale of scientific investment from 1917 onwards was certainly unprecedented, as was the intense ideological conflict. The Bolsheviks, who seized power in 1917, believed that proper investment in the sciences was crucial for the military and industrial development of the Soviet Union. 'Socialist reconstruction,' they called it. A number of prominent early Soviet politicians, including Nikolai Gorbunov, had a background in the sciences. Gorbunov, who worked as Vladimir Lenin's personal secretary, was a chemical engineer by training. He helped convince the Soviet authorities to establish a series of high-tech scientific institutes. These included the Physico-Technical Institute in Leningrad, which was founded in September 1918, less than a year after the October Revolution. At the same time, the Soviet government nationalized existing private laboratories, including the one run by the Moscow Physical Society. By 1930, the Soviets were spending over 100 million roubles a year on science. (This was roughly equivalent to the amount of money spent by the Soviets on munitions production for the military.)[15]

Initially, the Soviet government supported Russian scientists studying abroad. Following the First World War, the Bolsheviks recognized the importance of re-establishing links with scientists in Europe. Kapitza was one of a number of early Soviet scientists who studied at foreign universities. They were often sent with requests to purchase equipment and books, once again with the aim of building up scientific capacity back home. At the same time, the Soviets welcomed many foreign scientists to Russia. Between 1917 and 1930, the Russian Association of Physicists held a series of annual conferences, many of which were attended by leading European scientists. Paul Dirac and the German physicist Max Born both attended a meeting in Moscow in 1928, sharing the latest developments in quantum mechanics with their Russian counterparts. After the first day of the conference, all the participants – including Dirac and Born – boarded a steamship, where they continued their discussions as they sailed down the Volga River.[16]

There was, however, another side to Soviet science. As we've seen with Kapitza, scientists in the Soviet Union lived through a time of major ideological conflict. This was particularly the case in the period after Joseph Stalin came to power in 1922. By the early 1930s, Stalin had grown increasingly paranoid, turning inwards. He banned almost all foreign travel, particularly after George Gamow's defection in 1934. Scientists were also subject to political purges, particularly if they appeared insufficiently committed to Soviet ideology. This wasn't just a question of publicly supporting Marxism. Ideological disputes could easily extend to some of the most fundamental aspects of modern science. Many of the early Bolsheviks, including Lenin, saw a direct link between the recent revolution in the physical sciences and the political revolution taking place in Russia. Soviet scientists were therefore expected to be promoting revolutionary new ideas. Lenin himself even dedicated a chapter of his influential book *Materialism and Empirio-Criticism* (1909) to 'the recent revolution in natural science'. In it, Lenin described Albert Einstein's theory of relativity as revealing a 'crisis in modern physics', one mirrored by the recent crisis in society. According to Lenin, Einstein was one of 'the great reformers' of the modern age.[17]

With Lenin's seal of approval, Einstein was elected as a Foreign Member of the Russian Academy of Sciences. And by the early 1920s, students studying physics in Moscow and Leningrad were being taught about relativity, which was often pitched as a revolutionary antidote to the classical worldview of Isaac Newton. Still, not everyone agreed with Lenin about Einstein. Relativity might have been revolutionary, but for some it still smacked of 'bourgeois science'. Ernst Kolman, President of the Moscow Mathematical Society in the early 1930s, described Einstein as 'a great scientist but a poor philosopher'. For Kolman, the theories of special and general relativity were far too abstract, seemingly divorced from everyday experience. Kolman accused Einstein of 'replacing physical reality by mathematical symbols'. This was in fact an increasingly common criticism of both relativity and quantum mechanics in Soviet Russia, particularly under Stalin. It reflected a deep-seated belief – drawn from Marxist philosophy – in the importance of science as a practical undertaking, something grounded in the material world. In the Soviet Union, science had to be seen to serve the people.[18]

This was the world in which Russian scientists found themselves during the first decades of the twentieth century. A world in which leading Soviet politicians showed a deep interest in and willingness to support science. But also a world in which, at any moment, the political mood could change, sometimes with fatal consequences for those who found themselves on the wrong side of the Soviet leadership. Nonetheless, many early Soviet scientists supported the Russian Revolution. Some were even directly involved in it.

The physicist Yakov Frenkel was born into a family of radicals. During the 1880s, his own father had been exiled to Siberia after being identified as a member of a revolutionary organization. Frenkel shared his father's political sympathies. When the October Revolution brought an end to Tsarist rule, he made the most of the opportunities for ambitious young scientists. After studying physics at Petrograd University, he moved to Crimea in 1918 to take up a position at Tauride University. This was one of the new universities established by the Bolsheviks. In Crimea, Frenkel began to read about some of the latest ideas shaking modern physics, including Niels Bohr's work on the quantum model of the atom. At the same time, Frenkel kept up his political interests, joining the local Crimean Soviet and helping to reorganize education in the region along socialist lines.[19]

These were uncertain times. The Russian Revolution had brought about a full-blown civil war. Whilst the Bolsheviks controlled much of central Russia, the anti-communist White Army continued to fight in the south and west of the country. In July 1919, the White Army marched on Crimea. As a member of the local soviet, Frenkel was arrested and put in prison. Still, he didn't give up. Frenkel wrote to his mother from prison, reassuring her that 'I am not bored at all; I spend rather a lot of time reading'. He also played chess with his cellmates, at least until a prison guard confiscated the board. This turned out to be rather a productive time for Frenkel. Somewhat incredibly, it was actually in prison – during the middle of the Russian Civil War – that he began his most important theoretical work.[20]

Since around 1900, scientists had assumed that the flow of electricity in a metal could simply be explained by the free movement of electrons. These tiny negatively charged particles were imagined rather like a gas, freely moving in the space between atomic nuclei. Frenkel, however, realized that this couldn't be true. Quantum mechanics wouldn't allow it. Bohr, in his quantum model of the atom, had shown that electrons could only sit in particular orbits around an atomic nucleus. As Frenkel noted, electrons were therefore 'not free in the real sense of the word'. How then did electrons move to create electricity? Frenkel suggested a new model, based on quantum mechanics, in which electrons would effectively hop between adjacent atoms. This would create a flow of electricity, without having to imagine that electrons were entirely 'free' to move anywhere.[21]

Stuck in a prison in Crimea, Frenkel developed the first quantum mechanical explanation of electricity. Rather strikingly, this depended on rethinking the very notion of what it meant for an electron to be 'free' at a time when Frenkel himself was incarcerated. On top of this, Frenkel suggested a major theoretical concept that could be applied by physicists much more widely. He argued that the behaviour of electrons in a metal could be explained by imagining a new kind of particle, a 'collective excitation' as he called it. Frenkel was once again deliberate with his choice of words. This was a vision of quantum mechanics that fitted perfectly with Soviet ideology. There were no individuals, only the 'collective'. The identification of these new particles, which later became known as 'quasiparticles' in Europe and the United States, proved absolutely central to the development of quantum mechanics over the course of the twentieth century. The basic idea was that strange physical phenomena could be more easily explained if scientists imagined the existence of the collective action of as-yet-unidentified particles.[22]

Yakov Frenkel's article on quasiparticles, first published in 1924, represented the start of a major new research programme in fundamental physics. Soviet scientists led the way in this regard. Amongst those who followed up Frenkel's work were a number of women who, despite the important contributions they made to modern physics, are largely forgotten today, even in Russia. As we saw in the previous two chapters, women were typically excluded from higher education in nineteenth-century Russia, although a small number did manage to study science abroad. The Bolsheviks therefore prided themselves on the idea, if not always the reality, that women would contribute to the scientific and industrial development of the Soviet Union. As such, the October Revolution of 1917 brought about increased opportunities for bright young women to enter the world of science. Antonina Prikhot'ko was one of those women. Born in southern Russia in 1906, she was one of the first women to study physics at the Polytechnical Institute in Leningrad. Frenkel himself, who had been released from prison following the defeat of the White Army, was teaching there in the 1920s. As a student, Prikhot'ko sat in the freezing cold lecture hall in Leningrad, listening to Frenkel lecture on both relativity and quantum mechanics. She was therefore in a unique position, learning about quasiparticles before most physicists in Europe had even heard of them.[23]

Prikhot'ko graduated in 1929. The following year she took up a position at the Ukrainian Physico-Technical Institute in Kharkov. This was one of the new research centres established by the Bolsheviks with the aim of spreading scientific and industrial expertise across the Soviet Union. Over the next ten years, Prikhot'ko set to work, putting Frenkel's theoretical ideas into practice. She began a series of experiments in low-temperature physics, investigating the atomic structure of various crystals. As with the work of Peter Kapitza, these experiments relied upon enormous industrial machines, such as a helium liquefier. In the laboratory at Kharkov, Prikhot'ko could be found late at night, spanner in hand, tinkering with the liquefier. By measuring the amount of light absorbed by and released from different crystals at low temperatures, she was able to make inferences about atomic behaviour. Crucially, Prikhot'ko was the first to experimentally prove the existence of one of the quasiparticles predicted by Frenkel, the 'exciton'. Whilst all this might sound pretty abstract, Prikhot'ko's research actually had a much more practical side. Many of the crystals she was working on were used in the manufacture of industrial chemicals, including naphthalene (used as a pesticide) and benzene (used as a solvent in steel production). Prikhot'ko was therefore in many ways a model Soviet scientist. She used the latest scientific theories in quantum mechanics to conduct practical experiments that would contribute to real-world industrial development. On the basis of this work, Prikhot'ko was later awarded both the Lenin Prize and the Hero of Socialist Labour, two of the most prestigious civilian honours in Soviet Russia.[24]

Antonina Prikhot'ko worked in Kharkov at a particularly exciting time. The Ukrainian Physico-Technical Institute in the 1930s was full of aspiring young scientists, keen to make their mark on the modern world. Perhaps the most gifted of these was Lev Landau. Born in Baku in 1908, Landau was by all accounts a child prodigy, having mastered calculus by the age of thirteen. However, the rigid educational system under the Tsars did not suit him. Bored at school, Landau managed to get himself expelled after he insulted the headteacher. Luckily for Landau, the Russian Revolution reached Baku that same year. In an effort to open up education to the masses, the Bolsheviks removed all formal requirements for entry to local universities. Landau, aged just fourteen, jumped at the chance, and enrolled at Baku University to study physics. After a few years, he decided to transfer to Leningrad University to complete the remainder of his degree. In Leningrad, Landau met other young physicists, many of whom, like him, had revolutionary sympathies. Together they read the latest work on quantum mechanics and relativity, alongside the political writings of Vladimir Lenin and Leon Trotsky.[25]

In 1927, Landau graduated and began working as a researcher at the Physico-Technical Institute in Leningrad. He was then awarded a Rockefeller Fellowship to travel and study in Europe. Throughout the twentieth century, the Rockefeller Foundation, based in the United States, provided funds to support international collaboration between scientists. Although the Rockefeller Foundation had serious misgivings about the politics of the Soviet Union, it nonetheless saw scientific collaboration as a means to promote international peace. Landau was therefore able to spend over a year working with many of the leading scientists in Europe. He met Albert Einstein in Berlin and Werner Heisenberg in Leipzig, before travelling to Copenhagen to work with Niels Bohr. In 1931, Landau returned to Russia, more excited than ever about the new work being done in quantum mechanics. However, he started to get bored in Leningrad. Science there was still dominated by an older generation, even if some of the younger physicists like Frenkel were trying to push things forward. With this in mind, Landau decided to move and take up a new position at the Ukrainian Physico-Technical Institute in Kharkov. He arrived there in 1934, aged just twenty-six, and was immediately appointed as head of the theoretical department.[26]

Over the following years, Landau made a series of major theoretical breakthroughs. He ranged widely, working on everything from the physics behind the formation of stars to the fundamentals of magnetism. But his real passion was low-temperature physics. In Kharkov, Landau worked with a team of exceptional young scientists. Much of his research was conducted in partnership with Lev Shubnikov and his wife, Olga Trapeznikova, both of whom had studied physics in Leningrad in the 1920s. Landau's impressive work soon brought him to the attention of senior physicists in Moscow. In March 1937, Peter Kapitza wrote to Landau, inviting him to join the recently established Institute for Physical Problems. As we saw earlier, this was a new research centre, founded especially to support Kapitza's work on low-temperature physics. Landau understood that this was a great opportunity. The Institute for Physical Problems housed some of the best scientific equipment in the Soviet Union, most of it purchased from the Mond Laboratory in Cambridge. And unlike some of the old guard in Leningrad, Kapitza was also keen to support physicists doing genuinely novel theoretical work.[27]

In the spring of 1937, Landau travelled to Moscow to take up his new position. As it turned out, this was a good time to leave Kharkov. Between 1936 and 1938, Joseph Stalin orchestrated a major campaign of political repression, later known as the Great Terror. Anyone even vaguely suspected of 'counter-revolutionary' activity was arrested and either shot or sent to the Gulag. Up to one million people were killed during this period. In early 1937, the Great Terror reached Ukraine. Many of the scientists that Landau had worked with were never seen again. Lev Shubnikov was arrested at his laboratory. He was then taken to prison, tortured, and made to sign a confession stating that he was 'a member of a Trotskyist sabotage group working within the walls of the Ukrainian Physico-Technical Institute'. After a few months in prison, Shubnikov was executed by firing squad. His wife, the physicist Olga

Trapeznikova, was only spared because she had recently given birth to their only son.[28]

In the end, things caught up with Landau. On 28 April 1938, he was arrested in Moscow. Some of his colleagues in Ukraine, most likely under duress, had accused Landau of being a member of the same 'counter-revolutionary' group as Shubnikov. Landau spent the next year in prison. He was interrogated for hours, and made to take up stress positions, squatting on the ground with his arms tied behind his back. In all likelihood, Landau would have been executed if it weren't for the actions of his friend and mentor, Peter Kapitza. On the day that Landau was arrested, Kapitza wrote directly to Stalin himself. 'I beg you in view of his supreme talent to order very close attention to his case,' urged Kapitza. 'There is no doubt that the loss of Landau, both for our institute and for Soviet, and for world science, would not go unnoticed and will be very deeply felt,' he explained. Kapitza had only recently announced the discovery of superfluidity. At the time of his arrest, Landau was leading a team to try and explain this curious new phenomenon. Without Landau, Kapitza knew he would not be able to continue this line of research.[29]

The letter to Stalin seemed to work. Exactly a year after his arrest, Landau was released. He was severely malnourished, and could hardly walk. But within a few weeks, he was back at the Institute for Physical Problems. Kapitza was right about Landau. He really did have a 'supreme talent'. Three years later, Landau finally cracked the problem of superfluidity. In 1941, he published the first theoretical explanation of the behaviour of liquid helium at very low temperatures. Ever since Kapitza's discovery, physicists had assumed that the best way to think about superfluidity was to imagine that liquid helium started to behave rather like a gas, in which the atoms were free to move wherever they liked. Landau, however, showed that this was a really bad way of thinking about superfluidity. Instead, he drew on Yakov Frenkel's earlier work in quantum mechanics to show that the atoms in a superfluid were not entirely free. Rather, they moved in tiny swirling patterns. At the right temperature, these swirling atoms would reduce the friction in liquid helium to effectively zero. For his work on superfluidity, Landau was later awarded the Nobel Prize in Physics, one of the nine Soviet scientists to win a Nobel Prize in the twentieth century.[30]

The career of Landau is another reminder of the two different sides to science in the Soviet Union. On the one hand, it was only in the Soviet Union that a physicist like Landau could thrive. A radical intellectual, Landau was encouraged, particularly at the Institute for Physical Problems in Moscow, to develop revolutionary new scientific theories, pushing the boundaries of what was thought possible. The Soviet government also provided the equipment that Landau and others needed to conduct cutting-edge research, particularly in low-temperature physics. All this reflected the Bolsheviks' desire to use science as a means to further both the intellectual and industrial development of the Soviet Union in the years after the First World War. Yet at the same time, Landau – like many Soviet scientists – suffered in an atmosphere of intense ideological conflict. He was one of the lucky ones, arrested but later released on Stalin's orders. Many of those he worked with were not so fortunate. Even those who survived the Great Terror of 1936–8 lived under a shadow. Kapitza, the darling of Soviet science in the 1930s, was later dismissed from the Institute for Physical Problems after falling out with Stalin. And Landau spent the rest of his life under surveillance by the secret police. This ideological dimension to science was particularly strong in the Soviet Union. However, as we'll see, it was by no means unique.[31]

II. Einstein in China

On 4 May 1919, over 4,000 students took to the streets of Beijing. Although the Revolution of 1911 had brought an end to the Qing dynasty in China, many of the younger generation were still dissatisfied with the new national government. They held up banners, each painted with Chinese characters. 'Down with the militarists!' read one. 'Down with Confucius and his followers!' read another. This was the beginning of a mass protest, one that swept across China following the end of the First World War. It became known as the May Fourth Movement. The initial spark for the protests had been the perceived weakness of the Chinese government in response to the Treaty of Versailles, which transferred German-occupied territories in eastern China to Japan, despite the fact that China had also supported the Allies in the war.

However, the May Fourth Movement soon morphed into a wider criticism of traditional Chinese society. Many of the students felt that China was still stuck in the past. Alongside demands for new political institutions, including democracy, the protesters also called for greater investment in modern science. 'Science saves the nation!' some of the students could be heard shouting as they marched across Tiananmen Square. 'New science, new culture!' called out another.[32]

Albert Einstein's theory of relativity represented exactly the kind of modern science that many felt China lacked. The students studying at Peking University might well have learned about Einstein from one of the radical young professors there. Xia Yuanli was born into a family of political reformers. His father had been friends with many of the key figures involved in the foundation of the Republic of China following the Revolution of 1911. Xia himself had studied physics in the United States at the Sheffield Scientific School, part of Yale University. He had also spent time in Europe prior to the First World War, undertaking postgraduate study at the University of Berlin, where he had learned about relativity from the German physicist Max Planck. When the Qing dynasty collapsed in 1911, Xia returned to China, where he was quickly appointed Professor of Physics at Peking University.[33]

Lecturing just before the outbreak of the May Fourth Movement, Xia described the theory of relativity as 'the newest, most advanced, and most profound theory of today's physics'. Xia then explained some of the consequences of Einstein's work. 'The concept of absolute time cannot exist,' Xia told the students in Beijing. Not only that but 'time and space have lost their independence'. According to Xia, Einstein's theory of relativity was 'the most important achievement since Newton and Darwin'. It represented 'a great revolution in physics'. Xia soon found that the students were keen to learn more. In 1921, he decided to translate Einstein's own *Relativity: The Special and General Theory* (1916) from German into Chinese. This was the first book on relativity in China.[34]

Even before his trip to Shanghai in 1922, Albert Einstein was associated with revolution in China. One Chinese newspaper described the theories of relativity as nothing less than 'the starting point of the revolution of the whole world of science'. Another reported that 'the impact of the

Einsteinian revolution is even greater than Luther's Reformation in Germany or Marx's economic revolution in the past'.[35]

It was in fact Cai Yuanpei, one of the leaders of the May Fourth Movement, who had invited Einstein to China. Cai, who had been appointed Minister of Education following the Revolution of 1911, wanted to promote modern science in China. This in part was a continuation of one of the themes we saw in chapter 5. Since the middle of the nineteenth century, and particularly during the Self-Strengthening Movement in the last decades of the Qing dynasty, political reformers in China had been trying to replace ancient Confucian philosophy with modern science imported from Europe and the United States. With this in mind, Cai travelled to Germany in March 1921 in an effort to recruit leading European scientists to visit China. In Berlin, Cai met Einstein, offering him $1,000 – the equivalent of over $10,000 today – to give a series of lectures at Peking University. Einstein accepted the offer, although he ended up spending the majority of his time in Japan. Nonetheless, Cai was delighted to have this 'star of twentieth-century thought' set foot in China.[36]

The Revolution of 1911 brought about renewed interest and investment in modern science. Although this was not a communist revolution – that would come later with the founding of the People's Republic in 1949 – there are nonetheless many parallels with what was happening in the Soviet Union. Much like in Russia, leaders in China saw a close connection between political revolution and revolution in the sciences. Following the May Fourth Movement, the Chinese government established a series of new scientific institutions. In 1919, immediately after the student protests, Cai approved the foundation of a new physics laboratory at Peking University. And by 1930, China housed eleven new physics departments, including at universities in Wuhan and Shanghai.[37]

As well as building up capacity at home, China also began to make new connections abroad. Influential European physicists – including Einstein, Niels Bohr, and Paul Dirac – were all invited to give lectures in China. Thousands of Chinese students were also sent to study at universities in Europe, the United States, and Japan. This in some ways was part of a much longer trend. As we've seen throughout this book, China was far from isolated when it came to science. From the early modern period onwards, Chinese scholars had been exchanging ideas with people from across the world, and Chinese students had been studying science at European and American universities since the middle of the nineteenth century. Nonetheless, the Revolution of 1911 massively increased the scale of intellectual exchange. In the first four decades of the twentieth century, over 16,000 Chinese students travelled to the United States, the majority of whom studied science and engineering.[38]

Many Chinese students went to the United States under a new scheme created by the American government. In 1908, President Theodore Roosevelt approved the creation of the Boxer Indemnity Scholarship. At the time, China owed the United States government over $24 million. This was to be paid as compensation for the losses suffered in 1901 during an uprising against European and American troops stationed in China, known as the Boxer Rebellion. Roosevelt agreed that, instead of paying back the money directly, the Chinese government could use the funds to pay for scholarships at American universities. This, however, was no act of charity, but rather a shrewd piece of diplomacy. The aim was to pump cash into American universities at the same time as shaping the intellectual development of China. As one of Roosevelt's advisors put it, 'the nation which succeeds in educating the young Chinese of the present generation will be the nation which . . . will reap the largest possible returns in moral, intellectual, and commercial influence'.[39]

Zhou Peiyuan was one of the many Chinese students who studied abroad in the early decades of the twentieth century. He was born into a wealthy family in Jiangsu Province. However, following the Revolution of 1911, Zhou and his family moved around quite a bit, as different regions became overrun by warring factions. In the end, Zhou settled in Shanghai, where he began studying at a school run by American missionaries. But, like many of his generation, Zhou harboured deep misgivings about the state of Chinese society. When the May Fourth Movement broke out, he joined the students in protest. Outside his own school gates, Zhou could be found shouting 'down with imperialism!' The headmaster did not take kindly to this and Zhou was expelled from the school. His father was furious. What was he going to do with his life now?[40]

For a while, Zhou drifted. He spent some time at a Buddhist temple in the forests west of Shanghai. After meditating for days, Zhou finally decided to set out on a new path. He had heard about the opportunities for Chinese students abroad. He would travel to the United States and train to become, in his own words, a 'world-class physicist'. Zhou was clearly ambitious. However, before he could go to the United States, he first had to enrol at Tsinghua University in Beijing. This new institution had been established in 1911 especially to prepare Chinese students wishing to study on Boxer Indemnity Scholarships. Whilst studying in Beijing, Zhou also learned about relativity. He read reports in local newspapers about Albert Einstein's visit to Shanghai, and quickly purchased a copy of Xia Yuanli's translation of Einstein's book.[41]

Zhou graduated from Tsinghua University in 1924. That same year, he crossed the Pacific Ocean aboard a steamship and began his studies in the United States. He first spent two years at the University of Chicago, before moving to the California Institute of Technology, where he began studying for a PhD. In the PhD, which was later published in the *American Journal of Mathematics*, Zhou provided some of the first detailed solutions to the equations Einstein had proposed in his theory of general relativity. Ever since Einstein had published these 'field equations' in 1915, mathematicians had been looking for solutions which would describe actual physical systems: for example, what was the precise effect of the mass of a planet or a star on the curvature of space and the passage of time? Zhou, who had only turned to physics after getting expelled from school, provided an answer.[42]

Zhou returned to China in 1929 to take up a position as Professor of Physics at Tsinghua University. This was quite a turnaround: from student protestor to professor. Over the following years, Zhou continued to work on relativity. Then, in 1935, he received an extraordinary offer. He was invited to spend a year at the Institute for Advanced Study at Princeton University, where Einstein was based following his flight from Germany. At Princeton, Zhou and Einstein spent hours in conversation, discussing the wider implications of the theory of general relativity, particularly when it came to the structure of the universe. Was the universe static? Or was it expanding? This was one of the big questions in physics in the 1930s, and it was one that Einstein's equations held the key to. Zhou was one of those who argued, correctly, that the general theory of relativity implied the existence of an expanding universe. Alongside physics, Einstein chatted to Zhou about his time in

China, and his great appreciation for Chinese culture. 'When he talked to me alone, he showed deep sympathy to the Chinese working people . . . and entertained great expectations for us, a nation with a long history of civilization,' Zhou later recalled.[43]

In early 1937, Zhou returned to China once again. That summer, the Japanese army invaded Beijing. Zhou, along with all the staff and students at Tsinghua University, had to evacuate and relocate over 1,500 miles away to the southwestern province of Yunnan. Einstein did what he could to help, signing a public letter condemning the Japanese invasion. He was concerned that this might be the beginning of another major international conflict. Zhou appreciated the support and wrote to Einstein from his temporary office in Yunnan. 'We have to thank you for your sympathy for our cause and your effort in promoting the boycott of Japanese goods,' wrote Zhou. Once again, physics and politics were never far apart.[44]

Zhou Peiyuan was the first of a new generation of Chinese scientists, many of whom made important contributions to the development of modern physics. Alongside research in relativity, a number of Chinese scientists also worked on quantum mechanics. Like Zhou, they often trained abroad, returning to China to help establish new laboratories back home. Amongst these was a group of students who studied under the American physicist Robert Millikan.[45]

Ye Qisun and Zhao Zhongyao came from similar backgrounds. Like many Chinese scientists of this period, they were born into families of traditional scholars. Ye's father worked as a civil servant in the Qing bureaucracy, whilst Zhao's father was a schoolteacher. Both Ye and Zhao were expected to follow in the family tradition, either as bureaucrats or as teachers of Confucian philosophy. However, the Revolution of 1911 put an end to all that. Instead, Ye and Zhao – like many of their contemporaries – had to reinvent themselves. They chose to become modern scientists. Ye was the first to arrive in the United States, where he began studying with Millikan at the University of Chicago in 1918. Throughout this period, Millikan was working to test experimentally many of the new theories in quantum mechanics. Ye set to work on measuring the Planck constant. Named after the German physicist Max Planck, this constant represented the smallest possible amount of energy involved in any physical interaction – the very 'quantum' of 'quantum mechanics'. To determine such a tiny value, Ye had to devise a new experimental setup. This involved carefully measuring the energy of X-rays as they passed through an 'ionization chamber', a device filled with gas to detect radiation. When the X-rays passed through the ionization chamber, they collided with the gas particles, generating a tiny electrical current, which Ye measured and entered into his calculations. In 1921, he published an article, co-authored with the Harvard physicist William Duane, announcing the most accurate measurement of the Planck constant to date. This remained the standard value used by physicists right across the world for decades to come.[46]

In 1921, Millikan left the University of Chicago and moved to the California Institute of Technology. A few years later, Zhao Zhongyao arrived in the United States on a Boxer Indemnity Scholarship. He had travelled to California specifically to work with Millikan, who by this time had a reputation for supporting aspiring Chinese scientists. After some debate about suitable topics, Millikan agreed to take Zhao on as his PhD student. Zhao then began an exceptionally ambitious project in which he attempted to verify one of the latest theoretical breakthroughs in quantum mechanics. In 1929, two physicists working in Copenhagen had published an equation which they claimed could explain what happens when an electromagnetic wave, such as light, enters the atomic nucleus. Zhao decided to check whether the equation was any good. To do so, he exposed different chemical elements to gamma rays – a kind of high-energy electromagnetic wave. He then measured the amount of energy either absorbed or emitted by the atomic nucleus in each case. The results were surprising. For some atomic nuclei, the equation worked just fine. But for others, particularly heavy elements like lead, there was a significant excess of energy that didn't match the equation. Where had this energy come from? At first, Zhao wasn't sure. But in any case, his results were important enough to be published in the prestigious *Proceedings of the National Academy of Sciences*.[47]

It turned out that Zhao had been the first scientist in the world to observe the existence of a new fundamental particle: the 'positron'. The existence of this particle had been predicted in 1928 by the British physicist Paul Dirac. In certain circumstances, Dirac argued, it was possible for electrons to be produced with a positive rather than a negative

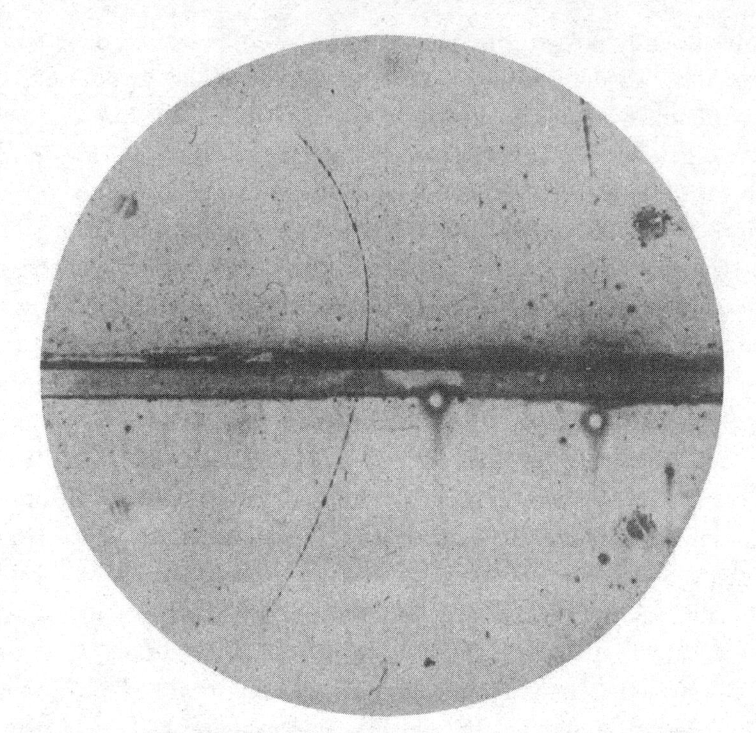

34. A photograph of a positron in a cloud chamber. The movement of the positron is indicated by the black curved line running from the lower left to the upper left.

charge. These strange particles represented an entirely new kind of matter – 'antimatter' as it came to be known. Dirac also realized that positrons could only exist for a tiny fraction of a second, as they would quickly be drawn towards a negatively charged electron, where the two would combine and 'annihilate' one another. Crucially, this reaction would give off a burst of energy. This was exactly what Zhao had detected in his experiments: the annihilation of an electron and a positron, hence the excess energy.

In 1930, Zhao received his PhD. He returned to China the following year to take up a position at Tsinghua University in Beijing. Despite having conducted such an important experiment, Zhao was never credited with the discovery of the positron. That accolade went to the American physicist Carl Anderson, one of Millikan's other students at the California Institute of Technology. Zhao and Anderson had worked on the same corridor, chatting about their experiments most days. 'His findings interested me greatly,' Anderson later wrote. Once Zhao returned to China, Anderson conducted further tests, confirming once and for all that Dirac was right: the positron was real. In 1936, Anderson was awarded the Nobel Prize in Physics for his discovery. At the time, he claimed that he had come across the positron 'by accident'. However, Anderson later acknowledged that he had been directly inspired by Zhao's earlier experiments.[48]

Zhao was one of the many Chinese physicists who studied abroad in the early decades of the twentieth century. Some went to the United States, others went to Europe and Japan. Together, they made a number of important contributions to the development of modern physics, working on everything from the mathematics behind general relativity to the existence of new fundamental particles. Many came from traditional scholarly backgrounds, looking for a new vocation following the Revolution of 1911 and the collapse of the Qing dynasty. They returned to China as modern scientists rather than Confucian scholars. Political leaders in China, much like in the Soviet Union, saw science as a means to modernize the nation. They associated the new theories of relativity and quantum mechanics with a better future. This was especially the case amongst those involved in the student protests which erupted across China in the spring of 1919. 'What China lacks and needs most is natural science,' argued Cai Yuanpei, one of the leaders of the May Fourth Movement. In the following section, we explore another side to the history of modern physics. In early twentieth-century Japan, there was no political revolution. Nonetheless, as we'll see, Japanese science was still transformed by that wider world of ideological conflict.[49]

III. Quantum Mechanics in Japan

Huddled around a pile of scientific journals, a group of Japanese students began to discuss the latest work in quantum mechanics. First, they went over Paul Dirac's new article on the hydrogen atom, then they moved on to Werner Heisenberg's paper on 'electron jumps'. For these young students, as with their counterparts in China, quantum mechanics represented the future. Japanese science needed to move beyond 'classical theory', argued one. 'Today's dead pedagogy is out of date,' declared another. This was the first meeting of the Physics Reading

Group, established in Tokyo in March 1926. The students who attended the weekly meetings had grown frustrated with the teaching at the University of Tokyo. At the time, the main course on physics only covered Isaac Newton's classical mechanics along with some of James Clerk Maxwell's work on electromagnetism. There was nothing on the new physics. No quantum mechanics. The students decided to take things into their own hands.[50]

Whilst it did not undergo a political revolution, for Japan the early twentieth century was nonetheless a period of major social change. When the Emperor Meiji passed away in 1912, many of the younger generation seized the opportunity. They demanded a new democratic politics and a new culture to go with it. Young Japanese men and women stopped attending traditional *kabuki* theatres, and instead started going to the cinema and listening to jazz. And at the same time that students in Tokyo were reading about quantum mechanics, others were joining the recently founded Marxist Society. In fact, a number of those who later went on to become leading physicists also joined the Japanese Communist Party, founded in 1922. Not everyone, however, was drawn to Marxism. Different political factions vied for influence. Nationalists called for greater military power in the wake of the First World War, liberals demanded the reform of parliament, whilst anarchists plotted to overthrow the government. Everyone, it seemed, had a vision for the future of Japan.[51]

Many saw science as the key to that future. Following the First World War, Japan began to invest more and more in science and technology. This in some ways was a continuation of a theme we saw in the previous two chapters. Following the Meiji Restoration of 1868, the Japanese government had started to send students to train abroad in the United States and Europe. This earlier generation of Japanese scientists then returned to help establish some of the first university departments in physics and biology. However, much like in China, the scale of scientific investment increased significantly following the First World War. Almost ten times as many students were studying at Japanese universities in 1930 compared to before the war. The Japanese government also established a number of new universities and scientific organizations in this period. These included the University of Osaka, founded in 1931, as well as the Japan Society for the Promotion of Science, founded in 1932.[52]

By far the most important of these new organizations was the Institute of Physical and Chemical Research, better known by its Japanese name, Riken. Established in Tokyo in 1917, Riken served both practical and intellectual ends. Following the First World War, Japan wanted to maintain its position of industrial and military supremacy in East Asia. 'The recent war . . . has taught us the urgent necessity of independence and self-sufficiency in military supplies and industrial materials,' explained the committee responsible for setting up Riken. At the same time, this new institution was intended as a centre for cutting-edge theoretical work in the sciences. These two aims actually complemented one another. Riken, which was partly funded by Jokichi Takamine, the wealthy industrial chemist we learned about in the previous chapter, quickly amassed patents for various chemical and industrial processes. These even included a special method for manufacturing Japanese rice wine, or sake. The money from these patents then helped to fund more theoretical research, particularly in physics. After all, the aim of Riken was not only to support industry, but also 'to contribute to world civilization, to enhance the status of our nation'. Japanese scientists were to become leaders in their respective fields. Riken soon developed a reputation as the place to go for ambitious young graduates who wanted to work on new scientific problems.[53]

Yoshio Nishina was one of those ambitious young graduates who came to work at Riken. He was born at the end of the nineteenth century into an influential family that had recently fallen on hard times. Nishina's grandfather was a samurai, but his father was of rather lower status, running a small farm on the outskirts of Okayama. Nonetheless, Nishina did well at school, and entered the Department of Electrical Engineering at Tokyo University in 1914, the year of the outbreak of the First World War. At university, Nishina excelled, graduating top of his class in 1918. This was an enviable position to be in. At the graduation ceremony, Nishina was given a silver watch by the Japanese emperor himself, whilst top engineering firms, which had done well out of the war, offered Nishina his pick of the most desirable jobs going. However, he had something else in mind. Despite his aptitude for engineering, Nishina wanted to become a scientist. His dream was to work in theoretical physics, and so instead of taking a well-paid position at an engineering firm, Nishina decided to go to work as a researcher in the Physics Department at Riken.[54]

By this time it was very common for Japanese scientists to undertake a period of study abroad. Nishina was no different, and in April 1921 he left Tokyo aboard a steamship bound for Europe. He was to spend a year studying physics at the University of Cambridge. Nishina's trip had been arranged by Hantaro Nagaoka, director of the Physics Department at Riken. As we saw in chapter 6, Nagaoka was one of the early pioneers of Japanese physics, having proposed a model of the atom strikingly similar to that of Ernest Rutherford, the director of the Cavendish Laboratory in Cambridge. By the 1920s, Nagaoka was well known amongst physicists in Europe. He had sent a number of promising students to work in Cambridge, and wrote personally to Rutherford to recommend Nishina. For a year, Nishina learned the basic experimental techniques of modern physics, toying with a special device – known as a cloud chamber – that could record the tracks made by subatomic particles. He also met a number of other international physicists who had come to work with Rutherford in Cambridge, including Peter Kapitza, who we encountered earlier. Still, Nishina yearned for more. Experimental physics was all well and good, but what he really wanted was to work on the theory behind quantum mechanics. Only then would he understand the fundamental nature of the universe itself.[55]

At the end of 1921, when Nishina was supposed to be preparing to return to Japan, he wrote to Niels Bohr. The two had met briefly during one of Bohr's visits to Cambridge. Nishina asked if he might join Bohr in Copenhagen, writing, 'if anyone wants assistance in the experiment or the calculation, I should do it with pleasure'. Bohr decided to take Nishina up on his offer, inviting the Japanese physicist to the Institute for Theoretical Physics in Copenhagen. With Bohr's backing, Nishina received a fellowship from the Rask-Ørsted Foundation, a Danish government organization which had been set up following the First World War to promote international scientific collaboration, and spent the next five years working in Copenhagen. What was supposed to be a one-year trip turned into nearly a decade abroad.[56]

In 1928, just before he left Copenhagen, Nishina made a major theoretical breakthrough. Earlier that year, the British physicist Paul Dirac had published an article which combined both relativity and quantum mechanics to describe the physics of the electron. Dirac, who knew Nishina from his time in Cambridge, forwarded a copy of the article to Copenhagen. It generated intense excitement. Dirac had shown that, at least in principle, it was possible to bring together relativity and quantum mechanics, which to that point had been treated almost completely separately. Nishina, however, wanted to go one further. Working with the Swedish physicist Oskar Klein, Nishina began to work on a formula which would extend Dirac's equation to describe an actual physical phenomenon: in this case, what happens when you fire an X-ray at an electron.[57]

Nishina and Klein worked intensely for the next few months, meeting every day in Copenhagen to discuss their results. The finished article was published in one of the leading German physics journals at the beginning of 1929. The mathematics was frankly mind-bending. But they had done it. What became known as the 'Klein–Nishina formula' was the first successful attempt to apply both relativity and quantum mechanics to a concrete physical phenomenon. This was in fact the very formula which inspired the experiments of Zhao Zhongyao, the Chinese physicist we met earlier. We can therefore start to get a sense of the remarkable international reach of physics in this period, in which a Japanese scientist working in Copenhagen might inspire a Chinese scientist working in California. For a brief moment in the early twentieth century, it looked like scientific collaboration might really lead to a more harmonious world.[58]

When Yoshio Nishina returned to Tokyo in 1928, he hardly recognized the city. During his absence, Japan had been rocked by the Great Kanto Earthquake of 1923. Over 100,000 people died, and many buildings were destroyed. Even the world of science had been affected. 'The main building of the Physics Department was cracked and on the verge of collapse. The Mathematics Department building was completely burned down,' noted one of the students at the University of Tokyo. Nishina returned to a nation still picking up the pieces. Nonetheless, many people saw this as an opportunity to start afresh. From the ashes, a new Japan would emerge.[59]

Nishina did more than anyone to promote quantum mechanics as part of the future of Japanese science. He invited both Paul Dirac and Werner Heisenberg to Japan, acting as their interpreter during a series of lectures the pair gave in 1929. These were later translated into Japanese by Nishina himself, providing students with an introduction to many of the basic concepts in quantum mechanics. Nishina also lectured across the country, helping to inspire a new generation of physicists, many of whom would go on to do important work in their own right. Possibly Nishina's most important contribution to Japanese science was in fact a lecture he gave in May 1931 at Kyoto University. In the audience was a young physics student named Hideki Yukawa. He sat transfixed as Nishina described the strange new world of quantum mechanics, quizzing the older professor after the lecture. Little did Nishina know that this young man would later become the first Japanese scientist to win a Nobel Prize.[60]

Yukawa was born in Tokyo in 1907. His father worked for the Geological Survey of Japan. Established in 1882, this was one of the new scientific institutions created during the Meiji Restoration, that period of Japanese history covered in the previous two chapters. Yukawa's father travelled the world, working with geologists from China and Europe, an exemplar of what it meant to be a modern Japanese scientist in the nineteenth century. There was, however, another major figure in Yukawa's early life. His grandfather was a respected classical scholar. Yukawa was therefore exposed to both modernity and tradition. His father taught him physics and chemistry, whilst his grandfather had him recite Chinese classics.[61]

In the end, Yukawa decided to follow his father and become a scientist. He entered Kyoto University in 1926 to study physics. However, like many young students across Japan, he found the teaching rather uninspiring. He was much more excited by the new physics of quantum mechanics than anything on the outdated syllabus. And so Yukawa decided to teach himself. He would spend hours in the physics library. 'I had no use for the old books that filled the shelves but wanted to learn, as quickly as possible, the articles that concerned the new quantum theory, those that had been published in foreign, especially German, journals, within the past two or three years,' he later wrote. For a nineteen-year-old student to teach himself quantum mechanics was quite a feat. But Yukawa was clearly hooked. He was joined in his studies by Sin-Itiro Tomonaga, another ambitious student at Kyoto University. (Tomonaga would also later win a Nobel Prize in Physics, only the second Japanese scientist to do so.) Together, the two young men would spend the evenings chatting about quantum mechanics, punctuated with the odd game of Go.[62]

Yukawa graduated right in the middle of a recession, the first signs of the Great Depression which would soon wreck the world economy. In late 1929, Yukawa weighed up his options. There were no graduate jobs, and he wondered whether he should become a priest? That would at least please his grandfather. But after attending Nishina's lecture in Kyoto, Yukawa decided to stick with his passion. He would become a theoretical physicist. After all, universities were just about the only places still hiring. In 1932, Yukawa took up a position as a lecturer in the Physics Department at Kyoto University. Much to the delight of the students, he immediately set up a new course on quantum mechanics. A year later, however, Yukawa was offered a position at Osaka University. This was one of the new universities established by the Japanese government in the 1930s, part of an effort to expand science across the country. It already had a reputation as the place to be for those conducting exciting new research. And it was ultimately in Osaka that Yukawa made the breakthrough for which he was awarded the Nobel Prize.[63]

On 17 November 1934, Yukawa presented his latest work at a meeting of the Physico-Mathematical Society of Japan. Apparently the audience took little notice, unaware they were listening to one of the most important theoretical developments in modern physics. In the paper, Yukawa solved a problem that had eluded some of the best scientists of the era. Two years earlier, the Cambridge physicist James Chadwick had discovered the existence of the neutron. This large uncharged particle could be found at the centre of the atomic nucleus, bound to the positively charged proton. There was, however, a problem here. It wasn't clear what was holding the atomic nucleus together. It couldn't be electrical charge, as the neutron didn't have a charge, and any positively charged protons would repel one another. Physicists therefore assumed there must be some other force holding the neutrons and protons in place. But what was it? Yukawa provided the answer. In his paper, which was published in early 1935, he predicted the existence of an entirely new fundamental particle, later known as the 'meson'. The meson, according to Yukawa, was the carrier of the strong nuclear force, binding protons and neutrons together.[64]

A few years later, Yukawa was proved right. Fittingly, it was his old mentor, Yoshio Nishina, who confirmed the existence of the meson. Nishina at this time was head of the Physics Department at Riken. He had also recently recruited Yukawa's university friend and fellow quantum mechanics enthusiast Sin-Itiro Tomonaga. Together, the two started to hunt for the meson. Yukawa had given a few clues. He predicted that the meson would only be detectable at very high energies and should have a mass about 200 times that of the electron. In late 1937, Nishina spotted a line in the cloud chamber that seemed to fit the bill. He had been experimenting with high-energy cosmic rays, watching what happened when they collided with other subatomic particles in the cloud chamber. In some cases, a new particle briefly appeared, visible as a thin white line on the photographs taken by Nishina. This was it, just as predicted. The 'Yukawa particle', as Nishina liked to call it. The meson was real.[65]

Unlike China and Russia, Japan did not undergo a revolution in the first decades of the twentieth century, yet science there was nonetheless shaped by the wider world of international politics. The death of the Emperor Meiji in 1912 had brought about calls to reform Japanese society. The younger generation wanted political and intellectual change: some joined the Japanese Communist Party, others started reading about quantum mechanics. Mitsuo Taketani, one of the physicists who worked with Yukawa at the University of Osaka, did both. At the same time, Japan was determined to cement its military and economic status in East Asia. New scientific institutions, such as Riken in Tokyo, served both political and intellectual goals. Riken would 'increase the nation's wealth' whilst also encouraging 'creative research in the disciplines of physics and chemistry'. By the 1930s, Japanese physicists working at new institutions like Riken had made a series of major breakthroughs. Like scientists in Russia and China, this new generation of scientists saw in modern physics a better future. For Yukawa, quantum mechanics represented the 'liberal spirit' of the age. This was a new science for a new Japan. In the following section, we explore how that idea of a brighter future shaped the development of modern physics in British India during the same period.[66]

IV. Physics and the Fight against Empire

Meghnad Saha was about to risk everything. Born into a poor Hindu family in Bengal, Saha had only recently been admitted to the prestigious Government Collegiate School in Dacca, in modern-day Bangladesh. He was getting on well, studying mathematics, physics, and chemistry. But in the summer of 1905, Saha took part in a protest which would define the rest of his life. At the time, Bengal was still under colonial rule, part of the British Empire in India. Since the late nineteenth century, many Indians had campaigned against the injustices of empire. However, the fight against colonialism was about to enter a new stage. In July 1905, the Viceroy of India announced his intention to divide Bengal into two new provinces: Hindu-majority West Bengal, and Muslim-majority East Bengal. Saha, like many other Bengalis, deeply resented the division of his homeland. When the Lieutenant-Governor of Bengal visited the Collegiate School in Dacca, the students decided to stage a boycott. They refused to turn up to class, instead standing outside the school gate jeering at the British colonial officer.[67]

Saha was expelled from the school the next day. As the first in his family to receive a secondary education, this must have been a tremendous blow. But Saha refused to give up. He didn't want to return to his tiny village, to become a shopkeeper like his father. Instead, Saha decided to stay in Dacca. He enrolled at a different school, one run by Bengalis rather than the British. And as his poor family could not support him, Saha had to earn his keep as a private tutor, cycling across the city on a rusty bicycle, teaching mathematics and physics at the homes of more wealthy students. In the end, all the hard work paid off. On the basis of his exam results, Saha was admitted to study at Presidency College, part of the prestigious University of Calcutta.[68]

In 1911, the poor boy from rural Bengal arrived in the beating heart of the British Empire in India. In Calcutta, Saha was taught by some of the finest scientists of the previous generation, including those we met in chapter 6. Prafulla Chandra Ray lectured on chemistry, whilst Jagadish Chandra Bose taught physics. Saha excelled at university, graduating with an MSc in 1915. He was particularly excited by the new work in quantum mechanics, even teaching himself German so he could read the original papers of Werner Heisenberg and Max Planck. However, the world of politics was about to catch up with him. At Presidency College, Saha studied alongside a number of student radicals, many of whom later went on to play a leading role in the fight against colonialism. These included Subhas Chandra Bose, an Indian nationalist who sided with Nazi Germany in the Second World War, as well as Atulkrishna Ghosh, one of the leaders of the revolutionary Jugantar Party.[69]

On graduating, Saha found his career choices severely limited. Most high-flying Indian graduates in mathematics and physics joined the Finance Department in Calcutta, helping to administer the colonial economy. But Saha was refused permission to take the relevant exams. The colonial government had learned about his boyhood protest against the Partition of Bengal, and was worried about Saha's more recent association with student revolutionaries in Calcutta. With few other options left open to him, Saha took up a position as a lecturer in physics at the University of Calcutta. Like many other student radicals, whether in India, China, or Japan, Saha ultimately chose to become a scientist when the political situation left him no other choice.[70]

The early twentieth century was a period of rising anticolonial activism in India. Whilst there had always been a degree of resistance to British rule, the Partition of Bengal in 1905 pushed many – including Meghnad Saha – to actively campaign for an end to empire. The British had completely disregarded the wishes of the Indian population, introducing the policy as part of a cynical attempt to divide and rule. As elsewhere, this political world shaped the development of science in India. Saha, like many Indian scientists in this period, was a committed anticolonial activist. He also believed that science itself had a role to play in ending colonialism. From the 1920s onwards, Saha argued that, to gain independence, India would need to industrialize and escape its reliance on Britain for scientific and technical support. This was a vision shared by many in the anticolonial movement. 'In the days to come India will again become the home of science, not only as a form of intellectual activity, but also as a means of furthering the progress of her people,' declared Jawaharlal Nehru at a meeting of the Indian Science Congress in Calcutta in 1938. Nehru, who had studied Natural Sciences himself at the University of Cambridge, would later go on to become the first prime minister of India following independence in 1947.[71]

Saha benefited immensely from this enthusiasm for science amongst Indian nationalists. In 1915, he began studying for a DSc at the University College of Science and Technology in Calcutta. This new institution had only been established the year before. It was the brainchild of two wealthy Bengali lawyers, both of whom were committed to the cause of Indian independence. One of the founders was even a member of the Indian National Congress, the main political organization committed to ending British rule. This new institution was to be 'an all-India College of Science', one 'to which students will flock from every corner of the Indian empire'. Shortly after completing his DSc, Saha was awarded a scholarship to undertake postgraduate study abroad. He hoped to travel to the United States, but in the end only got as far as Britain, arriving in London in early 1920. Saha, who was no friend of the British, now found himself in the imperial metropolis. He must have felt uneasy. Nonetheless, he made the most of it, taking a job working with the physicist Alfred Fowler at Imperial College, London. It was here that Saha made his first major breakthrough.[72]

At Imperial College, Saha began studying what happens when matter is heated up to extreme temperatures. Since the late nineteenth century, scientists had known that matter could enter a strange new state, known as plasma, at very high temperatures. In this state, electrons seemed to move freely between atoms, creating a cloud of electrical charge and emitting energy. But despite this basic understanding of high-temperature physics, no one really knew how to describe or explain what was happening in detail. Not until Saha published an article in the *Philosophical Magazine* in March 1920. In the article, Saha used his knowledge of quantum mechanics to describe the precise relationship between heat, pressure, and the degree of electrical energy – or 'ionization' – in the plasma. The formula he set out later became known as the 'Saha ionization equation'. It turned out to be incredibly useful, not just from a theoretical perspective, but also for explaining all kinds of physical phenomena. Saha's ideas were later used to identify the elements found in stars, as well as to describe what happens on the surface of the sun.[73]

Saha returned to India in 1921, where he took up a professorship at the University of Calcutta. Over the following years, he continued to combine his scientific and political interests. He worked closely with the Indian National Congress in the years running up to independence, serving on both the National Planning Committee and the Board of Scientific and Industrial Research in the 1930s. He also made regular contact with scientists in the Soviet Union. Saha even travelled to Russia in 1945, where he met Peter Kapitza at the Soviet Academy of Sciences in Moscow. He returned to India more enthusiastic than ever about the political power of modern science. According to Saha, Indian nationalists needed to focus on 'the monumental task of applying scientific and industrial methods . . . as has been done in the USSR'. When India finally did gain independence, Saha moved into politics and stood for election in the new parliament. In 1952, he was elected as a member of the Revolutionary Socialist Party in his beloved Bengal. Saha was certainly a radical in every sense of the word.[74]

Meghnad Saha was just one amongst a number of Indian physicists who mixed science with politics. Not all were as bold as Saha – some disagreed about how exactly to achieve independence, and few were quite so keen on following the path of the Soviet Union. Nonetheless, despite these differences, most Indian scientists in this period saw a connection between their work and the fight against empire. Satyendra Nath Bose was one of those who shared Saha's vision for both the future of physics and the future of India. In the summer of 1905, Bose was staging his own protest against the Partition of Bengal. Aged just eleven, the young Bose went door-to-door in Calcutta, collecting imported British textiles, before burning them in a pile in the street. This protest was part of the *swadeshi* movement sweeping through Bengal at the time, in which anticolonial nationalists called for a boycott of all British goods. This, they hoped, would reduce reliance on British imports whilst also stimulating Indian production. Unlike Saha, however, Bose didn't suffer any serious repercussions following his protest. He attended the Hindu School in Calcutta before entering Presidency College in 1909. After graduating, Bose then went on to join the new University College of Science and Technology as a lecturer in physics. It was here that he met Saha.[75]

Saha and Bose came from quite different backgrounds. Saha was born into a low-caste family in a rural village, whereas Bose was born into a high-caste family in Calcutta. But despite these differences, Saha and Bose became lifelong friends. They both believed that modern science would ultimately help Indians escape British rule. And the two also shared a deep interest in the new physics of relativity and quantum mechanics. Together, these young Bengali physicists taught themselves German, and began to buy up any imported copies of German scientific journals they could lay their hands on. There was an element of anticolonialism to this too, a boycott of British science in favour of the exciting new work coming out of Germany.[76]

Saha and Bose then did something remarkable. The pair translated Albert Einstein's original papers on special and general relativity from German into English. Published as *The Principle of Relativity* (1920) in Calcutta, this was in fact the very first English translation of Einstein's work available anywhere in the world. Students in Britain and the United States later purchased copies, learning about Einstein via India. It is worth pausing for a second to reflect on that. Two Bengalis brought German science to the English-speaking world. It is another reminder of the incredible international reach of science in this period, as well as the important role played by India in the development of modern physics.[77]

In 1921, Bose was hired as a lecturer at the University of Dacca, one of the new universities established following the First World War. He spent the next couple of years teaching relativity and quantum mechanics, whilst also working on his own research. Bose then plucked up the courage to write a letter to Einstein himself. 'Respected Sir, I have ventured to send you the accompanying article for your perusal and opinion. I am anxious to know what you think of it,' wrote Bose in June 1924. A few months earlier, Bose had sent an article to the *Philosophical Magazine* in London. It had been rejected by the editor. Undeterred, he sent the same article to Einstein, asking 'if you think the paper worth publication?'[78]

Einstein was amazed. Bose, who was pretty much unknown outside of India at this point, had come up with an entirely new way of thinking about the behaviour of fundamental particles, one based on quantum mechanics rather than classical physics. At the microscopic level, Bose realized, individual particles would often be indistinguishable from

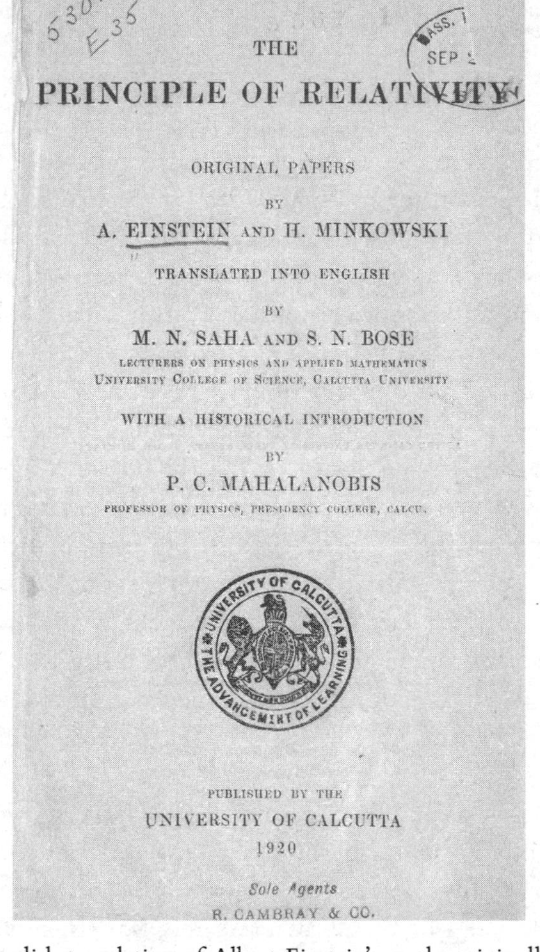

THE
PRINCIPLE OF RELATIVITY

ORIGINAL PAPERS
BY
A. EINSTEIN AND H. MINKOWSKI

TRANSLATED INTO ENGLISH
BY
M. N. SAHA AND S. N. BOSE
LECTURERS ON PHYSICS AND APPLIED MATHEMATICS
UNIVERSITY COLLEGE OF SCIENCE, CALCUTTA UNIVERSITY

WITH A HISTORICAL INTRODUCTION
BY
P. C. MAHALANOBIS
PROFESSOR OF PHYSICS, PRESIDENCY COLLEGE, CALCU.

UNIVERSITY OF CALCUTTA
THE ADVANCEMENT OF LEARNING

PUBLISHED BY THE
UNIVERSITY OF CALCUTTA
1920

Sole Agents
R. CAMBRAY & CO.

35. The first English translation of Albert Einstein's work, originally published in Calcutta in 1920.

one another. This made a mockery of the existing equations in thermodynamics. Instead, Bose developed a new statistical method for describing what was happening. This later became known as 'Bose–Einstein statistics'. Physicists also soon realized that only certain kinds of particles would follow this statistical pattern. Today, those particles are known as 'bosons' after Bose.[79]

This was 'an important step forward and pleases me very much', wrote Einstein in reply. In fact, Einstein was so impressed that he personally translated Bose's article from English into German. He then arranged for it to be published in one of the leading physics journals in Berlin. Einstein also suggested that Bose come to Europe, where the two could meet and discuss their ideas properly. At first, the University of Dacca was reluctant to let Bose go. But when Einstein wrote a letter of recommendation, that all changed. The university immediately granted Bose his request for leave. In September 1924, he boarded a steamship bound for Europe. 'I got a visa from the German Consulate just by showing them Einstein's card,' Bose later recalled. After spending some time in Paris, Bose then moved on to Berlin, where he finally met Albert Einstein himself. The two chatted about the future of quantum mechanics, which Einstein was less sure of by this point. They also talked politics. Einstein, as we saw earlier, was increasingly concerned with the state of society following the First World War. Naturally, the conversation turned to the British Empire in India. 'Do you really want that the British should quit your country?' asked Einstein. 'Of course,' replied Bose, 'we all want to determine our destinies.'[80]

After his meeting with Bose, Einstein took a much greater interest in both Indian science and Indian politics. He would often take the time to reply to letters sent by Indian scientists, even PhD students. Einstein also corresponded with many of the leading Indian political figures of the early twentieth century, including Mohandas Gandhi as well as Jawaharlal Nehru. As someone committed to international peace, Einstein found inspiration in Gandhi's philosophy of non-violence. 'You have shown by all that you have done that we can achieve the ideal even without resorting to violence,' wrote Einstein in a letter to Gandhi in September 1931. Gandhi was delighted that Einstein had taken an interest in the Indian independence movement. 'It is a great consolation for me that the work I am doing finds favour in your sight,' he replied. Gandhi's and Einstein's exchange of letters came less than a year after one of the most important moments in the history of science in India. In November 1930, an Indian scientist won the Nobel Prize in Physics for a discovery which changed how people understood the very nature of light.[81]

Sitting aboard the SS *Narkunda*, Chandrasekhara Venkata Raman looked out over the sparkling blue sea. He had recently attended a conference at the University of Oxford, and was now on his way back to India. In August 1921, as he crossed the Mediterranean, Raman's mind started to drift. Why was the sea blue, he wondered? Raman knew the standard answer. Since the middle of the nineteenth century, scientists had argued that the sea appears blue because it reflects the colour of the sky. That was what most physics textbooks still said at the time. 'The much admired dark blue of the deep sea has nothing to do with the colour of the sea, but is simply the blue of the sky seen by reflection,' wrote the British physicist Lord Rayleigh in 1910. But Raman wasn't so sure. He started playing around with a pocket light filter, checking the colour of the sea from different angles from the deck of the ship. Suddenly, Raman realized that Rayleigh was wrong. The sea wasn't simply reflecting the colour of the sky. Rather, the sea actually shifted the colour of light itself. This observation would ultimately lead Raman to win a Nobel Prize, making him the first Indian scientist to do so.[82]

Whilst still at sea, Raman dashed off a quick note to *Nature* in London. He hoped to convince scientists that 'the blue of the deep sea is a distinct phenomenon in itself', one that could only be explained using the latest theories in quantum mechanics. As the article was written by a relatively unknown Indian physicist, few people in Europe took much notice at first. But Raman was determined to show that he was right. In October 1921, Raman arrived back in Calcutta, where he had recently been appointed as the first Palit Professor of Physics at the University College of Science and Technology. He then began a series of experiments to show that water really did change the colour of light. The setup was pretty simple, but it did the job. Raman first placed a violet filter in front of an electric lamp. He then exposed a jar of water to the violet light and placed a green filter to one side of it. This would then allow Raman to detect, simply by eye, if the light had changed wavelength, and thus colour. If he was right, some of the violet light would increase in wavelength and turn green, passing through the filter.[83]

In 1928, Raman published his final results in the *Indian Journal of Physics*, a new periodical aimed at encouraging home-grown talent. In the article, he used his knowledge of quantum mechanics to describe what was happening when light interacts with a molecule of water. True, some of the light was simply reflected, as earlier physicists like Rayleigh had suggested. But crucially, some of the light was also absorbed by the water molecules. The remaining light would then have less energy, increasing in wavelength, and so changing colour. Raman called this 'a new radiation'. It soon became known as 'Raman scattering'. Two years later, he won the Nobel Prize in Physics. This was an immense personal achievement, but it was also a significant moment in the campaign for Indian independence. Raman had proved that an Indian scientist could make a major contribution to the development of modern physics, one that was recognized by the international scientific community.[84]

Shortly after winning the Nobel Prize, Raman left Calcutta to take up a new appointment. He had been offered the position of director of the Indian Institute of Science in Bangalore. Founded in 1909, the Indian Institute of Science had been set up by the wealthy industrialist Jamsetji Tata. As with many of the new institutions established in this period, the aim was to promote the growth of Indian industry through science and technology. Yet despite being largely funded by Indian donors, the Indian Institute of Science had long been dominated by British scientists. Every previous director of the institute had been British, as were most of the senior staff. Raman's appointment in 1933 was therefore met with much excitement amongst Indian nationalists. It marked the transition from British to Indian rule, at least in the world of science. Even Mohandas Gandhi, who was famously critical of the social impact of modern technology, visited the Indian Institute of Science to congratulate Raman.[85]

In Bangalore, Raman turned his theoretical discovery to practical use. He quickly realized that the scattering of light might reveal something about the structure of different materials. With this in mind, Raman started to make much more precise measurements, using photographic plates, of the degree to which different materials shifted the wavelength of light. This technique became known as 'Raman spectroscopy', and is still used by scientists today. Diamonds were a particular object of fascination for Raman, although not always easy to get hold of. He began by borrowing a friend's wedding ring, before finally convincing a local maharaja to loan him a much larger specimen. By measuring the degree of light scattering, Raman was able to explain how subtle differences in molecular structure affect the colour and brilliance of different types of diamond.[86]

Other researchers in Bangalore worked on more traditional industrial materials. Sunanda Bai, one of a small group of women employed at the Indian Institute of Science, conducted a series of experiments in the 1930s on the structure of different chemical compounds. Using Raman's method, Bai was able to identify the molecular structure and chemical properties of tetralin and nitrobenzene. Both of these chemicals were crucial to Indian industrial development at the time: tetralin was used to transform coal into liquid fuel, thus providing an alternative to importing petroleum, whereas nitrobenzene was used in the production of indigo dyes, a major Indian export. By better understanding the structure of these chemicals, Bai contributed to both Indian science and Indian industry.

Yet despite the fact that Indian women undertook such important research, working conditions were not always easy for them at the time. Most of the men involved in the Indian nationalist movement believed that women should remain in the home, where they could support the cause of independence as mothers, not scientists. And Raman himself disliked the idea of women working in the laboratory, stating in reply to one potential applicant that he did not want 'any girls in my institute'. Nonetheless, Bai was not entirely alone. Throughout the 1930s, a growing number of women in India rejected traditional gender roles, demanding an opportunity to work in previously male-dominated industries. Physics was no exception.

In Bangalore, Bai was joined by other pioneering Indian women. These included Anna Mani, who published a number of important articles on the molecular structure of precious stones. Born in 1918 in the southern city of Kerala, Mani came from a wealthy family. Her father owned a cardamom plantation, through which she would wander as a young girl. Mani's family expected her to marry and settle down as a dutiful wife, but she had other ideas. When Mani was just seven years old, she listened to Mohandas Gandhi speak at a rally in Kerala, and from that point on she was committed to anticolonialism. She soon decided that the best way she could support the cause of Indian independence was to become a scientist. At her next birthday, Mani rejected the traditional gift of a pair of diamond earrings and instead asked for a copy of *Encyclopaedia Britannica*. Studying hard, she later won a place on the physics course at Presidency College, part of the University of Madras. She then worked for nearly ten years with Bai and Raman in Bangalore. In a strange twist of fate, Mani spent much of her time at the Indian Institute

1. The skeleton of the *Megatherium* on display in nineteenth-century Madrid. The bones were originally excavated in 1788 close to the Luján River in Argentina.

2. The Russian zoologist and evolutionary thinker Ilya Mechnikov, winner of the 1908 Nobel Prize in Physiology or Medicine.

3. The Bengali physicist Jagadish Chandra Bose lecturing at the Royal Institution in London in 1897.

4. Postcard from the Paris Exposition of 1900, which coincided with the First International Congress of Physics.

5. The theoretical physicist Zhou Peiyuan (*on the far left*), who made a major contribution to the study of general relativity, alongside other leading Chinese intellectuals of the early twentieth century.

6. Albert Einstein and his wife Elsa at a reception in Japan during their visit in November 1922.

7. The Japanese physicist Aikitsu Tanakadate in his office at the University of Tokyo.

8. Chandrasekhara Venkata Raman, the first Indian scientist to win a Nobel Prize, examining the structure of a diamond at the Raman Research Institute in Bangalore.

9. The physicist Hideki Yukawa, who in 1949 became the first Japanese scientist to win a Nobel Prize.

10. A Chinese Communist Party propaganda poster from the 1960s showing students examining seeds in a laboratory. The caption reads 'Nurture Young Seedlings'.

11. The geneticists Obaid Siddiqi (*top row, first from left*) and Veronica Rodrigues (*top row, second from left*) alongside other scientists at the Tata Institute of Fundamental Research, Bombay, in 1976.

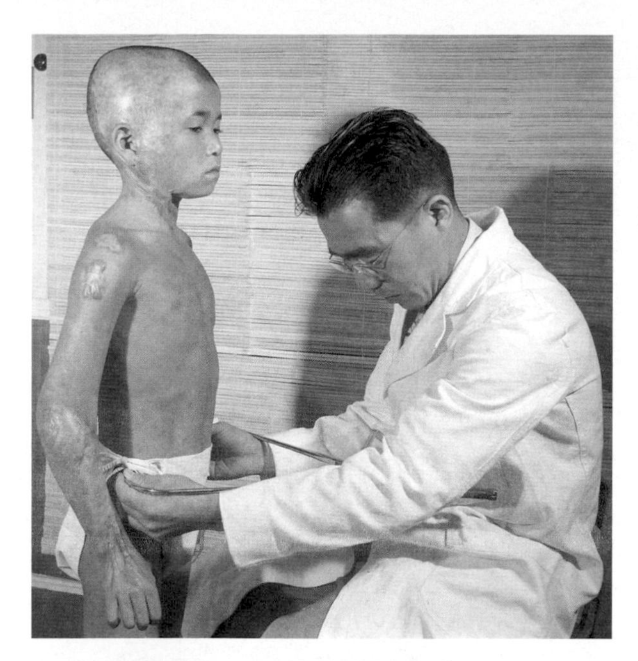

12. A Japanese doctor working for the Atomic Bomb Casualty Commission examining a young patient in Hiroshima in 1949. The victim has suffered radiation burns and the doctor is checking their growth rate by measuring the pelvis.

13. Yemeni Jewish families arriving at an immigration camp in Israel in 1949. Population geneticists undertook extensive studies of Jewish immigrants throughout the middle decades of the twentieth century.

14. The Israeli population geneticist Elisabeth Goldschmidt. Born to a Jewish family in Germany in 1912, Goldschmidt was forced to flee following the rise of Adolf Hitler and the Nazi Party in the 1930s.

15. Sarah Al Amiri, Chair of the United Arab Emirates Space Agency and Deputy Project Manager for the 2020 Emirates Mars Mission.

16. Moustapha Cissé, Staff Research Scientist and Director of the Google AI Center, Accra, Ghana.

of Science studying the molecular structure of diamonds and other precious stones, the very gifts she had refused as a young girl.[87]

Alongside Mani, Bai was joined by Kamala Sohonie. Born in 1911, Sohonie came from a family of scientists. Both her father and her uncle had studied chemistry, and they encouraged the young Sohonie to enter the same field. Following an undergraduate degree in physics and chemistry at the University of Bombay, she applied to work with Raman at the Indian Institute of Science in 1933. Sohonie had come top of her year at university, yet Raman chose to reject her application, another reminder of the discrimination women faced. 'Though Raman was a great scientist, he was very narrow-minded. I can never forget the way he treated me just because I was a woman,' Sohonie later wrote. Still, she wasn't going to take no for an answer. In a move of incredible bravery, Sohonie confronted Raman in his office in Bangalore, demanding he admit her. In the end, Raman backed down, and accepted Sohonie as a research student. She later went on to become the first Indian woman to receive a doctorate in the sciences, graduating with a PhD from the University of Cambridge in 1939, before returning to India to take up a professorship. Raman might not have wanted 'any girls' in his laboratory. But like it or not, women were here to stay.[88]

The Partition of Bengal in 1905 marked the beginning of the end of the British Empire in India. In attempting to divide and rule, the British in fact galvanized the Indian independence movement. As elsewhere, the great political changes taking place in early twentieth-century India had a profound effect on the development of modern science. Many Indian physicists saw their work as part of the fight against empire. Meghnad Saha was the boldest in this respect, described by one British intelligence officer in the 1920s as a 'rabid revolutionary'. But others, including Raman, shared Saha's scientific vision, if not his socialist politics. They believed that science would ultimately help India transform into an independent industrial economy. 'There is only one solution to India's economic problems and that is science, more science and still more science,' declared Raman in January 1948, just five months after independence. With this in mind, Indian nationalists helped establish a number of new scientific institutions. With proper financial support, scientists in India were able to make a number of important breakthroughs, particularly in fields linked to relativity and quantum mechanics. This enthusiasm for science was also shared by Indian political leaders, including Jawaharlal Nehru, the first prime minister of independent India. Like leaders in China and Japan, he saw in science a better world. 'The future belongs to science,' argued Nehru.[89]

V. Conclusion

A blinding light. A burning heat. And in an instant, the world changed. On 6 August 1945, a B-29 Superfortress dropped an atomic bomb on the Japanese city of Hiroshima. At least 50,000 people were killed, the vast majority civilians. Three days later, the Americans dropped a second atomic bomb, this time on Nagasaki. Estimates vary, but perhaps up to 200,000 people died in total across the two sites, either from the direct effect of the blasts or from subsequent radiation poisoning.

For the first few decades of the twentieth century, science seemed to hold the key to a better society. Many saw in relativity and quantum mechanics a chance to break with tradition, to forge a brighter future. Working with colleagues across the world, scientists from Russia, China, Japan, and India made a number of important contributions to the development of modern physics. Albert Einstein himself spent much of the 1920s and 1930s promoting international cooperation in the wake of the First World War. The outbreak of the Second World War in 1939, followed by the use of atomic weapons in 1945, put an end to all that. This marked the beginning of the Cold War, as international cooperation gave way to a new era of international conflict. There was a grim irony here. Many of the optimistic young scientists we've encountered in this chapter ended up working on nuclear weapons programmes in the 1950s and 1960s. After all, they understood better than anyone else how to harness the immense energy contained within the atom. Lev Landau begrudgingly did the calculations for the first Soviet nuclear weapon, whilst Ye Qisun trained many of the physicists who built the first Chinese atomic bomb. In the following chapter, we push forward into the second half of the twentieth century, exploring the growth of science during the Cold War and its aftermath. Ideological conflict continued to shape the development of modern science, but in new ways and in new places.[90]

8. Genetic States

Masao Tsuzuki had heard that things were bad, but nothing could prepare him for the scene of devastation that he encountered on arriving in the ruined city of Hiroshima. Disfigured faces, bodies lying under the rubble, and children vomiting blood – it must have been difficult to fathom how a single explosion could have caused so much suffering. Tsuzuki, a professor at the Imperial University of Tokyo, was one of the first scientists to enter Hiroshima after the dropping of the atomic bomb on 6 August 1945. Over the following days, he examined survivors and conducted autopsies, building up a detailed picture of the medical effects of the blast. 'The burn action was so violent and severe that the entire thickness of the skin was burned,' he reported. Tsuzuki also noted how many survivors seemed to be suffering from what he called 'atomic bomb radiation sickness'. Those who were not killed by the explosion nonetheless developed disturbing symptoms, including vomiting, blood loss, and fever. The most severely affected patients typically died within a week.[1]

In the immediate aftermath of the explosion, Tsuzuki understandably concentrated on the most direct and observable effects of the blast. However, attention soon turned towards the long-term consequences of the use of nuclear weapons. A year later, Tsuzuki noted that scientists did not fully understand how exposure to radiation might affect 'the coming foetus, children, and descendants' of atomic bomb survivors. Since the 1920s, it had been known that radiation could cause genetic mutations. However, no one had really considered what this meant for the future of humanity, not until August 1945. Could these mutations be passed on to future generations? Was it safe for those exposed to atomic radiation to have children? There was a need for 'hereditary studies', argued Tsuzuki. These concerns were in fact shared by many scientists, not just in Japan, but also in the United States. 'If they could foresee the results 1,000 years from now . . . they might consider themselves more fortunate if the bombs had killed them,' argued the

American geneticist Hermann Joseph Muller, who went on to win the 1946 Nobel Prize in Physiology or Medicine for his earlier discovery of the genetic effects of radiation. 'There have been planted hundreds of thousands of minute time-bombs in the survivors' germ cells,' warned Muller, referring to the risk that damaging genetic mutations might be passed on to the next generation.[2]

Given widespread public concern, both at home and abroad, the United States government decided it needed to do something. In November 1946, President Harry Truman authorized the creation of the Atomic Bomb Casualty Commission. By this point, Japan had surrendered, and the country was under American occupation. Organized by the National Academy of Sciences, the Atomic Bomb Casualty Commission was tasked with tracking both the short- and long-term health outcomes of atomic bomb survivors, known in Japanese as the *hibakusha* (literally, 'exposed one'). Much of this work concerned the genetic impact of the blasts. The 'unique possibility for demonstrating genetic effects caused by atomic radiation should not be lost', argued the National Academy of Sciences. The study was led by an American geneticist named James Neel, who was assisted by a number of Japanese scientists, doctors, and midwives. In fact, well over 90 per cent of the staff employed by the Atomic Bomb Casualty Commission were Japanese. Tsuzuki was quickly recruited, as he was one of the few scientists to have entered Hiroshima in the weeks following the explosion. He had also conducted some experiments on the biological effects of radiation prior to the Second World War, so understood better than most what the genetic consequences of the use of nuclear weapons was likely to be.[3]

The initial work of the Atomic Bomb Casualty Commission, conducted by Neel and Tsuzuki in collaboration with a Japanese doctor named Saburo Kitamura, focused on tracking the birth outcomes of survivors. Together, Tsuzuki and Kitamura would travel around Hiroshima, interviewing pregnant women and examining newborn babies for any signs of abnormalities. Early reports seemed to suggest that spontaneous abortion was more common in cases where the father had been exposed to a high dose of radiation, but that the actual children born to atomic bomb survivors were not obviously affected in terms of major birth defects. This was consistent with the idea that the most damaging genetic mutations probably caused the death of the embryo well before it had a chance to grow. And so, although it was not possible to demonstrate a conclusive link between radiation exposure and reproductive health, Neel nonetheless concluded that genetic mutations must have occurred in the bodies of survivors.[4]

Alongside birth outcomes, the commission also began studying the effects of radiation at the level of the chromosome. Masuo Kodani, a Japanese American geneticist who had been interned in the United States during the war, played a leading role in this work. After completing a PhD at the University of California, Berkeley, Kodani moved to Japan – in part because his Japanese wife had been declared an illegal immigrant by the American government – and began working for the commission in 1948. The focus of Kodani's research was on the number of chromosomes found in the cells of atomic bomb survivors. By this point, it was possible to identify individual chromosomes – which are the carriers of genetic information, made up of strands of DNA – under

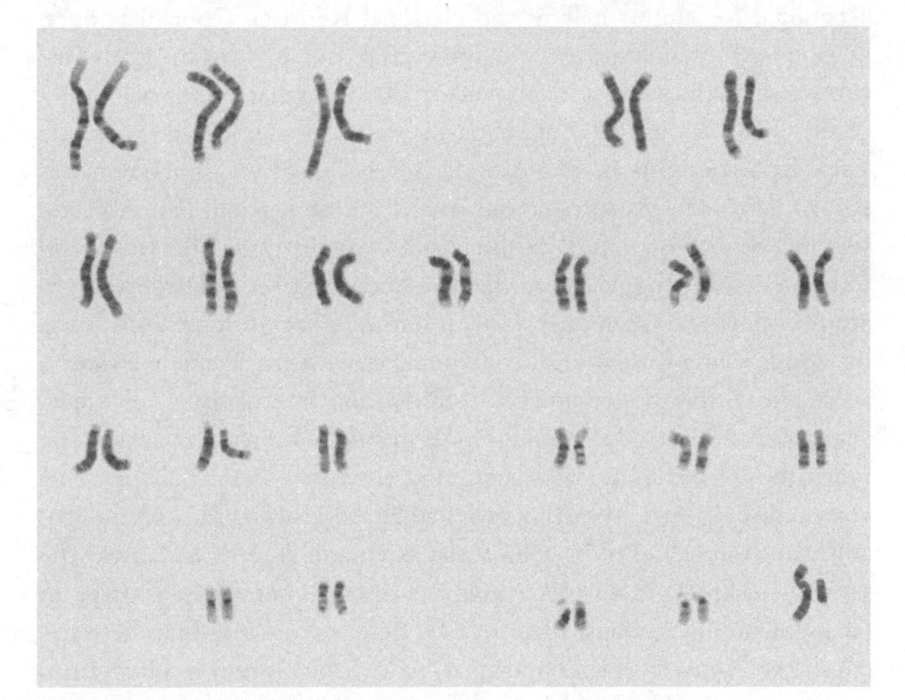

36. A typical set of human male chromosomes as observed under the microscope following staining. There are twenty-three pairs, forty-six in total.

the microscope. Kodani would take cell samples from patients, often from autopsies, stain them, and then carefully count the number of chromosomes he could see.[5]

In 1957, he published an important article documenting the existence of an additional chromosome in the testes of a number of male survivors of the atomic bombs. Whereas humans typically have 46 chromosomes – a fact that had only been confirmed the previous year by the Indonesian geneticist Joe Hin Tjio – Kodani found cases of atomic bomb survivors with either 47 or 48 chromosomes. Given that the presence of an additional chromosome can cause certain medical conditions, such as Down's syndrome and Klinefelter syndrome, and that these conditions can often be passed on to children, this was an incredibly significant discovery.[6]

The Atomic Bomb Casualty Commission was one of the largest scientific projects funded by the United States government in the immediate post-war period. At its height, the project employed over 1,000 staff and accounted for almost half of the National Research Council budget. This enormous investment was motivated, not just by medical concerns, but also by international politics. The 1940s marked the beginning of the Cold War, as the United States entered into an ideological battle with the Soviet Union. The Atomic Bomb Casualty Commission was part of a broader effort to extend American influence in East Asia and win the 'hearts and minds' of the Japanese population. This was going to be an uphill struggle, given that the United States had dropped two atomic bombs on Japan only a few months earlier. 'A long-term study of atomic bomb casualties in collaboration with the Japanese affords a most remarkable opportunity for cultivating international relations,' noted an American government report in 1947. This was just at the time when the United States was beginning to worry about the spread of communism in Asia – North Korea had already turned to communism, and was soon followed by China and Vietnam. And as we saw in the previous chapter, there was a long history of communist activity in Japan, including amongst scientists. By helping to rebuild Japanese science, the United States hoped to steer the country away from communism. It also hoped to allay fears about the continued testing of atomic weapons, something that was made much more difficult when a group of Japanese fishermen was inadvertently exposed to radioactive fallout following the detonation of an American hydrogen bomb on Bikini Atoll in March 1954.[7]

Throughout the 1950s, there continued to be major scientific disagreement over the effects of atomic radiation, particularly over the dose required to induce genetic mutations in humans. Some scientists believed that there was a minimum threshold dose, below which no genetic mutations could occur, and that it was therefore safe for humans to be exposed to relatively high doses of radiation, as might be encountered by workers in a nuclear power plant or those living close to nuclear weapons test sites. Others argued that this was wrong, and that even the smallest possible dose of radiation had the potential to induce a damaging genetic mutation. However, by the middle of the 1960s – thanks in part to the work of Japanese geneticists like Masuo Kodani – most scientists agreed that there was no threshold: exposure to radiation, no matter how small the dose, always had the potential to damage the genome.[8]

This, however, did not mark the end of the atomic age, but rather the start of it. Despite the knowledge of the damaging effects of radiation, governments around the world continued to invest in all kinds of nuclear technologies, particularly those related to energy and defence. This in turn created further demand for biological research into both the uses and effects of atomic radiation. The Atomic Bomb Casualty Commission, as we'll see, was just one of a number of institutions which brought together biological and nuclear science. At the international level, this work was supported by the United Nations, which throughout the 1950s and 1960s organized a series of conferences on the 'peaceful uses of atomic energy'. Every few years, scientists from around the world congregated in Geneva to discuss their research. Topics included the treatment of cancer using radiotherapy as well as the use of radiation to create new high-yield varieties of staple crops. 'I honestly believe that we are on the threshold of a new era in the study of . . . genetics,' wrote James Neel in 1957, echoing a widely held sentiment that the development of nuclear technologies, including atomic weapons, had brought about an unparalleled advance in the biological sciences.[9]

*

It is tempting to think that the history of modern genetics, based on molecular biology, began with the discovery of the structure of DNA. This is how the story is often told. Whilst the existence of DNA had been known since the late nineteenth century, it was only in 1953 that Francis Crick and James Watson, working together at the University of Cambridge, finally identified the famous 'double helix' structure of the molecule. Crick and Watson achieved this by examining X-ray photographs of DNA taken by Maurice Wilkins and Rosalind Franklin at Imperial College, London. This was a major breakthrough, one that helped scientists better understand how genetic inheritance works. Since the early twentieth century, scientists had known that chromosomes, which are made up of long strands of DNA, carry genetic information. Identifying the structure of DNA was therefore the first step in understanding how genes transmit biological characteristics. In fact, soon after Crick's and Watson's discovery, scientists proved that DNA codes for another molecule, called RNA, which in turn codes for proteins – the basic building blocks of life. In 1958, Crick referred to this process, in which DNA codes for RNA which then codes for proteins, as the 'Central Dogma' of modern molecular biology. Together, these discoveries eventually led to the development of new genetic technologies, such as gene editing and genome sequencing.[10]

It is true that the discovery of the structure of DNA was an important moment in the history of modern genetics. However, by focusing exclusively on this single discovery, we miss out on many other significant advances that were made in the biological sciences during the second half of the twentieth century. The emphasis on Crick and Watson also diverts attention away from scientists working in places outside of Europe and the United States, many of whom also played an important role in the development of the modern biological sciences. With this in mind, I want to suggest an alternative way of thinking about the history of modern genetics. Rather than starting in 1953 with the discovery of the structure of DNA in Cambridge, I think that we should instead start in 1945 with the dropping of the atomic bombs on Hiroshima and Nagasaki. This event marked the start of the Cold War. It also marked the beginning of the development of modern genetics. We've already seen how Japanese scientists working for the Atomic Bomb Casualty Commission did much of the early research into the genetic effects of radiation in humans. We've also seen how American investment in this research programme was motivated by Cold War fears over the spread of communism in Asia. In order to understand the history of modern genetics, we therefore need to look to the global conflict which defined the second half of the twentieth century – that is, the Cold War.[11]

Modern genetics was central to the process of state formation during the Cold War, not just in Europe and the United States, but right across Asia, the Middle East, and Latin America. This again is something that is often missed when focusing solely on the discovery of the structure of DNA. After all, most governments were not particularly interested in the structure of DNA – whether it was a double helix or not had no particular bearing on the future of the state. However, governments around the world were interested in the practical benefits offered by recent advances in genetics, particularly when it came to human health and food security.

For many states, the most immediate concern following the Second World War was how to feed the nation. The second half of the twentieth century was a period of massive population growth, with the world population increasing from a little over two billion in 1945 to five billion in 1990. This led to fears of what was referred to as the 'population bomb' – another allusion to the atomic age – in which millions of people might starve to death if the world's food supply did not dramatically increase. By the early 1960s, it was estimated that 80 per cent of the world's population suffered from malnutrition. Most governments recognized that the legitimacy of the state depended upon its ability to provide food for the population. This was particularly the case in Asia and Latin America, where many states had recently either gained independence or gone through a political revolution. With this in mind, governments around the world invested in plant genetics, hoping that it might be possible to engineer new high-yield varieties of crops such as rice and wheat. Much of this work was supported by the Rockefeller Foundation, which helped establish seed banks in countries ranging from Indonesia to Nigeria.[12]

Research in plant genetics was also encouraged by the United States government, which believed that the spread of world hunger would fuel the spread of communism. 'Communism makes attractive promises to underfed peoples,' wrote one prominent American geneticist in the early 1950s. The failure of states to provide enough food for the population was 'a threat to the peace of the world as well as our national security', warned the United States Agency of International Development, which had been set up in 1961 in order to provide scientific and technical assistance to various 'Third World' governments. By the end of the 1960s, there was talk of a 'Green Revolution', in which advances in plant genetics, chemical fertilizers, and irrigation techniques would solve the problem of world hunger. As the term suggests, this was imagined as an antidote to the 'Red Revolution' of the Soviet Union.[13]

Alongside plant genetics, governments invested in the developing area of human genetics. As we've already seen, there was widespread concern over the biological effects of atomic radiation following the bombing of Hiroshima and Nagasaki. These concerns only increased as more and more states developed nuclear weapons and constructed nuclear power stations. The relationship between atomic radiation and human genetics therefore became a national security issue for many governments, an important part of planning the response to any future nuclear war. At the same time, many states believed that by promoting the medical benefits of nuclear research – in both diagnosis and treatment – they could persuade a reluctant public of the advantages of living in an atomic age. Again, this idea was promoted by new international organizations, such as the World Health Organization (established in 1948) and the International Atomic Energy Agency (established in 1957), both of which provided funding for scientists from around the world to conduct research into the medical uses and effects of radiation.

More broadly, governments from Latin America to East Asia believed that modern genetics might bring about dramatic improvements in human health, particularly through a better understanding of inherited diseases. There was also an interest in the use of modern genetics to answer questions concerning national and ethnic identity, another major concern during a period of state formation and mass migration. Today, we know that race is not a meaningful biological category. Indeed, as early as 1950 the United Nations had issued a statement describing race as a 'social myth' rather than a 'biological fact'. Nonetheless, throughout the Cold War, governments around the world organized countless genetic surveys, hoping to distinguish different ethnic groups, such as 'Turks' and 'Arabs', by their genetic make-up, even if this ultimately proved impossible.[14]

As all this suggests, the development of modern genetics was inseparable from Cold War politics. However, whilst many historians have recognized the Cold War as a significant period for the development of modern science, they have tended to focus on scientific advances made in the United States, Europe, and the Soviet Union. In this chapter, I take a different approach, following the history of modern genetics as it developed across Latin America, Asia, and the Middle East. These, after all, were the regions in which the United States and Soviet Union battled for influence, hoping to shape the development, not only of science and technology, but also of world politics. Ultimately, in order to properly understand the history of science during the Cold War, we once again need to think in terms of global history. We begin with a Mexican geneticist on his way to the market.[15]

I. Mutations in Mexico

Efraím Hernández Xolocotzi had been driving for hours. It was an uncomfortable journey, rattling along in his old jeep through the Mexican countryside, but finally he arrived at his destination – a tiny market town in the southern state of Tabasco. Pulling up by the roadside, Hernández hopped out of his vehicle, and began chatting with people in the market. This was a relatively remote part of Mexico, and the locals didn't speak Spanish. Thankfully, Hernández was familiar with the Indigenous language of the region – one of the many dialects of Mayan – and was able to communicate without too much difficulty. He explained that he was looking to buy some maize, and the farmers in the market pointed him towards a stall piled high with cobs of corn. Hernández was delighted. He went over to the stall, examining each of the cobs closely, and then agreed to purchase the whole lot. The farmers must have wondered why he needed so much maize. Still, they didn't worry too much, as he paid a good price. Hernández then headed back to his jeep with the bags of corn, started the engine, and continued on his journey, winding his way towards the Yucatán Peninsula.[16]

Maize had been cultivated in Mexico for thousands of years, long before the arrival of Europeans in the sixteenth century. However, in the middle of the twentieth century, it became the focus of a major scientific investigation, one that formed the basis of the Green Revolution. Hernández was one of a number of geneticists employed to study maize by the Mexican Agricultural Program, which was established in 1943. It was located within the Mexican Ministry of Agriculture, but primarily funded by the Rockefeller Foundation, an American philanthropic organization. As we saw in the previous chapter, the Rockefeller Foundation played a major role in funding international science in the twentieth century. Alongside physics, the Rockefeller Foundation also invested in biology, particularly when there was an obvious practical application, as with plant genetics. In Mexico, the plan was to use the latest techniques of modern genetics in order to improve the yields of staple crops, such as wheat and maize.[17]

The Rockefeller Foundation certainly wanted to improve the lives of Mexican people. However, as with all philanthropy, there was an element of politics to this too. During the middle decades of the twentieth century, the United States grew increasingly concerned about the spread of communism, not just in Europe and Asia, but also closer to home. Following the Mexican Revolution of 1910–20, in which various armed groups fought for control following the overthrow of the president, Mexico seemed to be sliding towards radical socialism. Throughout the 1930s, the Mexican government redistributed large areas of farmland to impoverished peasants, and in 1938 the government appropriated a number of American-owned oil fields. These kinds of land seizures and collective ownership looked a lot like what was going on in the Soviet Union, and by the early 1940s, the American government was worried about the prospect of a communist state on its own border. The director of the Rockefeller Foundation shared these concerns, describing Mexico as 'tainted with Bolshevistic doctrines'. The Mexican Agricultural Program therefore served a number of overlapping political and scientific objectives. Chief amongst them was the notion that stopping the spread of hunger would help stop the spread of communism. By improving the yields of staple crops such as maize, the Rockefeller Foundation hoped to steer Mexico away from socialist politics. 'Hunger is a powerful enemy of peace,' wrote Paul Mangelsdorf, one of the American geneticists who worked for the Mexican Agricultural Program.[18]

Histories of the Green Revolution tend to focus on the contributions of American geneticists like Mangelsdorf. However, the Mexican Agricultural Program also employed a number of Mexican scientists who are today often forgotten. Hernández was one of those scientists. Born in 1913, he came from a humble background. His father was a peasant, possibly of Indigenous descent, and his mother was a teacher. Hernández knew the land well, and learned various Indigenous dialects as a boy, working in the fields with his father. However, Hernández moved around quite a bit, as his father looked for work and tried to stay out of trouble amidst ongoing conflict. In 1923, in the aftermath of the Mexican Revolution, ten-year-old Hernández emigrated with his mother to the United States. He studied at a local school in New Orleans, and later New York, before winning a scholarship to study biology at Cornell University, graduating in 1938. This was a great achievement, especially given that, much like today, Mexicans in the United States suffered from systematic racial discrimination, particularly when it came to education. After graduating from Cornell, Hernández was selected by the Rockefeller Foundation to undertake postgraduate study in genetics at Harvard University, spending two years learning the latest scientific techniques, before returning to Mexico in 1949. He was then hired as an 'Associate Geneticist' by the Mexican Agricultural Program, one of eighteen Mexican scientists who worked on the project.[19]

For two years, Hernández travelled across Latin America, sometimes by jeep, other times by train or boat. He reached as far south as Peru, and even crossed the Gulf of Mexico to collect specimens in Cuba. The son of an Indigenous farmer, Hernández knew more than anyone about the incredible variety of maize found in the region. 'The geographic distribution . . . was known only to E. Hernández Xolocotzi,' recalled one of the American scientists who worked for the programme. Hernández's fluency in a number of Indigenous dialects made tracking down different varieties of maize a lot easier. 'To collect the genetic variation of maize in a given community, one has to be persistent and to use a great deal of tact in dealing with the farmers,' explained Hernández. Even then, he sometimes struggled to convince people to sell him rare specimens, particularly the red varieties of corn used in certain rituals.

37. Different varieties of maize collected by geneticists in Latin America and the United States.

'I could not persuade the indigenous Hulchol population to sell me samples of their ceremonial maize varieties,' noted Hernández after returning empty-handed from a remote region of northwest Mexico. Nonetheless, after two years of intensive work, Hernández and his team had amassed a collection of over 2,000 different varieties of maize from across the Americas.[20]

Up to this point, the work undertaken by the Mexican Agricultural Program was not so different from the kind of natural history we saw in the eighteenth and nineteenth centuries. Hernández was collecting these different varieties in order to categorize them, with the ultimate aim of identifying those which might be crossed in order to increase yields. What was different, however, was the use of recent advances in genetics to guide this research. This was all explained in a book published by the Mexican Agricultural Program titled *Races of Maize in Mexico* (1952). Hernández was one of the co-authors, along with the American geneticists Edwin Wellhausen, Louis Roberts, and Paul Mangelsdorf. In *Races of Maize*, the team explained that the aim of the project was to combine an analysis of the 'vegetable characters of plants' with a study of 'genetic and cytological factors', meaning the examination of individual cells under the microscope. And so, as well as measuring the size of the leaves, tassels, and kernels of each specimen, the team also deployed the latest genetic techniques. One of the methods used by the programme was called Giemsa staining, which had been invented in the early twentieth century by the German chemist Gustav Giemsa. This staining technique made it possible to identify individual chromosomes, as well as distinctive bands of concentrated DNA, under the microscope, and so categorize different varieties of maize on this basis. Hernández himself was familiar with this technique, as it was the sort of thing he would have learned as part of his course on plant genetics at Harvard University in the 1940s.[21]

Through a combination of traditional natural history and modern genetics, the scientists built up a detailed picture of the 'extraordinary diversity of corn' in the Americas. Much of this work confirmed what Hernández had initially suspected, based on his existing knowledge of Mexican agriculture – that over the past 8,000 years the size of the corncob had increased through the hybridization of different varieties. More recent varieties, particularly those that had been bred after the Spanish conquest in the sixteenth century, tended to have larger cobs, whereas more ancient varieties, identified through archaeological remains, tended to have smaller cobs. The geneticists also found distinctive patterns of banding, referred to as 'chromosome knots', when examining the cells of more recent varieties under the microscope, which again seemed to confirm the long-term pattern of development. This genetic analysis then formed the basis of a major effort to increase food production in Mexico over the following decades. Different varieties of maize were selected according to their genetic characteristics. These hybrid varieties, which tended to have increased yields, were then sold to farmers for cultivation. By the late 1960s, improved varieties of maize made up 20 per cent of the annual crop.[22]

The Mexican Agricultural Program did not solve every problem, and not everyone supported its efforts to introduce improved varieties of maize. Food shortages continued in Mexico throughout the 1950s and 1960s, as did land seizures. At the same time, many Mexican scientists, including Hernández, worried that the Rockefeller Foundation was placing too much emphasis on industrial farming at the expense of smallholders and peasants. After all, the hybrid varieties produced by the programme were expensive to purchase. Mexican farmers were also encouraged to use more chemical fertilizers, which these varieties responded well to, despite the fact that overuse could cause long-term ecological damage. There was a related concern that a focus on improved varieties might end up destroying the very genetic diversity on which the Green Revolution depended. Some Mexican scientists even suggested that they would be better off seeking assistance from the Soviet Union, which promoted alternative 'socialist' farming methods, rather than the United States with its more industrial approach. Nonetheless, whatever people thought of it, the Mexican Agricultural Program marked an important moment in the history of modern genetics. The Green Revolution soon extended across Latin America, with the Rockefeller Foundation establishing similar programmes in Brazil and Columbia. As we'll see later in this chapter, the Mexican Agricultural Program in fact provided a model for many governments around the world, including those in Asia and the Middle East.[23]

Alongside work on plant genetics, Mexican scientists also made a number of important contributions to the development of human genetics. This was in part due to the efforts of the Rockefeller Foundation, which funded not only the Mexican Agricultural Program, but also a new Institute for Biomedical Research at the National Autonomous University of Mexico. The Mexican government itself also began investing more and more money in the biomedical sciences during this period. Much of the research in human genetics was undertaken by a team working for the Genetics and Radiobiology Program of the National Commission of Nuclear Energy. Just as we saw in Japan, the development of human genetics in Mexico was closely associated with the growth of nuclear science. Mexico in fact sits on substantial uranium deposits, located in the south of the country, which was one of the many reasons that the United States was so concerned about the future of its neighbour during the Cold War. However, unlike the United States, the Mexican government did not seek to develop nuclear weapons, and instead focused on the use of atomic energy for medical and scientific purposes.[24]

Established in 1960, the Genetics and Radiobiology Program was directed by a Mexican scientist named Alfonso León de Garay. Born in Puebla in 1920, de Garay studied medicine at a local university, before moving to Mexico City in 1947 to practise as a neurologist. It was during this period that the Mexican government started to look for ways to make use of the country's uranium deposits, establishing the National Commission of Nuclear Energy in 1953. De Garay himself became interested in radiobiology (the use of radiation to diagnose and treat medical conditions), as well as the long-term effects of radiation on the human body. In 1957, he was offered a fellowship to support postgraduate study in Europe by the International Atomic Energy Agency. De Garay chose to study at the Galton Laboratory at University College, London, spending three years there learning the latest techniques of genetic science. When de Garay returned to Mexico, he convinced the National Commission of Nuclear Energy to set up the Genetics and Radiobiology Program.[25]

De Garay quickly went about recruiting a team of promising young researchers to work with him. These included Rodolfo Félix Estrada, a graduate of the National Autonomous University who had initially worked as a geneticist for the Mexican Agricultural Program, as well as María Cristina Cortina Durán, who had studied at the National Autonomous University before completing a PhD at the University of Paris in the early 1960s. (Cortina Durán was also one of the first women to be employed by the Genetics and Radiobiology Program.) Together, the team conducted important research into the genetic effects of atomic radiation. Félix Estrada spent most days exposing fruit flies to radiation, and then seeing how long they survived afterwards in order to calculate the effect of different dosages. De Garay and Cortina Durán conducted similar experiments on human tissue, exposing cultured cells to radiation before examining them under the microscope. Through a series of very precise measurements, de Garay demonstrated that atomic radiation had the potential to shorten the length of human chromosomes, and so induce mutations. Cortina Durán focused on the relationship between radiation and cancer, helping to confirm earlier reports that exposure to radiation could induce a specific mutation on chromosome 22 which causes leukaemia. All this research fed into a series of major studies published throughout the 1960s by the United Nations Scientific Committee on the Effects of Atomic Radiation, of which de Garay was a leading member.[26]

*

In 1968, the Genetics and Radiobiology Program began its most ambitious project yet. That October, Mexico City hosted the Summer Olympics, in which over 5,000 athletes from around the world competed. This turned out to be one of the most contentious sporting events of the twentieth century. Just ten days before the opening ceremony, armed police opened fire on a crowd of protestors, in what became known as the Tlatelolco massacre. The crowd had been protesting against the Mexican government, which was widely considered antidemocratic and regularly resorted to police violence in order to maintain power. Political tensions continued throughout the games. South Africa was banned from taking part at the last minute, as other athletes threatened to pull out in protest against the Apartheid regime. And most famously, the African American sprinters Tommie Smith and John Carlos both wore a black glove and raised a fist on the podium following the men's 200 metres, a silent protest against racial injustice in the United States.

Amidst all this ongoing controversy, de Garay convinced the Mexican government to fund a major genetic study of Olympic athletes. The idea was to showcase the best of Mexican science on the world stage. The project would 'benefit all humanity by providing a better understanding of human excellence', explained de Garay. He even claimed that such research might prove useful in the 'early identification and selection of potential athletic types'. With the support of both national and international sporting committees, the scientists at the Genetics and Radiobiology Program set up a temporary laboratory in the Olympic Village, collecting blood samples from 1,256 athletes representing ninety-two different countries. These blood samples then underwent various kinds of genetic tests, including for sickle-cell disease as well as G6PD deficiency (a metabolic condition which causes the breakdown of red blood cells).[27]

This was also the first Summer Olympics in which all female athletes underwent genetic testing for sex. This was done by checking blood samples for the presence or absence of a Y chromosome, which is typically only found in men. (Transgender athletes were excluded from the Olympics until 2004, often on the basis of this kind of genetic testing.) Alongside this, the team of Mexican scientists took bodily measurements and photographs of each of the athletes, building up a detailed picture of what de Garay called 'their genetic and anthropological characteristics'. Those tested included some of the most famous athletes at the time, such as the Czechoslovak gymnast Věra Čáslavská, who turned away during the medal ceremonies in protest against the recent Soviet invasion of her home country, as well as John Carlos himself, who was even named in the final report published by de Garay.[28]

If all this sounds suspiciously like eugenics, that's because in many ways it was. After all, de Garay had studied at the Galton Laboratory in London, which was named after Francis Galton, the nineteenth-century founder of the eugenics movement. Galton infamously argued that human populations should be 'improved' through selective breeding. In the final report, de Garay cited Galton approvingly, as well as a more recent book published by the British Eugenics Society titled *Genetic and Environmental Factors in Human Ability* (1966). Today, many scientists like to think that eugenics simply disappeared after the Second World War, as it became associated with the atrocities committed by the Nazis during the Holocaust. Unfortunately, this is not the case. Cold War tensions reinforced concerns about the 'fitness' of competing human populations, leading many scientists to try and identify specific genes that might code for more or less desirable traits. There was even talk in the 1960s of a 'new eugenics', based on the latest techniques of molecular biology. It all turned out to be a false promise. De Garay himself admitted that 'there has been no good correlation found between any specific genes and any specific athletic achievement'. Nonetheless, the presence of widespread genetic testing at the 1968 Summer Olympics is an important reminder of the continued influence of eugenics during the second half of the twentieth century – a damaging legacy that the scientific world is still struggling to grapple with.[29]

By the early 1970s, Mexico was firmly established as a leading international centre for the study of genetics. This is a story which began with the Green Revolution. The hope was that by solving 'the problem of food', geneticists could divert Mexico away from socialism. The Mexican Agricultural Program, funded by the Rockefeller Foundation, also provided an opportunity for a new generation of Mexican scientists to undertake advanced training in genetics. A similar trend followed across Latin America, with leading scientists in Argentina and Brazil also training in the United States before returning home to set up new genetics laboratories. There was even a Latin American Society of Genetics, established in 1969, to help foster scientific links across the region. During the same period, Latin American governments invested in the field of human genetics. Mexican scientists often walked a fine line between genetics and eugenics. Such concerns about health and identity were not confined to Mexico. Throughout the Cold War, states around the world believed that genetics might unlock the key to a happier and healthier population. In the following section, we explore how similar concerns over food security and human health shaped the development of genetics in postcolonial India.[30]

II. Indian Genetics after Independence

Mankombu Sambasivan Swaminathan never forgot the photographs of starving children, emaciated bodies lying by the roadside. Between 1943 and 1944, three million Indians died in what became known as the Great Bengal Famine. Initially, the British colonial government tried to keep the news from getting out. But in August 1943, a newspaper in Calcutta printed a harrowing image of a Bengali girl stooped over the dead bodies of two young children. This photograph, along with continued reports of British mismanagement of the crisis, galvanized the anticolonial movement in India. Many recognized that the famine was not simply the result of a poor harvest or drought. Rather, the British had seized food supplies in order to support troops during the Second World War, leaving millions of Indians to starve. This was part of a long history of colonial mismanagement, dating back to the eighteenth century, which had caused multiple waves of famine.

Swaminathan lived in the Madras Presidency in the southeast of India. But he was nonetheless shocked and angered by the British government's response to the famine, particularly after seeing photographs of starving children in a local newspaper. The famine was a 'man-made problem', he declared. Swaminathan was in fact already committed to the cause of Indian independence. His father was a keen follower of Mohandas Gandhi, and the family all wore homespun cloth in support of the *swadeshi* movement to boycott British goods. Swaminathan also organized a student strike as part of Gandhi's Quit India campaign in

1942, walking out of class at the University of Travancore, where he was studying zoology. The Great Bengal Famine really just confirmed what Swaminathan had always believed. That the British only looked after themselves. Indians would not prosper until they were free from colonial rule.[31]

As we learned in the previous chapter, many Indian scientists in this period saw their work as part of the fight against colonialism. This was just as true of biology as it was of physics. Born in the small temple town of Kumbakonam in 1925, Swaminathan went on to become one of the world's leading plant geneticists, helping to bring the Green Revolution to India. His interest in plant genetics was directly motivated by his interest in Indian politics. Initially, Swaminathan had wanted to become a zoologist, but after hearing about the Great Bengal Famine in 1943, he decided to switch subjects and undertake a postgraduate degree in agricultural science. He hoped that by better understanding the genetics of staple crops such as rice and wheat, independent India could avoid the kind of devastating famines that were all too common under British rule. 'Man-made problems have to have man-made solutions,' he argued. In the summer of 1947, Swaminathan graduated with an MSc from the University of Madras. That same summer, on 15 August, India finally gained its independence from Britain. This marked the end of nearly 200 years of colonial rule, and Swaminathan celebrated in the street with his friends and family. Still, the festivities couldn't go on for too long. Swaminathan, like many Indian scientists, now turned to the practical task of building a new nation.[32]

Shortly after graduating, he joined the Indian Agricultural Research Institute in Delhi. Working alongside a team of committed Indian geneticists, Swaminathan began to tackle the problem of how to feed a nation of over 300 million people. Unsurprisingly, this was a major priority for the Indian government in the years immediately following independence. After all, anticolonial nationalists had spent the last few decades criticizing the British for failing to supply enough food. It was therefore essential for the legitimacy of the Indian state to avoid another famine. In fact, this research was considered so important that in 1948 the Prime Minister, Jawaharlal Nehru, personally visited the Indian Agricultural Research Institute in order to better understand the work being done there. Nehru himself had great faith in the power of modern science to support the new nation, particularly when it came to combating famine. 'Poverty has ceased to be inevitable now because of science,' he declared.[33]

Swaminathan soon realized that, in order to feed the nation, he would need to undertake further training in plant genetics. With this in mind, he travelled to Britain in 1950 and began studying for a PhD at the University of Cambridge. The focus of his research was on a phenomenon known as 'polyploidy', which is when a plant has double the usual number of chromosomes. This was a topic with a direct practical application, as plants with polyploidy often have higher yields. Swaminathan spent two years examining the cells of different plants under the microscope, carefully counting the number of chromosomes. He would then cross-reference this against the characteristics of each variety, particularly the yield, building up a detailed picture of the effects of polyploidy. In 1952, Swaminathan graduated from Cambridge, one of the first of a new generation of Indian scientists, no longer a colonial subject, but rather a citizen of an independent state. Swaminathan then spent a year doing postdoctoral work at the University of Wisconsin in the United States. He was even offered a job there. However, Swaminathan never forgot why he had become a scientist. 'I asked myself, why did I study genetics? It was to produce enough food in India. So I came back,' he later explained.[34]

It was around this time that he first learned about the work being done by the Mexican Agricultural Program. Excited by the potential of the Green Revolution, Swaminathan wrote to Norman Borlaug, one of the American geneticists working in Mexico, asking for assistance. This was part of a long and fruitful scientific exchange between India and Mexico, one that continues to this day. In March 1963, Borlaug visited the Indian Agricultural Research Institute in Delhi, bringing with him some samples of improved varieties of Mexican wheat in his suitcase. 'What Mexico did, your country can also do, except that yours should do it in half the time,' Borlaug told the Indian scientists in Delhi. Encouraged by Borlaug's enthusiasm, Swaminathan and his team began experimenting with these new varieties, planting seeds in test beds at the Indian Agricultural Research Institute. The Rockefeller Foundation also provided funding to allow a team of Indian geneticists to visit Mexico and learn more about the work being done by the Mexican

Agricultural Program. The results were very promising. Swaminathan found that by crossing the varieties of wheat used in Mexico with existing Indian varieties, he could produce new hybrids with increased yields that were also suitable for the local soil and climate.[35]

There was, however, a problem. These new hybrid varieties of wheat tended to produce a red-coloured flour. In Mexico, no one really minded. But in India, consumers preferred their flour to be much lighter, particularly for making traditional breads such as chapatis. This simple difference in colour threatened to derail the whole programme. That was until an Indian geneticist named Dilbagh Singh Athwal began a series of experiments using X-rays. Athwal, who had studied in Australia at the University of Sydney in the 1950s, knew that it was possible to induce genetic mutations by exposing plants to radiation. Perhaps, he thought, it might be possible to change the colour of wheat in this way? After a bit of trial and error, Athwal finally succeeded in inducing the mutation he hoped for – a variety of high-yield wheat that produced a light golden flour. With this problem solved, the Indian government began scaling up the agricultural programme in the late 1960s. By 1968, wheat production in India had increased by over 40 per cent. And by 1971, India was finally producing enough food to stop wheat imports from abroad. As elsewhere, the Green Revolution caused a fair amount of controversy in India. Smaller farmers were pushed out of the market, whilst the introduction of high-yield varieties went hand in hand with the overuse of chemical fertilizers, causing ecological damage. But for India's political leaders, if not its farmers, this was a price worth paying for food security.[36]

Much as we saw in Mexico, the development of modern genetics in India was closely associated with concerns over the supply of food. The Indian Agricultural Research Institute, originally founded in 1911 by the colonial government, soon emerged as a leading centre for the study of plant genetics in independent India. Scientists working there made a number of important breakthroughs, particularly in developing hybrid varieties of wheat suitable for the South Asian market. This work was only made possible thanks to a massive increase in science funding following independence. Between 1948 and 1958, the national science budget in India increased by close to a factor of ten. This reflected a conviction, promoted in particular by the Prime Minister, Jawaharlal Nehru, that India needed to invest in modern science and technology in order to escape from the problems of the past. Without the 'spirit of science', India was 'doomed to decay', Nehru warned. With this in mind, the Indian government initiated a series of 'Five-Year Plans' with the aim of building up scientific capacity. This initiative was directly inspired by the Soviet Union, which had run a series of Five-Year Plans since the late 1920s. Nehru himself was not a communist, but he was nonetheless sympathetic towards socialism, and believed that India had as much to learn from the Soviet Union as it did from the United States. In fact, during the 1950s a number of Indian geneticists were sent to Moscow, as well as Beijing, in order to learn about the agricultural science being done in communist states.[37]

The First Five-Year Plan of 1951 to 1956 saw the creation of a number of new scientific institutions. Amongst these was the Atomic Energy Establishment, set up in 1954 on the outskirts of Bombay. Following independence, the Indian government invested significantly in atomic research. The hope was that nuclear power might provide a secure source of energy for the new nation, and thus reduce reliance on imports of petroleum and gas. At the same time, the Indian government secretly initiated a nuclear weapons programme, conducting its first successful test in May 1974. Just as we've seen elsewhere, the development of atomic science in India went alongside the development of modern genetics. In 1958, Nehru himself ordered the Atomic Energy Establishment to undertake a study of 'the genetic effects of these explosions on the present and future generations'. This eventually led to the creation of a dedicated Molecular Biology Unit within the Atomic Energy Establishment.[38]

The new Molecular Biology Unit was directed by an outstanding Indian geneticist named Obaid Siddiqi. Born in 1932 in the northern state of Uttar Pradesh, Siddiqi came close to leaving India as a young man. In 1947, the British partitioned the Indian subcontinent into Muslim-majority Pakistan and Hindu-majority India. This resulted in one of the largest migration events in modern history, in which over fourteen million people moved from one country to the other. Religious violence broke out across the subcontinent, and hundreds of thousands of people lost their lives. Siddiqi was a Muslim and much of his extended family moved to Pakistan. It was a close call, but in the end Siddiqi decided to stay in India in order to complete his education. He enrolled at Aligarh Muslim University in Uttar Pradesh, and began studying for a degree in biology. During his time there, Siddiqi became involved in radical politics. In 1949, whilst still at university, he was arrested and held in a local prison along with a group of communist activists. Siddiqi later recalled being beaten by the guards. In the end, after two years, he was released without charge.[39]

Given his experience in prison, Siddiqi might well have been tempted to relocate to Pakistan. However, like many Indian Muslims, he ultimately considered India as his home, and saw no reason why he should move to a foreign country. Siddiqi was in fact rather patriotic. Through his scientific work, he hoped to contribute to the development of the new nation. And so, after graduating from Aligarh Muslim University in 1951, Siddiqi joined the Indian Agricultural Research Institute in Delhi. He was planning on dedicating his life to plant genetics. That was until 1954, when a freak hail storm destroyed the entire crop that Siddiqi had been working on. With his experiment ruined, he started to reflect on what he really wanted to do with his scientific career. He had just read about the discovery of the structure of DNA, which was announced in April 1953. Excited by this recent breakthrough, Siddiqi decided to retrain. In 1958, he moved to Scotland and began a PhD in molecular biology at the University of Glasgow.[40]

After receiving his PhD in 1961, Siddiqi was offered a position as a researcher at the University of Pennsylvania. By this time, it was becoming increasingly common for Indian scientists to undertake postdoctoral work in the United States. The American government was keen to support Indian scientific development, again in the hope of stemming the spread of communism in Asia. For their part, many Indian scientists saw the United States as an attractive alternative to Britain, which after all was a former colonial power. Siddiqi thrived amongst the American scientific community. He even got to meet his scientific hero, the American biologist James Watson, one of the co-authors of the original 1953 paper on the structure of DNA. It was also in the United States that Siddiqi made his first major breakthrough. Working with the American geneticist Alan Garen at the University of Pennsylvania, Siddiqi discovered a natural mechanism through which organisms are sometimes protected against certain genetic mutations. In some cases, a second mutation, known as a 'suppressor' mutation, cancels out the effect of an earlier more damaging one. Siddiqi and Garen worked on bacteria, but suppressor mutations occur in all organisms. Their findings therefore had broader implications for the study of human health, allowing scientists to pinpoint the effects of particular genetic mutations.[41]

By the early 1960s, Obaid Siddiqi was looking to return to India. However, at this time there were no laboratories in the country that were suitable for conducting cutting-edge research in molecular biology. With this in mind, he wrote to the director of the Atomic Energy Establishment in Bombay, a nuclear physicist named Homi Bhabha. 'I feel that in India, both from the point of view of facilities and the intellectual environment, the laboratories of the physical sciences would be more suitable places for developing molecular biology than the traditional biological institutions,' explained Siddiqi. The timing was just right. At Nehru's request, Bhabha had recently set up the Molecular Biology Unit within the Atomic Energy Establishment. In the summer of 1962, Bhabha invited Siddiqi to return to India to direct the new laboratory, which was soon relocated to the nearby Tata Institute of Fundamental Research. 'I am very interested personally in supporting the work in India in molecular biology and genetics,' wrote Bhabha, who at the time was helping to develop India's nuclear energy programme.[42]

Working in Bombay throughout the 1970s, Siddiqi made a series of major scientific breakthroughs. Much of this work was in the growing field of neurogenetics. During the Cold War, scientists worried about how genetic mutations, such as those caused by atomic radiation and chemical warfare, might impact on the function of the nervous system. This was a particularly pressing issue in the early 1970s, as the United States had recently deployed a devastating chemical weapon codenamed 'Agent Orange' in the Vietnam War. The chemical, sprayed from American helicopters across Vietnam, was used to destroy foliage and thus reduce cover for enemy soldiers. However, Agent Orange was later proved to cause cancer in humans alongside chronic inflammation of the skin. There was also a concern over the increased use of chemical fertilizers and pesticides following the Green Revolution. Some of these were also known to induce genetic mutations. In fact, Agent Orange itself was originally developed as a chemical herbicide.

Siddiqi began studying the effects of chemically induced mutations on the nervous system. Like many geneticists in this period, he chose to work on the fruit fly. These are easy to breed and have a small number of chromosomes, making genetic analysis more straightforward. In his laboratory in Bombay, he started exposing fruit fly larvae to a dangerous chemical called ethyl methane-sulphonate, or EMS. He also began exchanging letters with Seymour Benzer, an American geneticist based at the California Institute of Technology, where Siddiqi had spent a year as a visiting professor in 1968. Working together, Siddiqi and Benzer showed that it was possible to chemically induce a genetic mutation which causes paralysis in the fruit fly. The genes identified by Siddiqi and Benzer turned out to regulate the conduction of electrical signals within the nerves of the fly, hence the paralysis. This was an incredibly important discovery, one that opened up a whole new field of research. Up to this point, the fruit fly had mainly been used to study the genetics of relatively simple characteristics, such as the colour of the eye. Now, scientists began to study much more complex characteristics, such as the way genes regulate the development of the nervous system.[43]

Alongside his work with American geneticists, Obaid Siddiqi also conducted a number of important experiments in collaboration with an Indian geneticist named Veronica Rodrigues. Born in 1953, Rodrigues was one of a new cohort of Indian women to train in the sciences following independence. As we saw in the previous chapter, a small number of Indian women did enter the world of science in the early decades of the twentieth century. However, they faced significant barriers, not least the sexist attitudes of their male colleagues. Problems of sexual discrimination did not simply disappear following independence in 1947. Even by 1975, women still made up less than 25 per cent of those studying science at university in India. Nonetheless, thanks to the efforts of campaign groups such as the Indian Women Scientists' Association, this picture began to improve. Gradually, more and more Indian women were able to pursue a career in science. Some, like Rodrigues, went on to transform an entire field.[44]

Rodrigues is also another good example of how the wider world of international politics shaped the development of science during the Cold War. She had in fact spent the first twenty years of her life outside of India. Born in Kenya, she was the daughter of Goan immigrants who had travelled to East Africa in search of work. Rodrigues's parents most likely migrated to Kenya during the early decades of the twentieth century, when the British Empire recruited hundreds of thousands of Indian labourers to work in East Africa. The family was relatively poor, and Rodrigues's early years were tough. Thankfully, her mother and father managed to scrape together just enough money to send Rodrigues to a local school in Nairobi. It was here that she first developed a love of science. In 1971, Rodrigues went on to study at the University of East Africa in Uganda. However, shortly after arriving in the capital city of Kampala, Rodrigues was forced to flee. This was the year that Idi Amin launched a military coup in Uganda. Hundreds of thousands of people were killed in the violence that followed. Amongst other ethnic groups, Amin targeted the Asian population in Uganda. In August 1972, all Indians were ordered to leave the country. Rodrigues, however, didn't give up on her dream of studying science at university. Instead of returning to Nairobi, she decided to travel to Ireland, enrolling for a degree in biology at Trinity College, Dublin.[45]

Rodrigues graduated in 1976. By this time, she was technically stateless. Her student visa in Ireland had expired, and Rodrigues couldn't return to Uganda or Kenya. Britain had also recently tightened its immigration laws in order to prevent those from former colonies settling in the country. With nowhere else to go, Rodrigues started to think about moving to India. She wrote to the Tata Institute of Fundamental Research in Bombay, asking if there might be a place for her on the PhD programme there. Impressed by her determination to pursue a career in science, Siddiqi agreed to take Rodrigues on as a student in the Molecular Biology Unit. She arrived in Bombay at the end of 1976, aged twenty-three. This was the very first time that Rodrigues had ever set foot in India.[46]

Rodrigues's major breakthrough came in 1978, whilst she was still a PhD student. Through a series of careful experiments, she was able to isolate the particular genetic mutations that affect the sense of taste and smell in fruit flies. Like Siddiqi, Rodrigues used chemicals to induce genetic mutations in the flies. She then tested to see whether the flies had a preference for or against certain substances, such as sugar or quinine. Once she had done this, Rodrigues undertook a minute study of the anatomy of the mutant flies. This was the key bit of the research. Rodrigues was ultimately able to show that particular genes controlled for the development of certain sensors on the fly antennae. She was even able to map these genes to a particular region on one of the chromosomes. This was a foundational moment in the history of neurogenetics. Rodrigues proved that it was possible to trace the effect of a genetic mutation all the way through the nervous system, right down to the level of being able to detect a certain taste or smell.[47]

When India gained independence in 1947 it marked an important moment, not just in the political history of the nation, but also in the history of science. The Prime Minister of India, Jawaharlal Nehru, had himself studied Natural Sciences at the University of Cambridge, and was passionate about the possibility of science to transform the new nation. Through a series of Five-Year Plans, modelled on those of the Soviet Union, the Indian government began to build up scientific capacity, establishing new laboratories and institutions. These would be 'temples of science built for the service of our motherland', declared Nehru in 1954. The focus of much of this early scientific work was on solving the problem of hunger. By the early 1980s, India had emerged as an important research hub in the region, with scientists from Bangladesh, Sri Lanka, Burma, Vietnam, and Thailand all travelling to study plant genetics at the Indian Agricultural Research Institute.[48]

Decolonization fundamentally shaped the development of modern science in twentieth-century India. Obaid Siddiqi, an Indian Muslim, narrowly escaped the violence that followed the Partition of India in 1947. Veronica Rodrigues too lived through the end of empire. Her life represents a period in the history of science that I think we urgently need to remember today. This is a history of how the end of empire transformed promising young scientists into stateless migrants. But it is also a history of how those same scientists seized the opportunity of independence to forge a new path. In the following section, we explore another side to the history of science during the Cold War. Across the border, scientists in China were grappling with one of the most significant political events of the twentieth century – the rise of the Chinese Communist Party.[49]

III. Communist Genetics under Chairman Mao

Li Jingzhun had been planning his escape for months. Finally, in February 1950, he decided it was no longer safe to stay in China. Accompanied by his wife and four-year-old daughter, Li boarded a train from Beijing. It was around Chinese New Year, so he hoped that the authorities would not notice that he was gone before it was too late. Over the following weeks, Li and his family travelled south, before finally reaching Canton. Then, in the dead of night, they crossed over the border into Hong Kong, which at this time was still a British colony. Li's daughter was so exhausted that he had to carry her on his shoulders for the final part of the journey. And when Li arrived in Hong Kong, he collapsed, overwhelmed with tiredness and emotion. He was finally free. Free from political persecution. And free to carry out his scientific research in peace.[50]

One of the leading geneticists of the twentieth century, Li found himself an enemy of the state when the Chinese Communist Party came to power in 1949. Prior to the outbreak of the Second World War, Li had completed a PhD in plant genetics at Cornell University in the United States. He was one of the new generation of Chinese scientists that we met in the previous chapter, those who trained abroad in the first few decades of the twentieth century. However, when Li returned to China in the early 1940s, he found that the country had descended into civil war. Over the following years, the Chinese Communist Party, led by Mao Zedong, secured much of the mainland, whilst the Nationalist Party retreated to the island of Taiwan. On 1 October 1949, Mao declared the foundation of the People's Republic of China. The world's most populous country was now the world's largest communist state.[51]

At the time, Li was teaching genetics at Beijing Agricultural University. He soon learned that he was no longer welcome. At the end of October, the new dean of the university, a Chinese Communist Party official, called all the staff to a meeting. Li and the others were told that they needed to stop teaching Mendelian genetics. (This was the most widely accepted genetic theory of the time, in which characteristics are passed on exclusively through genetic material contained within the chromosomes.) Instead, the scientists at Beijing Agricultural University were ordered to teach an alternative genetic theory promoted by a Soviet scientist named Trofim Lysenko. This new theory was apparently 'a great achievement of the conscious, thorough application of Marxism and Leninism to the biological sciences'. Li was horrified. Lysenko was infamous. At a meeting of the Leningrad Academy of Agricultural Sciences in August 1948, Lysenko had given a speech denouncing the work of European and American geneticists. According to Lysenko, Mendelian genetics was completely incompatible with Marxism. It was an 'idealist doctrine', he claimed. The concept of the 'gene' was an abstraction from 'the real regularities of animate nature'. Instead, Lysenko tried to resurrect the old idea of the inheritance of acquired characteristics, which he believed was much more in keeping with Marxist philosophy, with its focus on materialism and collective action. Anyone who disagreed would be sent to the Gulag.[52]

Throughout the 1950s, Lysenko's theory, which turned out to be completely false, spread throughout China. The official Chinese Communist Party newspaper, the *People's Daily*, told readers that Lysenkoism represented 'a fundamental revolution in biology' and that 'the old genetics . . . must be thoroughly reformed'. Similarly, another newspaper proudly announced that 'the reactionary theories of heredity propounded by Mendel . . . have already been deleted from the biology textbooks'. During the same period, Soviet scientists were invited to lecture at Chinese universities, whilst Russian textbooks were translated into Chinese. A cinema in Beijing even screened a Soviet propaganda film, dubbed in Chinese, which explained the basics of Lysenko's theory. This was all part of Mao's attempt to forge an alliance with the Soviet Union during the early 1950s. China needed to 'learn from the advanced experience of the Soviet Union', declared Mao. This would help accelerate Chinese scientific development as well as 'strengthen our solidarity with the Soviet Union . . . [and] with all the socialist countries'.[53]

Li chose to leave China rather than be forced to teach the 'new genetics' promoted by the Chinese Communist Party. Shortly after escaping to Hong Kong, he wrote a brief letter describing his experiences. It was printed in the *Journal of Heredity*, the official publication of the American Genetic Association, under the title 'Genetics Dies in China'. This was the first time that the international scientific community had heard about the spread of Lysenkoism in China. Beijing Agricultural University had been 'completely taken over by the Communists . . . the courses on Mendelian genetics were suspended immediately', reported Li. He also described the strict ideological conformity imposed by the Chinese Communist Party, explaining that 'one must declare his allegiance to the Lysenko theory or leave. The latter has been my choice.' Li then ended the letter with an appeal for help. 'If I may be of any service to any of the American universities or institutions that you know of, I should be only too glad to offer them,' he wrote. The following year, Li was appointed as a professor at the University of Pittsburgh, where he remained for the rest of his career, conducting pioneering work on the use of new statistical methods in population genetics. He never returned to China.[54]

Li Jingzhun was just one of a number of scientists who fled China following the rise of Chairman Mao in 1949. The persecution he suffered is another reminder of how ideological conflict shaped the development of twentieth-century science, particularly during the Cold War. Throughout the 1950s, the United States government took great pride in helping scientists from around the world escape political repression. Li's experience was an example of the need to 'uphold scientific freedom and to challenge totalitarianism', argued one prominent American geneticist.[55]

It is important to remember, however, that this is only one side of the story. True, scientists in China faced exceptionally challenging conditions. Many were removed from their posts, never to be seen again. And even those that did follow the party line found themselves cut off from the wider world, with limited access to laboratory equipment and international scientific journals. Still, we should not assume that, simply because they were working in a communist state, Chinese scientists in this period were unable to do any worthwhile research. Such a view simply reinforces a Cold War narrative which portrayed China as a backward nation, opposed to modernization. This narrative also does a disservice to the many Chinese scientists who, despite the extraordinary circumstances, managed to make a number of important contributions to the development of modern science. Ultimately, in order to properly understand the history of science in twentieth-century China, we need to get the balance right. We need to acknowledge the oppressive nature of the communist regime, particularly under Chairman Mao. But we also need to recognize the achievements of Chinese scientists, rather than simply discounting them.[56]

Contrary to popular belief, Mao himself was not opposed to modern science. In fact, like many socialist leaders around the world, Mao believed that science would flourish under communism. 'We can assuredly build a socialist state with modern industry, modern agriculture, and modern science,' declared Mao in 1957. He repeated this claim a few years later, arguing that 'scientific experiment' was one of the 'three great revolutionary movements for building a mighty socialist country'. With this in mind, the Chinese government invested a significant amount of money in the development of new scientific institutions, tripling the national science budget during the First Five-Year Plan of 1953–7. In 1959, Mao even authorized the creation of a new Institute of Genetics, affiliated with the Chinese Academy of Sciences in Beijing. And in 1967, China conducted its first successful nuclear weapons test, astounding many American policymakers who had assumed that the country was simply incapable of producing any kind of advanced technology.[57]

During the same period, the Chinese Communist Party moved away from its commitment to Lysenkoism. This was in part due to the changing geopolitical situation. In 1956, Mao began to break with the Soviet Union, which he believed was insufficiently committed to the cause of world revolution. That same year, Mao gave an influential speech in which he recognized the need for greater intellectual diversity, particularly when it came to science. 'Let a hundred flowers bloom, and a hundred schools of thought contend,' he declared. This prompted a group of Chinese scientists to organize a major conference on the future of genetics. During the opening session, a Chinese Communist Party official clearly signalled that Lysenkoism was no longer state policy. 'Our Party does not want to interfere in the debate on genetics like the Soviet Party,' he explained. The official even put a Marxist spin on the recent discovery of the structure of DNA, pointing out that this proved that the concept of the gene had a material basis. (At the core of Marxist philosophy was the idea that everything, even scientific concepts like the 'gene', was a product of the material conditions of life. As Marx put it, 'it is not the consciousness of men that determines their existence, but their social existence that determines their consciousness'.) The official then concluded with a reference to Mao's speech, stating that in science, as elsewhere, the policy of the Chinese Communist Party was to 'let a hundred flowers bloom'.[58]

Much as we've seen elsewhere, renewed interest in modern genetics in China was largely motivated by concerns over the supply of food. During the Second World War, China had suffered a major famine, in which over two million people died. This was then followed by the Great Chinese Famine of 1959–61. Over the course of three years, well over fifteen million people died in what turned out to be one of the worst famines in human history. The famine was caused by a number of different factors, but chief amongst them was the Chinese Communist Party's policy of redirecting rural farmers towards the production of iron and steel rather than food. This was then exacerbated by the adoption of Lysenkoism, as Chinese agricultural scientists spent much of the 1950s wasting their time on futile experiments. Naturally, Mao was unwilling to admit responsibility. Still, the Chinese Communist Party recognized that it could not afford a repeat of such a disaster, investing significantly in the development of agricultural science and modern genetics from the 1960s onwards.[59]

Yuan Longping was haunted by memories of the Great Chinese Famine. He later recalled seeing bodies lying by the roadside and children eating soil in a desperate attempt to survive. It was this grim experience that motivated Yuan to search for a new way to increase crop yields in China. Today, he is remembered for having developed the first varieties of hybrid rice, an important breakthrough that many scientists in Europe and the United States thought was impossible. Born in 1930 in Beijing, Yuan represents the other side of the history of genetics in China. Unlike most of the previous generation of Chinese scientists, Yuan was not educated in the United States. Instead, he studied plant genetics at Southwestern Agricultural University in the early 1950s, one of the new institutions established by the Chinese Communist Party.

Yuan was studying at a time when Lysenkoism still dominated the teaching of genetics in China. He was even required to learn Russian whilst at university. However, one of Yuan's lecturers secretly introduced him to Mendelian genetics, sharing an old Chinese translation of a popular American textbook with him. This was a risky thing to be involved in, and the lecturer was later removed from his post, never to be seen again. Yuan soon learned to keep his head down, although he kept reading about Mendel, hiding his copy of the textbook by wrapping it up in a recent edition of the *People's Daily*.[60]

After graduating in 1953, Yuan was assigned to work at the Anjiang Agricultural School, located in an old Buddhist temple in the far west of Hunan Province. Even in this remote part of China, Lysenkoism influenced how geneticists conducted their research. Yuan was asked to conduct bizarre experiments, grafting a tomato plant onto a sweet potato, in the hope of producing a new hybrid. Needless to say, the experiments failed. A few years later, the Great Chinese Famine reached Hunan. Yuan witnessed the devastation first hand. 'I saw five people fall down dying, on the roadside, at the ridge of the fields or under a bridge,' he later recollected. Following the great famine of 1959–61, Yuan was finally able to start teaching Mendelian genetics at the Anjiang Agricultural School. As mentioned earlier, by this time China had split from the Soviet Union, and so it was once again safe to criticize Lysenkoism. Nonetheless, Yuan was still expected to follow a socialist model of scientific research. The Chinese Communist Party promoted the idea of 'mass science', in which 'old peasants' and 'educated youths' would learn from one another. 'To a large extent, inventions come not from experts or scholars but from the working people,' explained the *People's Daily*. University-educated scientists like Yuan were therefore expected to spend time in the fields, learning from rural farmers. Chairman Mao referred to this as the 'rural scientific experiment movement'.[61]

Yuan therefore spent much of his time in the surrounding fields, speaking with farmers and instructing peasants in the basics of Mendelian genetics. This, as it turned out, proved rather useful. In the summer of 1964, whilst walking through the local paddy fields, Yuan came across an unusual variety of rice plant, with strangely shaped flowers. Intrigued, he took the specimen back to the Anjiang Agricultural

School. Flowers naturally have both male and female reproductive organs. The male organs, known as the anthers, produce the pollen, and the female organs, known as the carpels, receive the pollen. Examining the strange specimen of rice plant under the microscope, Yuan quickly noticed that the anthers were all shrivelled up and not producing any pollen. This suggested that the plant was what is known as a 'male sterile'.[62]

Yuan immediately recognized the importance of what he had discovered. Rice is a naturally self-pollinating plant. Scientists had therefore assumed that it was practically impossible to breed hybrid rice, as the plant would always pollinate itself before there was a chance to cross it with a different variety. This is one of the reasons why geneticists in the United States and Mexico had focused their efforts on maize, which cross-pollinates naturally. Yuan, however, suddenly realized that it might be possible to breed hybrid rice after all. In the fields of Hunan, he had discovered a rice plant that, simply because of a random genetic mutation, was unable to pollinate itself. Crucially, the female reproductive organs of the plant were still intact and capable of being pollinated by another rice plant. In theory, it would therefore be feasible to select a different variety of rice and cross it with this male sterile specimen, thus creating what many thought was impossible – an improved hybrid variety of rice.[63]

In 1966, Yuan reported his discovery in the *Chinese Science Bulletin*, the main periodical published by the Chinese Academy of Sciences in Beijing. This marked the beginning of a major programme to breed hybrid rice in China. In many ways, this was an example of Mao's 'mass science' in action. Yuan had made his discovery whilst working alongside peasant farmers in rural China. And in order to scale up the programme, he needed to train those same peasant farmers to identify and collect more examples of the male sterile rice plant. Over the following years, Yuan and his team collected over 14,000 specimens, of which just five turned out to be suitable for cultivation. This was genetic science, but not as we often think of it. There was no high-tech laboratory, no X-rays, and no chemicals. Instead, Yuan brought genetics back into the field.[64]

Despite his apparent commitment to socialist science, Yuan was not immune from political persecution. One day in 1969 he arrived at work to find a handmade poster pasted to the wall. It read, 'Down with Yuan Longping, active counter-revolutionary!' This was at the height of a movement known as the Cultural Revolution in which Chairman Mao led a campaign against what he saw as the remaining elements of bourgeois society. Intellectuals in particular were targeted, as well as those who came from more middle-class backgrounds. Students at universities across China were encouraged to identify potential 'counter-revolutionaries' and report them to the authorities. Yuan's university education, as well as his interest in European and American genetics, marked him out. A few weeks later, the head of the Anjiang Agricultural School ordered Yuan to resign from his post. He was told that he had been reassigned to work in a nearby coalmine.[65]

During the Cultural Revolution, thousands of Chinese scientists were 'sent down' to work in similar labour camps. Many were never seen again. Yuan, however, was one of the lucky ones. After two months of backbreaking work, he was suddenly released and told to return to the Anjiang Agricultural School. It was his science that had saved him. An official working in the State Science and Technology Commission had read Yuan's article in the *Chinese Science Bulletin* and recognized its importance for the future of agriculture in China. The official then wrote a telegram to the authorities in Anjiang, ordering Yuan's release. With the approval of the Chinese Communist Party, Yuan was finally able to continue his research in peace. It took a bit of trial and error, crossing different varieties, but in 1973, Yuan successfully developed the world's first hybrid rice plant that could be used in agricultural production, something that many scientists had previously thought was impossible.[66]

In many ways, the development of modern genetics in the People's Republic of China was exceptional. During the early 1950s, the Chinese Communist Party promoted the discredited theories of the Soviet biologist Trofim Lysenko, causing a number of leading geneticists to flee the country. Even after the Chinese Communist Party rejected Lysenkoism, genetics still proved a source of deep ideological conflict. The geneticist Yuan Longping, who otherwise acted as a model socialist scientist, only narrowly escaped the ideological purges of the Cultural Revolution. All this was certainly extraordinary, matched only by the experience of the Soviet Union. Yet in many other ways, the history of modern genetics in China followed a very similar pattern to that which we've seen elsewhere. Rather than regarding China as an aberration, we should therefore try and understand how it fits into a broader history of Cold War science.

In China, just like in Mexico and India, the development of modern genetics was closely tied to the practical demands of the state, particularly the demand for increased food production. Somewhat ironically, then, the Green Revolution – which the United States promoted as part of its fight against communism – ultimately found one of its biggest supporters in none other than Chairman Mao. Throughout the 1960s, Mao endorsed what he called 'scientific farming'. His hope was that the development of improved varieties of staple crops, along with the use of chemical fertilizers and pesticides, would help modernize Chinese agriculture and feed the nation. It seemed to work. Today, the most recent version of Yuan's hybrid rice is grown, not only in China, but also in India, Vietnam, and the Philippines, helping to feed hundreds of millions of people across Asia.[67]

IV. Genetics and the State of Israel

Every morning, Joseph Gurevitch would get in his car and drive down to one of the immigration camps on the outskirts of Jerusalem. Once there, he would begin his medical rounds – examining patients, administering vaccinations, and taking blood samples. Between 1949 and 1951, over 600,000 Jewish immigrants arrived in Israel. The vast majority passed through one of the camps set up by the government following the foundation of the State of Israel in 1948. Many of the immigrants came from Europe, often survivors of the Holocaust. Others came from Jewish communities in the Middle East, Africa, and Asia. All travelled to Israel in the hope of starting a new life, free from antisemitism, in what had long been promised as 'a national home for the Jewish people'. Gurevitch was one of the hundreds of doctors employed to both examine and care for the newly arrived immigrants. Born to an Orthodox Jewish family in Germany at the end of the nineteenth century, he studied medicine in Czechoslovakia following the First World War, before migrating to

Mandatory Palestine in the early 1920s. By the time of the foundation of the State of Israel, Gurevitch was working as a physician at the Hadassah Hospital in Jerusalem. And it was during this period that he started to become interested in 'the genetics of the Jewish people'.[68]

Walking around the immigration camps, Gurevitch was struck by the physical diversity of the different Jewish populations arriving in Israel. Yemeni Jews, for example, looked very different from Ashkenazi Jews, who in turn looked different from Persian Jews. Yet according to the Torah, all these different Jewish groups shared a common ancestry, dating back some 3,000 years. Gurevitch started to wonder whether it might be possible to trace this ancestry using the latest techniques of modern science. With this in mind, he began collecting thousands of blood samples from Jewish immigrants in the camps around Jerusalem, storing them in the blood bank at the Hadassah Hospital. Each of the blood samples was carefully labelled in order to identify the specific ethnic group from which it had been taken, before being tested to determine the blood type – either A, B, AB, or O – of the individual in question. Once all this was complete, Gurevitch began comparing the ratios of the different blood types found amongst various Jewish communities.[69]

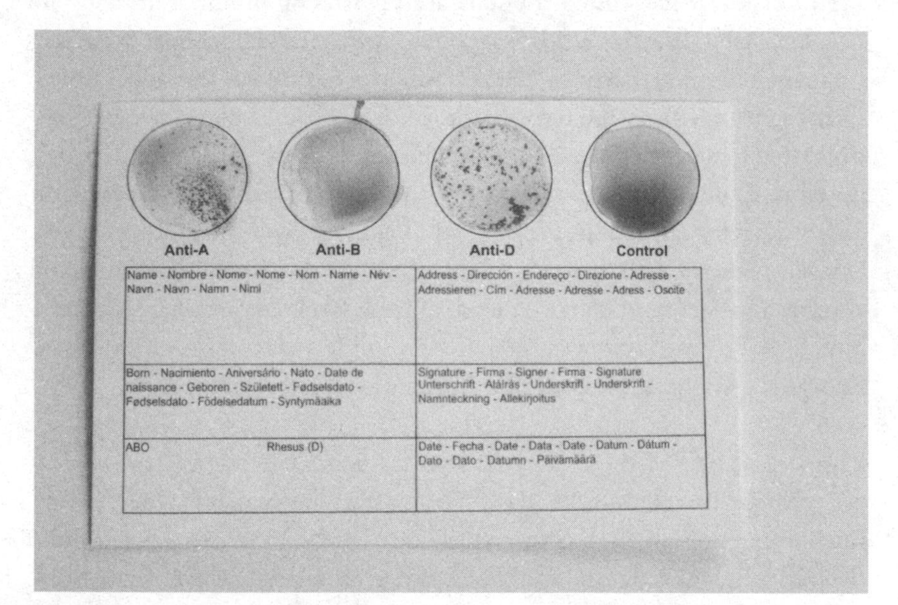

38. A test kit for identifying the ABO and rhesus factor blood groups. Blood tests were widely used by population geneticists in the twentieth century.

The ABO blood group system had been discovered around 1900, and so Gurevitch would have first learned about it during his medical training back in Europe. During the 1920s and 1930s, a number of other blood group systems were also discovered, such as the rhesus system and the MN system, each of which plays a different role in human health. For example, the ABO system helps regulate the coagulation of the blood. That is why it is so important to receive the correct blood type in a transfusion, as mixing the wrong types causes the blood to coagulate. During the First World War, states around the world began setting up blood banks in order to provide the right blood for transfusions, particularly for soldiers injured in combat. These blood banks, whilst primarily intended for use in medical care, also provided a new opportunity for genetic research. For the first time, geneticists had access to large collections of blood samples which could easily be cross-referenced against individual patient records. Like many other scientists in this period, Gurevitch believed that blood tests might provide the key to tracing the genetic history of humankind.[70]

Throughout the 1950s, Gurevitch published a series of articles on Jewish genetics. By comparing the frequencies of different blood types, he tried to show what united the separate Jewish communities arriving in Israel, as well as what made particular groups distinct. For example, Gurevitch claimed that both 'Kurdistani Jews' and 'Baghdad Jews' tended to have around the same frequencies of A, B, and O blood types. This suggested that they possessed a shared heritage. However, Gurevitch also noted that the relative frequencies of the M and N antigens were quite different, with around 40 per cent of 'Baghdad Jews' possessing the M antigen compared to around 30 per cent of the 'Kurdistani Jews'. In another article, Gurevitch even went as far as to claim that a certain combination of rhesus antigens was shared by 'all Jewish communities'. This, he argued, 'suggests the common origin of the Jewish people'.[71]

The second half of the twentieth century was a period of major political change in the Middle East. Following the Second World War, European colonial empires were forced to withdraw from the region – the British from Egypt and Palestine, the French from Syria and Lebanon. This led to the creation of a number of new states, including the State of Israel in 1948. In Israel, as elsewhere, modern science was widely understood to be essential for the success of the new nation. 'Israel is a small country, lacking material wealth and poor in natural resources. The importance of scientific research for its development cannot be overstressed,' argued the President of the Hebrew University of Jerusalem in 1960. This view was shared by many political leaders, including the first prime minister of Israel, David Ben-Gurion, who authorized the creation of a number of new scientific institutions, such as the Institute for Biomedical Research, established in 1952. The Israeli government also increased funding for existing scientific institutions, many of which dated from the period of the British Mandate for Palestine, such as the Hebrew University of Jerusalem.[72]

State investment in modern science was in fact common across the Middle East in this period. Following the Egyptian Revolution of 1952, Gamal Abdel Nasser approved the creation of the Egyptian National Research Centre, whilst in Turkey, the government established the Scientific and Technological Research Council shortly after the 1960 military coup. Both the Egyptian and Turkish governments also invested in genetic research, often in the hope of improving agriculture and human health. Egyptian and Turkish doctors, like their Israeli counterparts, were similarly interested in the genetic make-up of Middle Eastern populations. They too wrestled with questions of national identity. The Republic of Turkey tried to distinguish Turks from other ethnic groups, such as Arabs and Jews, who had long lived in the lands formerly occupied by the Ottoman Empire before its collapse in 1922. Similarly, the Egyptian government under Nasser promoted the idea of a shared Arab identity as the basis of regional cooperation following decolonization, hence the investment in genetic studies of the population.[73]

We've already seen how, during the Cold War, modern science – and genetics in particular – could be turned to a variety of different political uses. This was certainly true in Israel, especially when it came to the question of national identity. The Israeli Declaration of Independence explicitly identified 'the land of Israel' as 'the birthplace of the Jewish people', whilst the 1950 Law of Return declared that 'every Jew has a right to come to this country'. The question of who was and was not Jewish therefore became a key political issue in the middle decades of the twentieth century. Joseph Gurevitch was just one amongst a number of Israeli doctors who believed that modern genetics might provide a way to tackle this problem. During the same period, Israeli political leaders also discussed whether there needed to be some kind of 'regulation of immigration', perhaps even selection based on medical criteria. Indeed, the 1950 Law of Return actually included a clause which allowed the Israeli government to reject anyone who might 'endanger public health'. That was in part why the government set up the immigration camps, in order to medically assess new arrivals, as well as administer vaccinations and antimalarials. These two concerns – over national identity and public health – played a key role in shaping the development of modern genetics in the Middle East.[74]

In September 1961, the Hebrew University of Jerusalem hosted a major international conference on population genetics. Those attending included the American geneticist James Neel, who as we saw earlier worked as part of the Atomic Bomb Casualty Commission in Japan, as well as the British geneticist Arthur Mourant, who had recently published an influential book titled *The Distribution of the Human Blood Groups* (1954). Others travelled from India, Brazil, and Turkey in order to share their latest work on the origins of different human populations. There were, however, no representatives from neighbouring Arab states, despite the fact that scientists in these countries were also working on similar problems in population genetics at this time. For example, Munib Shahid, a Lebanese doctor based at the American University of Beirut, had recently published a series of articles on the prevalence of sickle cell anaemia in the Arab population, whilst Karima Ibrahim, an Egyptian doctor based at the State Serum Institute in Cairo, had actually co-authored an article with Mourant on 'The Blood Groups of the People of Egypt'. However, given the recent Arab–Israeli War of 1948 and the Suez Crisis of 1956, in which Israeli troops occupied the Sinai Peninsula, it is perhaps unsurprising that neither Shahid nor Ibrahim attended the conference in Jerusalem.[75]

The conference had been organized by an Israeli geneticist named Elisabeth Goldschmidt. Like many other Jewish scientists in this period, Goldschmidt was a refugee from Nazi Germany. Born to a Jewish family in 1912, Goldschmidt began studying medicine at Frankfurt University in the early 1930s, but was forced to flee following the rise of the Nazi Party. After escaping to Britain, Goldschmidt enrolled at the University of London, studying zoology and graduating in 1936. She then emigrated to Mandatory Palestine and began a PhD on the genetics of the mosquito at the Hebrew University of Jerusalem. Following a year in the United States, Goldschmidt returned to Israel in 1951, helping to set up the first dedicated genetics course at the Hebrew University. Goldschmidt also established the Genetics Society of Israel in 1958, serving as its first president.[76]

The other major figure behind the 1961 conference was an Israeli doctor named Chaim Sheba. Much like Goldschmidt, Sheba grew up in Europe during a period of rising antisemitism. Born in Austria–Hungary in 1908, he attended a series of local Jewish schools before studying medicine in Vienna in the early 1930s. Sheba then emigrated to Mandatory Palestine in 1933, deciding it was best to leave Austria given the recent electoral success of the Nazi Party in neighbouring Germany. By the early 1950s, he was working at the Tel-Hashomer Hospital, located just outside of Tel Aviv. Like Gurevitch, Sheba spent much of his time in the nearby immigration camps, collecting blood samples and tending to patients. And it was during this time that he too started to become interested in 'the genetic differentiation among the Jewish groups of Israel'.[77]

By the early 1960s, Israel was widely recognized as an important site for the study of population genetics. 'Israel, with its diverse population, drawn from so many parts of the world and so many different environments presents a unique laboratory for the geneticist,' announced the rector of the Hebrew University of Jerusalem during the opening address of the 1961 conference. And whilst the papers presented covered a wide variety of subjects, the majority focused on the relationship between population genetics and disease. Goldschmidt, for example, presented her recent research on the prevalence of Tay–Sachs disease – an inherited condition affecting the nervous system – in Ashkenazi Jews, whilst Sheba discussed his work on the prevalence of G6PD deficiency – a kind of metabolic disorder – in different Jewish groups.[78]

To be clear, this kind of research was not unique to Israel, but was in fact common across the world throughout the Cold War. Other scientists at the meeting presented their work on different regions and ethnic groups. A Japanese geneticist described his recent study of the 'differences between Caucasians and Japanese', whilst a Brazilian geneticist presented his research on mutations amongst those he referred to as 'Whites' and 'non-Whites'. As might be expected, the Israeli participants made sure to clearly distinguish their research from the kind of eugenics practised by the Nazis. Throughout the 1960s, Goldschmidt in particular campaigned vigorously against the continued influence of eugenics in modern science, reminding the international community that 'pseudo-genetic argumentation served as a pretext to the extermination of millions'. Another scientist at the conference also urged participants to remember that 'population genetics is a field in whose name great outrages have been committed'.[79]

The Cold War was a period in which scientific understandings of race and identity underwent a significant shift. Prior to the Second World War, most scientists understood race as a straightforward biological fact. However, in the aftermath of the Holocaust, this view came under increasing attack. 'For all practical social purposes, "race" is not so much a biological phenomenon as a social myth,' argued the United Nations in its influential 'Statement on Race', published in 1950. Rather than seeing race as a fixed biological concept, geneticists started thinking about it as something in constant flux. The focus of modern population genetics was not therefore on identifying fixed racial groups, but rather tracing the migration and mixing of different communities over time. This was one of the reasons why blood groups proved such a popular topic of research. 'A study of blood groups show[s] a heterogeneity in the proudest nations and support[s] the view that the races of the present day are but temporary integrations,' explained the British geneticist Arthur Mourant. Within any given ethnic group, there was in fact a great deal of genetic diversity. 'We must disavow any mystic notion of blood as a racial factor,' he concluded.[80]

This view of race, however, was much easier to maintain in principle rather than in practice. During a period in which many new states were in the process of formation, the political demand for a strong sense of national identity often took precedence. We've already seen how, shortly after the formation of the State of Israel in 1948, Joseph Gurevitch claimed to have identified the 'common origin of the Jewish people' through his study of ABO blood groups. Sheba made a similar claim, arguing that the prevalence of G6PD deficiency – which was known to be genetically inherited – could be used to trace the 'ethnic origin' of different Jewish groups. Others were more sceptical. Goldschmidt, for example, denied that Tay–Sachs disease was a good marker of Jewish identity, whilst Mourant argued that 'the genetical constitution of modern Jewish communities shows a wide range of variation'. In the end, most scientists tried to find a balance, arguing that whilst there was no single 'Jewish gene', it was nonetheless possible to trace the migration of different Jewish groups through their genetic history.[81]

At the same time as Chaim Sheba and Arthur Mourant were discussing the genetic history of humankind, another group was exploring the origins of agriculture. Historians had long believed that the earliest farming communities, dating to around 10,000 years ago, were located in the region between Palestine and Persia, an area commonly referred to as the 'Fertile Crescent'. In the early 1960s, a team of scientists at the Hebrew University of Jerusalem began testing this hypothesis. They were led by a plant geneticist named Daniel Zohary. Born in Jerusalem in 1926, Zohary was the son of an eminent botanist who had emigrated to Mandatory Palestine from Austria following the First World War. As a young boy, he would accompany his father on botanical field trips, particularly around the Sea of Galilee, learning the basics of plant taxonomy. In 1946, Zohary entered the Hebrew University of Jerusalem, studying botany in the hope of following in his father's footsteps. His degree, however, was interrupted by the outbreak of the Arab–Israeli War in 1948. The original campus of the Hebrew University of Jerusalem, located on Mount Scopus, had to be evacuated, as it was overrun by Jordanian troops. Zohary himself managed to escape, and went on to serve in the war, but one of his best friends was killed. Once the fighting was over, Zohary returned to complete his degree at the new university campus, located at Givat Ram.[82]

At this point, Zohary's scientific knowledge was not so different from his father's. That all changed following a visit to the United States in the early 1950s. Between 1952 and 1956, Zohary studied for a PhD in genetics at the University of California, Berkeley. It was here that he learned the techniques that would later prove so useful in identifying the origins of domesticated crops. Zohary would spend his days examining plant chromosomes under the microscope, staining them and comparing banding patterns. It was also in California that Zohary met his lifelong friend and collaborator, an American geneticist named Jack Harlan, who later worked for the United States Department of

Agriculture. Together, Zohary and Harlan hoped 'to discover when, where, and under what circumstances [the] early domestication of cereals came about'. However, Zohary quickly realized that, if he really wanted to grapple with this problem, he would need to return to the 'Fertile Crescent' itself. And so, after completing his PhD, Zohary moved back to Israel, taking up a position in the Department of Genetics at the Hebrew University of Jerusalem in 1956.[83]

Zohary's approach to the history of agriculture had a lot in common with the work conducted by the Mexican Agricultural Program that we encountered earlier. Zohary first went out collecting different varieties of wild plants, particularly those he thought might be related to staple crops such as wheat and barley. This was actually easier said than done, particularly as the region covered by the 'Fertile Crescent' extended well beyond Israeli territory. Zohary had to call in some favours, writing to Harlan in the United States as well as to botanists in Britain, Iran, and the Soviet Union, requesting that they send samples from local seed banks. This work was made easier thanks to the recent establishment of a major regional seed bank, supported by the United Nations Food and Agricultural Organization, located in Izmir, western Turkey. Having amassed a vast collection, Zohary then began comparing the different varieties of wild plants. In the 1950s, he focused on what he called 'chromosome analysis', which meant staining plant chromosomes and comparing them under the microscope, a technique he had learned in California. However, following a series of technological breakthroughs in the 1970s, Zohary was also able to analyse actual sequences of DNA, extracted directly from the plants he wanted to compare. It was then possible to accurately calculate the 'genetic distances' between different plants, determining which were close relations and which were distant cousins. 'The impact of these new molecular techniques is just starting to be felt in solving [the] problems of the origin of cultivated plants,' noted Zohary.[84]

After nearly three decades of intensive research, Zohary published his major work, titled *Domestication of Plants in the Old World* (1988). In this book, which was co-authored with the German archaeologist Maria Hopf, Zohary confirmed that staple crops such as wheat and barley were indeed first domesticated in the ancient Middle East, around 10,000 years ago. Crucially, he was also able to identify the wild ancestors of many contemporary crops, demonstrating their exact 'genetic relationship' to one another. This was a considerable intellectual achievement, but there was also a practical side to Zohary's work. The discovery of 'the original wild ancestors of cultivated cereals . . . opens a possibility for their utilization as genetic material for further crop improvement', noted one of Zohary's colleagues at the Hebrew University of Jerusalem. This was a simple idea, but it turned out to be very effective. By crossing existing varieties of wheat and barley with their wild ancestors, agricultural scientists were able to significantly increase crop yields. Zohary himself recognized the implications of his research, helping to develop, not only improved varieties of wheat and barley, but also vegetables and fruit. This was all part of a major drive towards achieving self-sufficiency in food production in Israel, something that was all the more pressing given the sharp increase in population that followed the arrival of hundreds of thousands of Jewish immigrants from the late 1940s onwards.[85]

During the second half of the twentieth century, scientists presented the Middle East as at the 'crossroads' of human history. Whether it was the migration of different ethnic groups, or the origins of agriculture, the lands around Palestine were widely understood to be the location of some of the most important events of the past 10,000 years. In this section, we've seen how Israeli scientists deployed the latest advances in modern genetics in order to better understand this history. Much as we've seen elsewhere, the development of modern genetics in Israel was closely tied to the process of state formation. Scientific interest in Jewish genetics was motivated by concerns over unrestricted immigration, whilst research into the deep history of agriculture was part of a broader programme to increase food production.[86]

Israeli scientists, many of whom were refugees from Nazi Germany, or survivors of the Holocaust, also played an important role in fighting antisemitism in science. Elisabeth Goldschmidt, the founder of the Genetics Society of Israel, did much to combat the continued influence of eugenics in post-war population genetics. At the same time, however, other Israeli scientists believed that modern genetics might provide a way to trace the ethnic origins of different Jewish communities. This somewhat contradictory approach to human genetics was not unique to Israel, but was in fact characteristic of the post-war period. In Turkey, geneticists used blood samples to distinguish between 'Arabs' and 'Turks', whilst in Iran, the same technique was used to trace the origins of the Zoroastrian population. Similar studies were conducted across Asia and the Americas. Officially, the scientific community rejected the concept of race as a meaningful biological category. Yet this often proved difficult to balance with the political demand for a strong sense of national identity, both in the Middle East and elsewhere. Today, we are still living with the legacies of this unresolved tension between genetics, race, and nationalism.[87]

V. Conclusion

On 26 June 2000, President Bill Clinton held a press conference in the East Room of the White House. He was joined by the German, French, and Japanese ambassadors to the United States, as well as the British Prime Minister, Tony Blair, via video-link. With the world's press watching, Clinton began his speech. 'We are here to celebrate the completion of the first survey of the entire human genome,' he announced. He then went on to explain how 'more than 1,000 researchers across six nations have revealed nearly all three billion letters of our miraculous genetic code'. Ten years earlier, the United States had launched the Human Genome Project. It cost $3 billion, but by the summer of 2000, scientists had finally completed the draft sequence of the entire human genome. The hope was that a map of the human genome would help scientists better understand the cause of diseases, such as cancer and Parkinson's. Medicine could then be personalized right down to the level of the individual, with those more at risk due to genetic factors identified before they developed symptoms. And although the project was led by the United States, it was a truly international effort, with geneticists working in Britain, France, Germany, Japan, and China all contributing to the sequencing. Different teams in different countries were assigned particular sections of the human genome, such as a particular chromosome. The results were then combined to give the complete genetic sequence.[88]

For many, including Clinton, the Human Genome Project was a symbol of the end of the Cold War. The project had launched just as the

Soviet Union was beginning to collapse, and the researchers involved spanned continents, even including scientists working in China, which since the death of Chairman Mao in 1976 had started to liberalize its economy and develop diplomatic relations with the United States. The Human Genome Project, Clinton claimed, would 'be directed towards making life better for all citizens of the world'. This view was shared by Blair, who spoke of a 'global community . . . now working across national frontiers to safeguard our shared values and put this remarkable scientific achievement at the service of all mankind'.[89]

As we've seen in this chapter, the development of modern genetics was fundamentally shaped by Cold War politics, particularly the process of state formation. It is therefore tempting to think of the Human Genome Project as a moment of transition, in which the era of Cold War rivalry gave way to a new era of globalization. That is certainly how both Bill Clinton and Tony Blair, perhaps the two politicians most associated with the wave of globalization that followed the collapse of the Soviet Union, understood the Human Genome Project. The idea that 'in genetic terms, all human beings, regardless of race, are more than 99.9 percent the same' proved exceptionally appealing to those looking to promote a vision of 'shared humanity'. The Human Genome Project was imagined as part of a future without racial discrimination.[90]

It would be a mistake, however, to finish the story here. The end of the Cold War was not the end of history, and the expansion of globalization during the 1990s did not bring about a more harmonious world. The Human Genome Project certainly did not put an end to racism. As we are now all too aware, globalization – in science, as in society more generally – in fact led to even greater fragmentation, dividing people more than ever and reinforcing existing inequalities. Even the promise of personalized medicine largely failed to materialize, whilst scientists continue to debate the ethics of gene editing.

All this was reflected in the field of genetics as it developed throughout the 2000s. Almost as soon as the Human Genome Project was complete, scientists and political leaders began to challenge the idea that a single reference genome could stand in for the whole of humanity. After all, the vast majority of the genetic material sequenced by the Human Genome Project came from a single male donor – almost certainly white – living in Buffalo, New York. With this in mind, states around the world began setting up their own national genome projects. These included the Iranian Human Genome Project (launched in 2000), the Indian Genome Variation Consortium (launched in 2003), the Turkey Genome Project (launched in 2010), the Genome Russia Project (launched in 2015), and the Han Chinese Genome Initiative (launched in 2017). All these projects had the effect of promoting ethnic nationalism, in which nations were once again seen in racial terms. This was most obviously the case with the Chinese example, which focused exclusively on the Han majority ethnic group, ignoring the genetic and ethnic diversity of the wider Chinese population. The Cold War might have been over, but genetics was just as much a tool of state formation in the 2000s as it was during the 1950s.[91]

At the same time, governments began to target minority ethnic groups, which came to be blamed for all kinds of social and political problems. The Genome Russia Project, for example, explicitly distinguished between what it called 'Ethnical Russian Groups' and 'Ethnical Non-Russian Groups'. The latter included a number of ethnic minorities which the government considered a threat to national security, such as the Chechens, who had fought against Russian troops in Chechnya throughout the 1990s in a bid for independence. The United States government made similar use of genetic testing to target minority ethnic groups. At the beginning of 2020, the Department of Homeland Security started collecting DNA samples from migrants crossing the US–Mexico border, with the results fed back into a massive criminal database. The use of genetics as a tool of state surveillance also became increasingly common in China throughout the 2000s. In 2016, the Chinese government began collecting DNA samples from the Uyghur minority ethnic group, most of whom are Muslim. This was all part of a broader effort to track and subdue the Uyghur population, culminating in the forced removal of over one million Uyghurs to detention camps across Xinjiang in northwest China. Today, the 'shared humanity' promised by modern genetics seems further away than ever.[92]

Epilogue: The Future of Science

On the morning of 28 January 2020, Charles Lieber, Chair of the Department of Chemistry and Chemical Biology at Harvard University, was arrested by special agents working for the Federal Bureau of Investigation (FBI). A world-renowned expert in nanoscience, Lieber was accused of 'aiding the People's Republic of China'. In court documents, the FBI alleged that Lieber was a 'contractual participant' in China's Thousand Talents Plan. According to the FBI, this had been set up in 2008 in order to 'lure Chinese overseas talent and foreign experts to bring their knowledge and experience to China and reward individuals for stealing proprietary information'. FBI officers alleged that Lieber had been recruited by the Wuhan University of Technology in 2011 and paid $50,000 a month. Lieber, the FBI alleged, had then 'repeatedly lied about his involvement in the Thousand Talents Plan'. This, the FBI claimed, amounted to fraud. At the time of writing, the trial is ongoing. Lieber denies all charges, but if found guilty, he faces up to five years in prison and a fine of up to $250,000.[1]

That same day, the FBI charged two Chinese nationals with similar crimes. Ye Yangqing, who had been working as a researcher at the Department of Physics, Chemistry, and Biomedical Engineering at Boston University, was accused of 'acting as an agent of a foreign government'. After intercepting a number of WeChat messages, the FBI concluded that Ye had been 'completing numerous assignments from PLA [People's Liberation Army] officers such as conducting research, assessing U.S. military websites and sending U.S. documents and information to China'. Even more dramatically, Zheng Zaosong, a researcher based at the Beth Israel Deaconess Medical Center in Boston, was charged with 'attempting to smuggle 21 vials of biological research to China'. In December 2019, Zheng had been arrested whilst trying to board a flight from Boston to China after customs officials 'discovered the vials hidden in a sock inside one of Zheng's bags'. He was promptly handed over to the FBI for questioning.[2]

Announcing the charges, Special Agent Joseph Bonavolonta made clear the geopolitical motivation behind the FBI's investigations. 'No country poses a greater, more severe, and long-term threat to our national security and economic prosperity than China,' Bonavolonta told reporters. 'China's goal, simply put, is to replace the United States as the world's leading power, and they're breaking the law to get there,' he claimed. These investigations were all part of a broader FBI programme which, beginning in 2018, sought to root out Chinese spies within American scientific institutions. In recent years, a number of Chinese and Chinese American scientists have been arrested and charged with failing to disclose financial or institutional links to China. Universities are also beginning to cut ties with Chinese technology companies, such as Huawei, which are increasingly seen as a security threat. And in December 2018, Meng Wanzhou, the Chief Financial Officer of Huawei, and daughter of the company's founder, was arrested in Canada following an extradition request from the United States. Meng was accused of stealing trade secrets, allegations which she denies. But if found guilty by an American court, she faces up to ten years in prison.[3]

Throughout this book, I have argued that the best way to understand the history of modern science is to think in terms of key moments in global history. We began in the fifteenth century, with the colonization of the Americas, before exploring the growth of trading and religious networks across Asia and Africa in the sixteenth and seventeenth centuries. We then moved on to the eighteenth century, a period in which European empires and the transatlantic slave trade expanded significantly. In the nineteenth century, we witnessed an age of capitalism, nationalism, and industrial warfare. And finally, in the twentieth century, we uncovered a world of ideological conflict, a world of anticolonial nationalists and communist revolutionaries. Each of these four periods of world historical change shaped the development of modern science. Global connections brought different people and scientific cultures together, sometimes out of choice, often by force.

Today, we are living through another key moment in global history. Scientists around the world find themselves at the centre of a geopolitical conflict fought between China and the United States. From the late 2000s onwards, the world entered what is best described as a 'New Cold

War'. At its heart, this is a struggle between China and the United States for economic, political, and military dominance. Following the 2007–8 financial crisis, the economic gap between the United States and China dramatically narrowed, and in 2010, China overtook Japan to become the second largest economy in the world. In order to secure continued economic growth, as well as access to natural resources and energy, China began to expand internationally throughout the early 2010s. This culminated in the launch of China's Belt and Road Initiative in 2013, an international financing and infrastructure project which is funding everything from new ports in Sri Lanka to railways in Kazakhstan. And so, whilst most analysts focus only on the United States and China, we need to recognize that this New Cold War – much like the original Cold War of the twentieth century – is global. What happens in Latin America, Africa, South Asia, and the Middle East matters in fundamental ways to both the future of science and the future of politics.[4]

In order to understand the world of science today, we need to pay close attention to the relationship between globalization and nationalism. In the 1990s, politicians and scientists tended to be rather naïve about globalization, thinking that it would lead to a more harmonious and productive world, sweeping away past inequalities in the process. By connecting people together, globalization was supposed to make us both more wealthy and more cosmopolitan. This turned out to be a false promise. Globalization in fact increased inequality within most nations, even as it reduced some of the inequality between them. The overall economic gap between China and the United States might have narrowed, but the richest 10 per cent of people in the United States both own and earn comparatively more today than they did in 1990. The same is true of China, which now has more billionaires than any other country in the world except the United States. This growth in inequality led to a resurgence of nationalism – exactly the opposite of the kind of cosmopolitan future imagined by champions of globalization. Over the past decade, we've seen Britain leave the European Union, Donald Trump elected as President of the United States, the revival of Hindu nationalism in India, and the rise of right-wing political leaders across Latin America.[5]

It is this strange combination of globalization and nationalism which really characterizes the New Cold War. States around the world see their participation in the globalized world of science as a means to assert national and regional authority. This is exactly why the United States is so concerned about Chinese influence at American universities. As we'll see, it is also why China is investing, not just in sending students to the United States, but also in forging scientific links across Asia and Africa.

In this epilogue, I reveal how our own global historical moment is shaping the development of modern science. We follow recent trends in three major fields of scientific research: artificial intelligence (AI), space exploration, and climate science. The future of each of these fields will depend on how both scientists and politicians confront the twin forces of globalization and nationalism. The future of science and the future of the world are inextricably linked.

In July 2017, the Chinese Communist Party unveiled its 'New Generation of Artificial Intelligence Development Plan'. This plan set out a timetable to transform China into the world-leader in AI research by 2030. It is already well on its way to reaching this goal. China publishes more AI papers than any other nation, including the United States, and the Chinese Communist Party is investing in expensive new research facilities, including a brand new Academy of Artificial Intelligence in Beijing. According to the 2017 plan, AI will provide 'a new engine for economic development' in China, with AI industries projected to contribute $146 billion to the Chinese economy by 2030. AI promises to aid in the 'great rejuvenation of the nation'.[6]

As it stands, computers are nowhere near being able to match a human when it comes to general intelligence, meaning the ability to perform multiple complex and interrelated intellectual tasks. However, it is possible to train a computer to become very good at a specific task, such as identifying a person from a photograph. This is what modern AI, or 'machine learning' as it is often referred to, is all about. Scientists write algorithms, essentially a set of instructions, which allow a computer to train itself at a given task. The algorithm is then fed a lot of data – for example, hundreds of thousands of digitized photographs of human faces. Analysing the photographs, the algorithm gradually learns to distinguish different facial features, and perhaps other characteristics. The more data you give the algorithm, the more it learns, and the better it gets at the task at hand. Facial recognition is one of the big areas of AI research, but there are many others. AI is already being used to make investment decisions, identify military targets, diagnose diseases, and translate foreign languages. With such a wide variety of uses, the economic and geopolitical benefits of advanced AI research are potentially enormous.

The recent explosion of interest in AI, in both China and beyond, neatly illustrates how the New Cold War is shaping the development of science today. The Chinese Communist Party itself describes AI research as 'a new focus of international competition'. And the former head of Google China, Kai-Fu Lee, has even gone as far as to suggest that China and the United States are engaged in an arms race, as each seeks to become the next 'AI superpower'. For countries like China and the United States, AI has the potential to transform the economy, disrupting existing patterns of employment and creating whole new fields of work. At the same time, AI is seen as the key to national security. States around the world are increasingly turning to AI technologies in order to enhance surveillance and military hardware, for example, through facial-recognition software. Competition between nations, alongside the globally connected nature of research in computer science, is fuelling massive increases in funding for AI. This has resulted in a number of significant breakthroughs in recent years.[7]

Some of these breakthroughs have the potential to transform our lives in positive ways. In 2019, a team of researchers at the Guangzhou Medical University published an article describing the use of AI to scan through millions of patient records in order to spot the early signs of common childhood illnesses. This was achieved by matching up patterns of symptoms and cross-referencing these against medical test results. The researchers found that their algorithm was able to accurately diagnose everything from gastroenteritis to meningitis, even in cases where these diseases had been missed by doctors. AI is increasingly deployed in hospitals around the world. Scientists have developed algorithms which can analyse an X-ray or MRI scan and identify signs of disease. These algorithms are already operating at the level of a trained radiologist, resulting in cheaper and faster diagnoses of diseases such as cancer.[8]

Whilst all this might sound relatively benign, there are still reasons to be sceptical about the positive impact of AI. Recent breakthroughs are a consequence of a massive increase in the amount of personal data being collected by both private companies and national governments. After all, the basic ideas behind modern AI have been around for decades, with proof of concept for tasks such as facial recognition accomplished in the 1960s. But without the volume of data required to train the algorithms on which AI is based, scientists had been unable to make much progress. Now, with companies like Facebook and states like China collecting personal data from hundreds of millions of people, it is possible to train algorithms to do things that were previously thought impossible.

This is one of the reasons that China has such a competitive advantage when it comes to AI. The Chinese state collects an incredible amount of personal data from its citizens, ranging from medical records and spending habits to energy usage and online activity. This data then acts as the raw material for training a new generation of AI algorithms. No wonder that China has some of the most advanced facial-recognition software in the world, far surpassing the capability of major American companies. This software is regularly used to monitor Chinese citizens' movements. Even more disturbingly, facial-recognition software deployed by Huawei is allegedly able to identify the ethnicity of an individual, and then alert the authorities if it spots a member of the Uyghur minority ethnic group, over a million of whom are currently held in detention camps across Xinjiang.[9]

Recent developments in AI are the product of a New Cold War, and this is a conflict which is already extending across the globe. Both Chinese and American companies are investing in AI research facilities in Africa. Once again, there are positives and negatives. On the one hand, increased investment is allowing African scientists to push AI research in directions which matter to them. The Google AI Center in Ghana, which opened in 2019, is a good example of this. Researchers there have been working on training algorithms to identify the onset of disease in African staple crops, such as cassava. This software will then help African farmers respond more quickly to outbreaks. There is also a project to improve algorithms for processing and translating African languages, which up to now have largely been neglected by researchers in the United States and Europe. Moustapha Cissé, Professor of Machine Learning at the African Institute for Mathematical Sciences, and

Director of the Google AI Center in Ghana, is optimistic. 'The future of machine-learning research is in Africa,' he recently told a reporter.[10]

Yet at the same time, foreign investment in African AI research can be exploitative. This is particularly the case when it comes to China's Belt and Road Initiative. In 2018, a Chinese company called CloudWalk signed a deal, as part of the initiative, to supply facial-recognition software to the Zimbabwean government. CloudWalk promised to 'help build a national facial database in Zimbabwe'. The potential introduction of a mass-surveillance system in a country with a poor record for human rights was met with widespread criticism. As elsewhere, it is likely that the Zimbabwean government will use the technology to crack down on political dissent. We also need to be clear about exactly what is motivating Chinese investment in African science. In order to make advances in AI, scientists need large volumes of data. China is already collecting about as much as it can from its own citizens. In order to expand further, it needs to collect personal data globally. This is of course exactly what American companies have been doing for the past decade or so. Facebook in particular has expanded aggressively into Africa in recent years. Why? It has a lot to do with the relatively poor state of data protection legislation and enforcement across Africa. This is partly what is driving major foreign investment in African AI. Indeed, as part of the deal signed between CloudWalk and the Zimbabwean government, Chinese researchers will be able to access African facial data remotely. This will then be used to further improve facial-recognition algorithms back in China.[11]

AI is also booming in the Middle East. Once again, the research agenda is being shaped by global politics. In September 2020, the United Arab Emirates (UAE) and Israel signed a peace agreement. The deal, which was brokered by the United States, was a major diplomatic breakthrough, one that was rightly celebrated by all those who hope for long-term peace in the Middle East. In signing the agreement, the UAE became only the third Arab state to recognize Israel's sovereignty. As part of the deal, the UAE and Israel also agreed to begin collaborating on AI research. Scientists at the Mohamed bin Zayed University of Artificial Intelligence in Abu Dhabi would now take part in a series of workshops with their counterparts at the Weizmann Institute of Science in Israel. Prior to the peace deal, such scientific collaboration was practically impossible, as Emiratis could not travel to Israel, and Israelis could not travel to the UAE.[12]

In this deal, we see how scientific collaboration can help foster international peace. But at the same time, we shouldn't lose sight of what is motivating Middle Eastern states to invest in AI. In both Israel and the UAE, the core concern is national security. The Israel Defense Forces already use AI to identify potential military targets in Palestine. According to one Israeli military engineer, their software is able to predict 'the most likely areas [rocket] launchers will be set up and at what hours. That enables us to know in advance what will happen and what areas should be attacked.' Security is also a major driver of AI investment in the UAE. The Emirati security services are already using facial-recognition software to track the population and stifle political dissent. During the COVID-19 pandemic, police in Dubai even used the same software to monitor whether individuals were adhering to social distancing guidelines.[13]

The forces of globalization and nationalism are shaping the growth of AI as the New Cold War plays out across Asia, Africa, and the Middle East. The same is true of another major field of scientific research, one that has echoes of the original Cold War of the twentieth century. In recent years, there has been growing interest in space exploration. Countries ranging from China and Japan to India and Turkey are all investing in space programmes, marking the beginning of a new space race. These programmes often require international collaboration, another reminder of the way in which globalization continues to shape scientific research. The establishment of the United Arab Emirates Space Agency in 2014 is a good example of this. In order to build up capacity, the UAE recruited scientists and engineers from the United States and South Korea to advise on the design of satellites and help plan future space missions. After six years of hard work, the UAE launched an unmanned mission to Mars in the summer of 2020. Once again, there was an international element to this. The UAE Mars probe was launched aboard a Japanese rocket from the Tanegashima Space Center in Japan. Through international collaboration, the UAE became the first Arab state in space. 'This is the future of the UAE,' announced Sarah Al Amiri, the Deputy Project Manager of the Emirates Mars Mission.[14]

But what kind of future is this? There is certainly much to be optimistic about. For a start, the Emirates Mars Mission is part of a drive to improve the representation of women in science and technology in the UAE. Al Amiri and her team, a third of whom are women, will undoubtedly inspire a new generation of female scientists in the region. More of course needs to be done, particularly in a country where the law still discriminates against women, but progress is nonetheless welcome. Alongside this, investment in space science is part of a broader effort to transition away from an oil-based economy, as the UAE seeks to become a hub for scientific and technological development in the Middle East. There are even plans to build a space port in Abu Dhabi to support commercial space tourism.[15]

As with all space exploration, there is an element of nationalism to this too. The UAE wants to maintain a leadership role within the Middle East, and it sees the prestige associated with space science as a means to do that. Shortly after the launch of the Mars probe, the UAE government Twitter account shared a message 'of pride, hope, and peace to the Arab region'. The UAE then promised to lead the way in a new 'golden age of Arab and Islamic discoveries'. The Emirates Mars Mission was also carefully planned to coincide with the fiftieth anniversary of the founding of the UAE. When the mission finally reached Mars in early 2021, it formed a major part of the national celebrations. According to the Emirati government, this was 'a defining moment in our history and marks the UAE joining advanced nations involved in space exploration'.[16]

A similar kind of nationalism is driving Chinese investment in space exploration. In November 2020, China launched an unmanned mission to the Moon. The official scientific purpose was to collect lunar rocks and return these to Earth for analysis. But the Chinese Communist Party couldn't resist a publicity stunt. Whilst collecting rocks, the spacecraft also planted a Chinese flag on the lunar surface. Up to that point, only five other flags had been planted on the Moon, all by the United States. That same year, China also launched an unmanned mission to Mars. As with the UAE, this was planned to coincide with a major political anniversary. The mission reached Mars in early 2021, coinciding with the hundredth anniversary of the founding of the Chinese Communist Party. State media even started referring to the Mars mission as a '100-year anniversary gift'.[17]

Governments around the world clearly see the development of a space programme as a marker of national prestige. However, space science also serves more practical ends, particularly when it comes to security and defence. The Turkish government was upfront about this following the foundation of the Turkish Space Agency in 2018. 'Turkey has proven its technological capabilities in the defence industry,' explained the Minister of Industry and Technology at the time. 'Space technologies will allow us to expand in novel and unique dimensions,' he argued, just as Turkey was beginning to design and manufacture its own military drones and rockets. India too sees space science as a means to assert its military strength. Alongside a series of unmanned lunar missions, India is also investing heavily in related military technologies. In March 2019, Prime Minister Narendra Modi announced that India had conducted its first successful anti-satellite test, shooting down one of its own satellites with a surface-to-space missile. This was widely seen as a counter to the threat posed by India's neighbour, China, which has launched multiple military and surveillance satellites in recent years. Modi, the leader of the nationalist Bharatiya Janata Party, celebrated the missile test, arguing that India was now one of the world's leading 'space powers'.[18]

We've seen how globalization and nationalism are shaping the development of AI and space exploration. Rivalry between states, particularly between China and the United States but also regional powers like the United Arab Emirates and Israel, is fuelling both collaboration and competition in scientific research. This is the New Cold War which characterizes our own global historical moment. There is one final area of scientific research which, more than any other, is being shaped by these twin impulses of globalization and nationalism.

We are living through a climate emergency. This is clearly a global problem, with greenhouse gas emissions causing irreversible damage to our shared environment. Climate change will wreak havoc on the world, as livelihoods are destroyed and hundreds of millions of people are forced to become climate refugees. The basic facts about climate change have been known for decades, as was made clear in the first Intergovernmental Panel on Climate Change (IPCC) assessment in 1990. Set up by the World Meteorological Organization and the United

Nations Environment Programme, the IPCC brought together experts from across the globe to assess the scientific evidence for climate change and suggest potential solutions. In its first assessment, the IPCC concluded that an increase in greenhouse gas emissions had likely caused an increase in the average global temperature over the previous hundred years. At this rate, the average global temperature was predicted to increase by a further 3°C over the next hundred years. The IPCC also emphasized that climate change needed to be studied by a global community of scientists in order to coordinate a global response. This was part of an attempt to rebalance scientific power in the wake of the Cold War and decolonization. Scientists from Communist China and the former Soviet Union worked alongside those from Europe, the United States, Latin America, South Asia, Africa, and the Middle East.[19]

Yet despite the best efforts of the IPCC, little concrete action was taken on combating climate change in the 1990s and 2000s. Various international agreements were signed, such as the Kyoto Protocol in 1997, which required states to reduce greenhouse gas emissions, but global warming continued apace. The mood, however, is starting to change. Many of the world's biggest polluters now realize that climate change is a major threat to national security and economic prosperity. China is a good example here. Every year, China emits more carbon dioxide than any other country in the world. China is also responsible for causing massive environmental damage through its Belt and Road Initiative, which has seen the expansion of unsustainable infrastructure projects across Asia, Africa, and the Middle East.

More recently, however, the Chinese Communist Party has started to acknowledge the threat posed by climate change, if not for the world, then for the Chinese state. After all, with its large coastal cities, major river deltas, and expansive deserts, China is especially vulnerable to climate change. Even a small rise in sea levels could devastate economic centres on the coast, such as Shanghai and Guangzhou. A major drought or flood could also dramatically impact food supplies, and so reduce public support for the ruling Chinese Communist Party. In response to this threat, China has invested significantly in climate science and green energy research. Tsinghua University in Beijing, for example, has set up a dedicated research group working on 'new energy' technologies, such as batteries suitable for storing renewable sources of energy. China is also the biggest producer of solar power in the world, as well as the biggest manufacturer of electric cars. At the end of 2020, the Chinese president, Xi Jinping, even announced plans for China to become carbon neutral by 2060.[20]

National self-interest is clearly the impetus for the Chinese response to climate change. Yet China also recognizes that it cannot hope to combat climate change on its own. In 2016, China set up the 'Digital Belt and Road Program' as part of its global climate strategy. Based at the Chinese Academy of Sciences in Beijing, this project brings together international expertise in order to monitor environmental and climatic change throughout Asia, Africa, and the Middle East. As part of the Digital Belt and Road Program, China is installing climate monitoring equipment in remote locations, particularly in countries with relatively poor meteorological services. China has also set up a number of joint research stations. For example, in 2019 the University of Ruhuna in Sri Lanka opened a new oceanography station, at which both Sri Lankan and Chinese scientists are working on monitoring climatic changes in the Indian Ocean. Finally, China is providing satellite imagery to go with this data, and using advanced AI algorithms to analyse and model changes in the climate. Scientists involved in the Digital Belt and Road Program include those from China, Russia, India, Pakistan, Malaysia, and Tunisia, amongst others. This, then, is a prime example of the way in which science today is shaped by both nationalism and globalization. China is just one amongst a number of states trying to find a way between a narrow nationalist response to climate change, and the need to work together as part of a global scientific community.[21]

Regional cooperation, as with the Digital Belt and Road Program, is a major theme in recent climate science. Global climate models, after all, are not particularly helpful for individual states hoping to plan for the future. Instead, scientists and politicians have started to worry more about how climate change will affect particular regions, and in what ways. This has led to the creation of a number of regional institutions. For example, in 2012 a group of African states formed the Southern African Science Service Centre for Climate Change and Adaptive Land Management (SASSCAL). The centre is based in Namibia and directed by the Rwandan–South African climate scientist Jane Olwoch. Scientists involved in the project include those from Angola, Botswana,

South Africa, Zambia, Germany, and Namibia itself. Much like the Digital Belt and Road Program, the idea is to pool resources and data in order to generate more accurate regional climate models. There is also a focus on the particular energy and climate challenges faced by Africa, such as desertification, in which previously fertile land dries up and becomes infertile.[22]

Latin America is another major player when it comes to the future of climate research. Once again, the current focus is on more regional studies. Carolina Vera, an IPCC climate scientist based at the University of Buenos Aires, is conducting pioneering work in this regard. She works closely with local maize farmers along the Matanza River in Argentina in order to produce flood-risk maps. In the past, climate scientists tended to ignore the knowledge of Indigenous people and local farmers. Vera, however, is combining modern climate science with local knowledge. Her team collects rainfall data using scientific equipment and then interviews local farmers in order to better understand the timing and impact of flooding in the region. 'I needed a dialogue with those who might use or benefit from my research, and to work with them as equals,' explained Vera in a recent article. By bringing together scientific and local knowledge, Vera and her team are able to produce more accurate flood-risk maps. These regional results are then fed back into the global climate models produced by the IPCC.[23]

Science is not, and has never been, a uniquely European endeavour. Throughout this book, we've seen how people and cultures from around the world contributed to the making of modern science. From Aztec naturalists and Ottoman astronomers to African botanists and Japanese chemists, the history of modern science needs to be told as a global story. The same is true when it comes to the future of science. Indeed, there is no reason to think that the next big scientific discovery will come out of a laboratory in Europe or the United States. Exciting new work in artificial intelligence, space exploration, and climate science is already taking place in Asia, Africa, the Middle East, and Latin America. Chinese computer scientists are making breakthroughs in machine learning, Emirati engineers are sending spacecraft to Mars, and Argentine environmental scientists are helping to produce new climate models.

There is much to celebrate, but science also faces serious problems. Private companies and national governments are collecting massive amounts of personal data in a bid to become the next 'AI superpower'. Countries like Turkey and India are investing huge sums of money in costly space programmes, the value of which – beyond military and nationalist posturing – isn't always clear. And whilst the world is slowly waking up to the climate emergency, states typically only act when it is in their own national interest.[24]

Scientists now find themselves on the frontline of a New Cold War. In the United States, this has led to what one group of scientists has accurately described as 'racial profiling', whereby those of Chinese descent are increasingly targeted by the FBI for investigation. In China, a number of Uyghur scientists have disappeared over the past few years, whilst in Turkey, critics of President Recep Tayyip Erdoğan, including many leading scientists, have been detained. It is a similar story in Africa and Latin America. In Sudan, the geneticist Muntaser Ibrahim, an expert on African genetic diversity based at the University of Khartoum, was arrested during a peaceful protest in February 2019, although he was, thankfully, released following a military coup later that year. In Brazil, many climate scientists have started to publish anonymously. They fear retaliation from Jair Bolsonaro, the right-wing president who since being elected in 2019 has frozen the science budget and pursued a policy of deforestation in the Amazon.[25]

The challenges are great, but the future of science depends on finding a way between the twin forces of globalization and nationalism. How might we do this? We need to begin by getting the history right. The myth that modern science was invented in Europe is not only false, it is also deeply damaging. There is little hope of working together as a global scientific community when most of the world is excluded from the story. Narratives about a medieval 'golden age' of Islamic, Chinese, or Hindu science are equally unhelpful, popular as they are amongst nationalist politicians today. These narratives simply serve to relegate the scientific achievements of the world beyond Europe to the distant past. Yet as we saw throughout this book, Muslim, Chinese, and Hindu scientists continued to contribute to the development of modern science long after the medieval period.

At the same time, we need to move beyond a naïve view of

globalization and its history. Modern science was undoubtedly the product of global cultural exchange. However, this cultural exchange took place in the context of deeply uneven power relations. The histories of slavery, empire, war, and ideological conflict are at the heart of the story of the origins of modern science. Seventeenth-century astronomers travelled aboard slave ships, eighteenth-century naturalists worked for colonial trading companies, nineteenth-century evolutionary thinkers fought in industrial wars, and twentieth-century geneticists continued to promote racial science throughout the Cold War. We need to actively engage with the legacies of these histories, rather than simply ignoring them. The future of science ultimately depends on a better understanding of its global past.

Notes

Given the scope of this book, references are restricted to works on which I directly relied in the writing. For a similar reason, I have kept points of discussion in the notes to an absolute minimum.

Introduction: The Origins of Modern Science

1 This story is repeated, more or less explicitly, in almost all surveys of the history of science written from the middle of the twentieth century onwards. Examples include Herbert Butterfield, *The Origins of Modern Science* (London: G. Bell and Sons, 1949), Alfred Rupert Hall, *The Scientific Revolution* (London: Longmans, 1954), Richard Westfall, *The Construction of Modern Science: Mechanisms and Mechanics* (Cambridge: Cambridge University Press, 1977), Steven Shapin, *The Scientific Revolution* (Chicago: University of Chicago Press, 1996), John Gribbin, *Science: A History, 1543–2001* (London: Allen Lane, 2002), Peter Bowler and Iwan Rhys Morus, *Making Modern Science: A Historical Survey* (Chicago: University of Chicago Press, 2005), and David Wootton, *The Invention of Science: A New History of the Scientific Revolution* (London: Allen Lane, 2015).

2 Kapil Raj, *Relocating Modern Science: Circulation and the Construction of Knowledge in South Asia and Europe, 1650–1900* (Basingstoke: Palgrave, 2007) is the closest to my own work in terms of argument, but is confined to a particular region (South Asia) and a particular time period (pre-1900). Arun Bala, *The Dialogue of Civilizations in the Birth of Modern Science* (Basingstoke: Palgrave, 2006) also makes a similar argument, although again is confined to the earlier period. Other existing works that cover a broader range of regions tend to simply reinforce European exceptionalism, for example, H. Floris Cohen, *The Rise of Modern Science Explained: A Comparative History* (Cambridge: Cambridge University Press, 2015), Toby Huff, *Intellectual Curiosity and the Scientific Revolution: A Global Perspective* (Cambridge: Cambridge University Press, 2010), and James E. McClellan III and Harold Dorn, *Science and Technology in World History: An Introduction*, 3rd edn (Baltimore: Johns Hopkins University Press, 2006).

3 On the need for a global history of science, see Sujit Sivasundaram, 'Sciences and the Global: On Methods, Questions, and Theory', *Isis* 101 (2010).

4 Jeffrey Mervis, 'NSF Rolls Out Huge Makeover of Science Statistics', Science, accessed 22 November 2020, https://www.sciencemag.org/news/2020/01/nsf-rolls-out-huge-makeover-science-statistics, Jeff Tollefson, 'China Declared World's Largest Producer of Scientific Articles', *Nature* 553 (2018), Elizabeth Gibney, 'Arab

World's First Mars Probe Takes to the Skies', *Nature* 583 (2020), and Karen Hao, 'The Future of AI is in Africa', MIT Technology Review, accessed 22 November 2020, https://www.technologyreview.com/2019/06/21/134820/ai-africa-machine-learning-ibm-google/.

5 David Cyranoski and Heidi Ledford, 'Genome-Edited Baby Claim Provokes International Outcry', *Nature* 563 (2018), David Cyranoski, 'Russian Biologist Plans More CRISPR-Edited Babies', *Nature* 570 (2019), Michael Le Page, 'Russian Biologist Still Aims to Make CRISPR Babies Despite the Risks', New Scientist, accessed 13 February 2021, https://www.newscientist.com/article/2253688-russian-biologist-still-aims-to-make-crispr-babies-despite-the-risks/, David Cyranoski, 'What CRISPR-Baby Prison Sentences Mean for Research', *Nature* 577 (2020), Connie Nshemereirwe, 'Tear Down Visa Barriers That Block Scholarship', *Nature* 563 (2018), *A Picture of the UK Workforce: Diversity Data Analysis for the Royal Society* (London: The Royal Society, 2014), and 'Challenge Anti-Semitism', *Nature* 556 (2018).

6 Joseph Needham's multivolume *Science and Civilisation in China* (Cambridge: Cambridge University Press, 1954 to present) is the most famous work celebrating ancient Chinese science, largely at the expense of the modern. Seyyed Hossein Nasr, *Science and Civilization in Islam* (Cambridge, MA: Harvard University Press, 1968) provided a single-volume equivalent for the Islamic world. See also Jim Al-Khalili, *Pathfinders: The Golden Age of Arabic Science* (London: Allen Lane, 2010) for a popular introduction to medieval Islamic science. For the history and politics of the 'golden age', see Marwa Elshakry, 'When Science Became Western: Historiographical Reflections', *Isis* 101 (2010).

7 'President Erdoğan Addresses 2nd Turkish–Arab Congress on Higher Education', Presidency of the Republic of Turkey, accessed 14 December 2019, https://tccb.gov.tr/en/news/542/43797/president-erdogan-addresses-2nd-turkish-arab-congress-on-higher-education.

8 Butterfield, *Origins of Modern Science*, 191, James Poskett, 'Science in History', *The Historical Journal* 62 (2020), Roger Hart, 'Beyond Science and Civilization: A Post-Needham Critique', *East Asian Science, Technology, and Medicine* 16 (1999): 93, and George Basalla, 'The Spread of Western Science', *Science* 156 (1967): 611. Twentieth-century historians of science were drawing on an earlier Orientalist tradition, dating from the late eighteenth century, of equating 'Europe' with 'modernity', one that was significantly reinforced during the Cold War, and particularly following decolonization, see Elshakry, 'When Science Became Western'.

9 Elshakry, 'When Science Became Western', Poskett, 'Science in History', and Nathan Rosenberg and L. E. Birdzell Jr, 'Science, Technology and the Western Miracle', *Scientific American* 263 (1990): 42.

10 David Joravsky, 'Soviet Views on the History of Science', *Isis* 46 (1955): 7.

11 Elshakry, 'When Science Became Western', Benjamin Elman, ' "Universal Science" Versus "Chinese Science": The Changing Identity of Natural Studies in China, 1850–1930', *Historiography East and West* 1 (2003), and Dhruv Raina, *Images and Contexts: The Historiography of Science and Modernity in India* (New Delhi: Oxford University Press, 2003), particularly 19–48 and 105–38.

Part One: Scientific Revolution, c.1450–1700

1. New Worlds

1 I have opted to use the term 'Aztec' rather than the more precise 'Mexica' in this chapter. Similarly, I use 'Tenochtitlan' rather than 'Mexico-Tenochtitlan'. On the history of this terminology, see Alfredo López Austin, 'Aztec', in *The Oxford Encyclopaedia of Mesoamerican Cultures*, ed. Davíd Carrasco (Oxford: Oxford University Press, 2001), 1:68–72.

2 Davíd Carrasco and Scott Sessions, *Daily Life of the Aztecs*, 2nd edn (Santa Barbara: Greenwood Press, 2011), 1–5, 38, 80, 92, 164, 168, and 219, James McClellan III and Harold Dorn, *Science and Technology in World History: An Introduction*, 3rd edn (Baltimore: Johns Hopkins University Press, 2006), 155–64, Miguel de Asúa and Roger French, *A New World of Animals: Early Modern Europeans on the Creatures of Iberian America* (Aldershot: Ashgate, 2005), 27–8, Jan Elferink, 'Ethnobotany of the Aztecs', in *Encyclopaedia of the History of Science, Technology, and Medicine in Non-Western Cultures*, ed. Helaine Selin, 2nd edn (New York: Springer, 2008), 827–8, and Ian Mursell, 'Aztec Pleasure Gardens', Mexicolore, accessed 12 April 2019, http://www.mexicolore.co.uk/aztecs/aztefacts/aztec–pleasure–gardens/.

3 Francisco Guerra, 'Aztec Science and Technology', *History of Science* 8 (1969): 43, Carrasco and Sessions, *Daily Life*, 1–11, 38, 42, 72, and 92, and McClellan III and Dorn, *Science and Technology in World History*, 155–64.

4 Frances Berdan, 'Aztec Science', in Selin, ed., *Encyclopaedia of the History of Science*, 382, Francisco Guerra, 'Aztec Medicine', *Medical History* 10 (1966): 320–32, E. C. del Pozo, 'Aztec Pharmacology', *Annual Review of Pharmacology* 6 (1966): 9–18, Carrasco and Sessions, *Daily Life*, 59–60, 113–5, 173, and McClellan III and Dorn, *Science and Technology in World History*, 155–64.

5 Carrasco and Sessions, *Daily Life*, 72 and 80.

6 Iris Montero Sobrevilla, 'Indigenous Naturalists', in *Worlds of Natural History*, eds. Helen Curry, Nicholas Jardine, James Secord, and Emma Spary (Cambridge: Cambridge University Press, 2018), 116–8, and Carrasco and Sessions, *Daily Life*, 88 and 230–7.

7 Peter Dear, *Revolutionizing the Sciences: European Knowledge and Its Ambitions, 1500–1700* (Basingstoke: Palgrave, 2001), and John Henry, *The Scientific Revolution and the Origins of Modern Science* (Basingstoke: Palgrave, 1997).

8 Herbert Butterfield, *The Origins of Modern Science* (London: Bell, 1949), David Wootton, *The Invention of Science: A New History of the Scientific Revolution* (London: Penguin Books, 2015), Robert Merton, 'Science, Technology and Society in Seventeenth-Century England', *Osiris* 4 (1938), Dorothy Stimson, 'Puritanism and the New Philosophy in 17th Century England', *Bulletin of the Institute of the History of Medicine* 3 (1935), Christopher Hill, *Intellectual Origins of the English Revolution* (Oxford: Clarendon Press, 1965), Steven Shapin and Simon Schaffer, *Leviathan and the Air-Pump* (Princeton: Princeton University Press, 1985), Elizabeth Eisenstein, *The Printing Press as an Agent of Change: Communications and Cultural Transformations in Early Modern Europe* (Cambridge: Cambridge University Press, 1997), and Steven Shapin, *The Scientific Revolution* (Chicago: University of Chicago Press, 1998).

9 Toby Huff, *Intellectual Curiosity and the Scientific Revolution: A Global Perspective* (Cambridge: Cambridge University Press, 2010), Antonio Barrera-Osorio, *Experiencing Nature: The Spanish American Empire and the Early Scientific Revolution* (Austin: University of Texas Press), Jorge Cañizares-Esguerra, *Nature, Empire, and Nation: Explorations of the History of Science in the Iberian World* (Stanford: Stanford University Press, 2006), William Burns, *The Scientific Revolution in Global Perspective* (New York: Oxford University Press, 2016), Klaus Vogel, 'European Expansion and Self-Definition', in *The Cambridge History of Science: Early Modern Science*, eds. Katharine Park and Lorraine Daston (Cambridge: Cambridge University Press, 2006), and McClellan III and Dorn, *Science and Technology in World History*, 99–176.

10 Alfred Crosby, *The Columbian Exchange: Biological and Cultural Consequences of 1492* (Westport: Praeger, 2003), 1–22, and J. Worth Estes, 'The European Reception of the First Drugs from the New World', *Pharmacy in History* 37 (1995): 3.

11 Katharine Park and Lorraine Daston, 'Introduction: The Age of the New', in Park and Daston, eds., *Cambridge History of Science: Early Modern Science*, Dear, *Revolutionizing the Sciences*, 10–48, and Shapin, *Scientific Revolution*, 15–118.

12 Anthony Grafton with April Shelford and Nancy Siraisi, *New Worlds, Ancient Texts: The Power of Tradition and the Shock of Discovery* (Cambridge, MA: The Belknap Press, 1992), 1–10, Paula Findlen, 'Natural History', in Park and Daston, eds., *The Cambridge History of Science: Early Modern Science*, 435–58, and Barrera-Osorio, *Experiencing Nature*, 1–13 and 101–27.

13 Crosby, *Columbian Exchange*, 24, Grafton, *New Worlds, Ancient Texts*, 84, and Asúa and French, *A New World of Animals*, 2.

14 Andres Prieto, *Missionary Scientists: Jesuit Science in Spanish South America, 1570–1810* (Nashville: Vanderbilt University Press), 18–34, and Thayne Ford, 'Stranger in a Foreign Land: José de Acosta's Scientific Realizations in Sixteenth-Century Peru', *The Sixteenth Century Journal* 29 (1998): 19–22.

15 Prieto, *Missionary Scientists*, 151–69, Grafton, *New Worlds, Ancient Texts*, 1, and Ford, 'Stranger in a Foreign Land', 31–2.

16 José de Acosta, *Natural and Moral History of the Indies*, trans. Frances López-Morillas (Durham, NC: Duke University Press, 2002), 37 and 88–9, Prieto, *Missionary Scientists*, 151–69, Grafton, *New Worlds, Ancient Texts*, 1, and Ford, 'Stranger in a Foreign Land', 31–2.

17 Acosta, *Natural and Moral History of the Indies*, 236–7.

18 Grafton, *New Worlds, Ancient Texts*, 1–10, Park and Daston, 'Introduction: The Age of the New', 8, and Ford, 'Stranger in a Foreign Land', 26–8.

19 Arthur Anderson and Charles Dibble, 'Introductions', in *Florentine Codex: Introduction and Indices*, eds. Arthur Anderson and Charles Dibble (Salt Lake City: University of Utah Press, 1961), 9–15, Arthur Anderson, 'Sahagún: Career and Character', in Anderson and Dibble, eds., *Florentine Codex: Introduction and Indices*, 29, and Henry Reeves, 'Sahagún's "Florentine Codex", a Little Known Aztecan Natural History of the Valley of Mexico', *Archives of Natural History* 33 (2006).

20 Diana Magaloni Kerpel, *The Colors of the New World: Artists, Materials, and the Creation of the Florentine Codex* (Los Angeles: The Getty Research Institute), 1–3,

Marina Garone Gravier, 'Sahagún's Codex and Book Design in the Indigenous Context', in *Colors between Two Worlds: The Florentine Codex of Bernardino de Sahagún*, eds. Gerhard Wolf, Joseph Connors, and Louis Waldman (Florence: Kunsthistorisches Institut in Florenz, 2011), 163–6, Elizabeth Boone, *Stories in Red and Black: Pictorial Histories of the Aztecs and Mixtecs* (Austin: University of Texas Press, 2000), 4, and Anderson and Dibble, 'Introductions', 9–10.

21 Victoria Ríos Castaño, 'From the "Memoriales con Escolios" to the Florentine Codex: Sahagún and His Nahua Assistants' Co-Authorship of the Spanish Translation', *Journal of Iberian and Latin American Research* 20 (2014), Kerpel, *Colors of the New World*, 1–27, Anderson and Dibble, 'Introductions', 9–13, and Carrasco and Sessions, *Daily Life*, 20.

22 Anderson and Dibble, 'Introductions', 11, Reeves, 'Sahagún's "Florentine Codex"', 307–16, and Kerpel, *Colors of the New World*, 1–3.

23 Bernardino de Sahagún, *Florentine Codex. Book 11: Earthly Things*, trans. Arthur Anderson and Charles Dibble (Santa Fe: School of American Research, 1963), 163–4 and 205, Guerra, 'Aztec Science', 41, and Corrinne Burns, 'Four Hundred Flowers: The Aztec Herbal Pharmacopoeia', Mexicolore, accessed 12 April 2019, http://www.mexicolore.co.uk/aztecs/health/aztec-herbal-pharmacopoeia-part-1.

24 Sahagún, *Florentine Codex. Book 11: Earthly Things*, 24.

25 Sobrevilla, 'Indigenous Naturalists', 112–30, and Asúa and French, *A New World of Animals*, 44–5.

26 Benjamin Keen, *The Aztec Image in Western Thought* (New Brunswick: Rutgers University Press, 1971), 204–5, Lia Markey, *Imagining the Americas in Medici Florence* (University Park: Pennsylvania State University Press, 2016), 214, and Kerpel, *Colors of the New World*, 6 and 13.

27 Andrew Cunningham, 'The Culture of Gardens', in *Cultures of Natural History*, eds. Nicholas Jardine, James Secord, and Emma Spary (Cambridge: Cambridge University Press, 1996), 42–7, Paula Findlen, 'Anatomy Theaters, Botanical Gardens, and Natural History Collections', in Park and Daston, eds., *The Cambridge History of Science: Early Modern Science*, 282, Paula Findlen, *Possessing Nature: Museums, Collecting, and Scientific Culture in Early Modern Italy* (Berkeley: University of California Press, 1996), 97–154, and Barrera-Osorio, *Experiencing Nature*, 122.

28 Dora Weiner, 'The World of Dr. Francisco Hernández', in *Searching for the Secrets of Nature: The Life and Works of Dr. Francisco Hernández*, eds. Simon Varey, Rafael Chabrán, and Dora Weiner (Stanford: Stanford University Press, 2000), Jose López Piñero, 'The Pomar Codex (ca. 1590): Plants and Animals of the Old World and the Hernandez Expedition to America', *Nuncius* 7 (1992): 40–2, and Barrera-Osorio, *Experiencing Nature*, 17.

29 Harold Cook, 'Medicine', in Park and Daston, eds., *The Cambridge History of Science: Early Modern Science*, 407–23, and López Piñero, 'The Pomar Codex', 40–4.

30 Weiner, 'The World of Dr. Francisco Hernández', 3–6, and Harold Cook, 'Medicine', 416–23.

31 Simon Varey, 'Francisco Hernández, Renaissance Man', in Varey, Chabrán, and Weiner, eds., *Searching for the Secrets of Nature*, 33–8, Weiner, 'The World of Dr. Francisco Hernández', 3–6, and Pinero, 'The Pomar Codex', 40–4.

32 Simon Varey, ed., *The Mexican Treasury: The Writings of Dr. Francisco Hernández* (Stanford: Stanford University Press, 2001), 149, 212, and 219, Jose López Pinero and Jose Pardo Tomás, 'The Contribution of Hernández to European Botany and Materia Medica', in Varey, Chabrán, and Weiner, eds., *Searching for the Secrets of Nature*, J. Worth Estes, 'The Reception of American Drugs in Europe, 1500–1650', in Varey, Chabrán, and Weiner, eds., *Searching for the Secrets of Nature*, 113, Arup Maiti, Muriel Cuendet, Tamara Kondratyuk, Vicki L. Croy, John M. Pezzuto, and Mark Cushman, 'Synthesis and Cancer Chemopreventive Activity of Zapotin, a Natural Product from *Casimiroa Edulis*', *Journal of Medicinal Chemistry* 50 (2007): 350–5, Ian Mursell, 'Aztec Advances (1): Treating Arthritic Pain', Mexicolore, accessed 24 January 2021, https://www.mexicolore.co.uk/aztecs/health/aztec-advances-4-arthritis-treatment, Varey, 'Francisco Hernández, Renaissance Man', 35–7, and del Pozo, 'Aztec Pharmacology', 13–17.

33 David Freedberg, *The Eye of the Lynx: Galileo, His Friends, and the Beginnings of Modern Natural History* (Chicago: University of Chicago Press, 2003), 246–55, Pinero, 'The Pomar Codex', 42, Vogel, 'European Expansion and Self-Definition', 826, and Asúa and French, *A New World of Animals*, 98–100.

34 Millie Gimmel, 'Reading Medicine in the Codex de la Cruz Badiano', *Journal of the History of Ideas* 69 (2008), Sandra Zetina, 'The Encoded Language of Herbs: Material Insights into the de la Cruz–Badiano Codex', in Wolf, Connors, and Waldman, eds., *Colors between Two Worlds*, and Vogel, 'European Expansion and Self-Definition', 826.

35 William Gates, 'Introduction to the Mexican Botanical System', in Martín de la Cruz, *The de la Cruz–Badiano Aztec Herbal of 1552*, trans. William Gates (Baltimore: The Maya Society, 1939), vi–xvi, and Gimmel, 'Reading Medicine', 176–9.

36 Martín de la Cruz, *The de la Cruz-Badiano Aztec Herbal of 1552*, trans. William Gates (Baltimore: The Maya Society, 1939), 14–15.

37 Gimmel, 'Reading Medicine', 176–9.

38 Raymond Stearns, *Science in the British Colonies of America* (Urbana: University of Illinois Press, 1970), 65, Paula Findlen, 'Courting Nature', in Jardine, Secord, and Spary, eds., *Cultures of Natural History*, Cook, 'Medicine', 416–23, Barrera-Osorio, *Experiencing Nature*, 122, Grafton, *New Worlds, Ancient Texts*, 67, and Worth Estes, 'The Reception of American Drugs in Europe, 1500–1650', 111–9.

39 Gimmel, 'Reading Medicine', 189, and Freedberg, *Eye of the Lynx*, 252–6.

40 Surekha Davies, *Renaissance Ethnography and the Invention of the Human: New Worlds, Maps and Monsters* (Cambridge: Cambridge University Press, 2016), 149–70, Laurence Bergreen, *Over the Edge of the World: Magellan's Terrifying Circumnavigation of the Globe* (New York: Morrow, 2003), 160–3, and Antonio Pigafetta, *The First Voyage around the World*, ed. Theodore J. Cachey Jr (Toronto: University of Toronto Press, 2007), 12–17.

41 Alden Vaughan, *Transatlantic Encounters: American Indians in Britain, 1500–1776* (Cambridge: Cambridge University Press, 2006), xi–xii and 12–13, and Elizabeth Boone, 'Seeking Indianness: Christoph Weiditz, the Aztecs, and Feathered Amerindians', *Colonial Latin American Review* 26 (2017): 40–7.

42 Anthony Pagden, *The Fall of Natural Man: The American Indian and the Origins of Comparative Ethnology* (Cambridge: Cambridge University Press, 1982), Joan-Pau Rubiés, 'New Worlds and Renaissance Ethnology', *History of Anthropology* 6 (1993), and J. H.

Eliot, 'The Discovery of America and the Discovery of Man', in Anthony Pagden, ed., *Facing Each Other: The World's Perception of Europe and Europe's Perception of the World* (Aldershot: Ashgate, 2000), David Abulafia, *The Discovery of Mankind: Atlantic Encounters in the Age of Columbus* (New Haven: Yale University Press, 2009), and Rebecca Earle, *The Body of the Conquistador: Food, Race and the Colonial Experience in Spanish America, 1492–1700* (Cambridge: Cambridge University Press, 2012), 23–4.

43 Cecil Clough, 'The New World and the Italian Renaissance', in *The European Outthrust and Encounter*, eds. Cecil Clough and P. Hair (Liverpool: Liverpool University Press, 1994), 301, Davies, *Renaissance Ethnography*, 30 and 70, Acosta, *Natural and Moral History of the Indies*, 71, and Crosby, *Columbian Exchange*, 28.

44 Saul Jarcho, 'Origin of the American Indian as Suggested by Fray Joseph de Acosta (1589)', *Isis* 50 (1959), Acosta, *Natural and Moral History of the Indies*, 51, and Pagden, *Fall of Natural Man*, 150.

45 Acosta, *Natural and Moral History of the Indies*, 51–3 and 63–71.

46 Diego von Vacano, 'Las Casas and the Birth of Race', *History of Political Thought* 33 (2012), Manuel Giménez Fernández, 'Fray Bartolomé de las Casas: A Biographical Sketch', in *Bartolomé de las Casas in History: Towards an Understanding of the Man and His Work*, eds. Juan Friede and Benjamin Keen (DeKalb: Illinois University Press, 1971), 67–73, and Pagden, *Fall of Natural Man*, 45–6, 90, and 121–2.

47 G. L. Huxley, 'Aristotle, Las Casas and the American Indians', *Proceedings of the Royal Irish Academy* 80 (1980): 57–9, Vacano, 'Las Casas', 401–10, and Giménez Fernández, 'Fray Bartolomé de las Casas', 67–73.

48 Bartolomé de las Casas, *Bartolomé de las Casas: A Selection of His Writings*, trans. George Sanderlin (New York: Alfred Knopf, 1971), 114–5, and Christian Johns, *The Origins of Violence in Mexican Society* (Westport: Praeger, 1995), 156–7.

49 Earle, *Body of the Conquistador*, 19–23.

50 Earle, *Body of the Conquistador*, 21–3.

51 Jorge Cañizares-Esguerra, 'New World, New Stars: Patriotic Astrology and the Invention of Indian and Creole Bodies in Colonial Spanish America, 1600–1650', *American Historical Review* 104 (1999), and Earle, *Body of the Conquistador*, 22.

52 Karen Spalding, 'Introduction', in Inca Garcilaso de la Vega, *Royal Commentaries of the Incas and General History of Peru*, trans. Harold Livermore (Indianapolis: Hackett Publishing Company, 2006), xi–xxii.

53 Inca Garcilaso de la Vega, *Royal Commentaries of the Incas and General History of Peru*, trans. Harold Livermore (Indianapolis: Hackett Publishing Company, 2006), 1–11.

54 Inca Garcilaso de la Vega, *First Part of the Royal Commentaries of the Yncas*, trans. Clements Markham (Cambridge: Cambridge University Press, 1869), 1:v–vi, 2:87, and 2:236–7.

55 Barbara Mundy, *The Mapping of New Spain: Indigenous Cartography and the Maps of the Relaciones Geográficas* (Chicago: University of Chicago Press, 1996), 14, and Hans Wolff, 'America – Early Images of the New World', in *America: Early Maps of the New World*, ed. Hans Wolff (Munich: Prestel, 1992), 45.

56 Hans Wolff, 'The Conception of the World on the Eve of the Discovery of America – Introduction', in Wolff, ed., *America*, 10–15, and Klaus Vogel, 'Cosmography', in Park and Daston, eds., *The Cambridge History of Science: Early Modern Science*, 474–8.

57 Vogel, 'Cosmography', 478.
58 Wolff, 'America', 27 and 45.
59 Rüdiger Finsterwalder, 'The Round Earth on a Flat Surface: World Map Projections before 1550', in Wolff, ed., *America*, and Wolff, 'America', 80.
60 María Portuondo, 'Cosmography at the *Casa*, *Consejo*, and *Corte* during the Century of Discovery', in *Science in the Spanish and Portuguese Empires, 1500–1800*, eds. Daniela Bleichmar, Paula De Vos, Kristin Huffine, and Kevin Sheehan (Stanford: Stanford University Press, 2009), and Barrera-Osorio, *Experiencing Nature*, 1–60.
61 Vogel, 'Cosmography', 484, and Mundy, *Mapping of New Spain*, 1–23 and 227–30.
62 Felipe Fernández-Armesto, 'Maps and Exploration in the Sixteenth and Early Seventeenth Centuries', in *The History of Cartography: Cartography in the European Renaissance*, ed. David Woodward (Chicago: University of Chicago Press, 2007), 745, G. Malcolm Lewis, 'Maps, Mapmaking, and Map Use by Native North Americans', in *The History of Cartography: Cartography in the Traditional African, American, Arctic, Australian, and Pacific Societies*, eds. David Woodward and G. Malcolm Lewis (Chicago: University of Chicago Press, 1998), and Brian Harley, 'New England Cartography and Native Americans', in *American Beginnings: Exploration, Culture, and Cartography in the Land of Norumbega*, eds. Emerson Baker, Edwin Churchill, Richard D'Abate, Kristine Jones, Victor Konrad, and Harald Prins (Lincoln, NE: University of Nebraska Press, 1994), 288.
63 Juan López de Velasco, 'Instruction and Memorandum for Preparing the Reports', in *Handbook of Middle American Indians: Guide to Ethnohistorical Sources*, ed. Howard Cline (Austin: University of Texas Press, 1972), 1:234, Guerra, 'Aztec Science and Technology', 40, and Mundy, *Mapping of New Spain*, xii and 30.
64 Mundy, *Mapping of New Spain*, 63–4 and 96.
65 Mundy, *Mapping of New Spain*, 135–8.
66 Christopher Columbus, *The Four Voyages of Christopher Columbus*, trans. J. M. Cohen (London: Penguin Books, 1969), 224.
67 Wootton, *The Invention of Science*, 57–108, makes the same point, but without recognizing the role of Indigenous Amerindian knowledge in this process.

2. Heaven and Earth

1 Aydın Sayılı, *The Observatory in Islam and Its Place in the General History of the Observatory* (Ankara: Türk Tarih Kurumu Basımevi, 1960), 259–88, Stephen Blake, *Astronomy and Astrology in the Islamic World* (Edinburgh: Edinburgh University Press, 2016), 82–8, and Toby Huff, *Intellectual Curiosity and the Scientific Revolution: A Global Perspective* (Cambridge: Cambridge University Press, 2010), 138.
2 Sayılı, *Observatory in Islam*, 213 and 259–88, Vasiliĭ Vladimirovich Barthold, *Four Studies on the History of Central Asia* (Leiden: E. J. Brill, 1958), 1–48 and 119–24, and Benno van Dalen, 'Ulugh Beg', in *The Biographical Encyclopedia of Astronomers*, ed. Thomas Hockey (New York: Springer, 2007).
3 Stephen Blake, *Time in Early Modern Islam* (Cambridge: Cambridge University Press, 2013), 8–10, and Sayılı, *Observatory in Islam*, 13–14 and 259–88.

4 See Seyyed Hossein Nasr, *Science and Civilization in Islam* (Cambridge, MA: Harvard University Press, 1968), and Jim Al-Khalili, *Pathfinders: The Golden Age of Arabic Science* (London: Allen Lane, 2010), for an overview.

5 Marwa Elshakry, 'When Science Became Western: Historiographical Reflections', *Isis* 101 (2010). Histories of Islamic astronomy often end with Ulugh Beg, hence my choice to start with him.

6 Sayılı, *Observatory in Islam*, 262–90.

7 Huff, *Intellectual Curiosity*, 138, and İhsan Fazlıoğlu, 'Qūshjī', in Hockey, ed., *The Biographical Encyclopedia of Astronomers*.

8 Sayılı, *Observatory in Islam*, 272, Huff, *Intellectual Curiosity*, 135, and Blake, *Astronomy and Astrology*, 90.

9 David King, 'The Astronomy of the Mamluks', *Muqarnas* 2 (1984): 74, and Huff, *Intellectual Curiosity*, 123.

10 Barthold, *Four Studies*, 144–77.

11 Jack Goody, *Renaissances: The One or the Many?* (Cambridge: Cambridge University Press, 2009), and Peter Burke, Luke Clossey, and Felipe Fernández-Armesto, 'The Global Renaissance', *Journal of World History* 28 (2017).

12 Michael Hoskin, 'Astronomy in Antiquity', in *The Cambridge Illustrated History of Astronomy*, ed. Michael Hoskin (Cambridge: Cambridge University Press, 1997), and Michael Hoskin and Owen Gingerich, 'Islamic Astronomy', in Hoskin, ed., *The Cambridge Illustrated History of Astronomy*.

13 Hoskin, 'Astronomy in Antiquity', 42–5.

14 Abdelhamid I. Sabra, 'An Eleventh-Century Refutation of Ptolemy's Planetary Theory', in *Science and History: Studies in Honor of Edward Rosen*, eds. Erna Hilfstein, Paweł Czartoryski, and Frank Grande (Wrocław: Polish Academy of Sciences Press, 1978), 117–31, F. Jamil Ragep, 'Ṭūsī', in Hockey, ed., *The Biographical Encyclopedia of Astronomers*, and Sayılı, *Observatory in Islam*, 187–223.

15 John North, *The Fontana History of Astronomy and Cosmology* (London: Fontana Press, 1994), 192–5, F. Jamil Ragep, 'Nasir al-Din al-Tusi', in *Naṣīr al-Dīn al-Ṭūsī's Memoir on Astronomy*, trans. F. Jamil Ragep (New York: Springer-Verlag, 1993), F. Jamil Ragep, 'The *Tadhkira*', in *Naṣīr al-Dīn al-Ṭūsī's Memoir*, and Nasir al-Din al-Tusi, *Naṣīr al-Dīn al-Ṭūsī's Memoir*, 130–42.

16 Michael Hoskin and Owen Gingerich, 'Medieval Latin Astronomy', in Hoskin, ed., *The Cambridge Illustrated History of Astronomy*, 72–3.

17 Avner Ben-Zaken, *Cross-Cultural Scientific Exchanges in the Eastern Mediterranean, 1560–1660* (Baltimore: Johns Hopkins University Press, 2010), 2, and North, *Fontana History of Astronomy*, 255.

18 George Saliba, *Islamic Science and the Making of the European Renaissance* (Cambridge, MA: The MIT Press, 2007), and George Saliba, 'Whose Science is Arabic Science in Renaissance Europe?', Columbia University, accessed 20 November 2018, http://www.columbia.edu/~gas1/project/visions/case1/sci.1.html.

19 Ernst Zinner, *Regiomontanus: His Life and Work*, trans. Ezra Brown (Amsterdam: Elsevier, 1990), 1–33, and North, *Fontana History of Astronomy*, 253–9.

20 Zinner, *Regiomontanus*, 1–33, and North, *Fontana History of Astronomy*, 253–9.

21 Noel Swerdlow, 'The Recovery of the Exact Sciences of Antiquity: Mathematics, Astronomy, Geography', in *Rome Reborn: The Vatican Library and Renaissance Culture*, ed. Anthony Grafton (Washington, DC: Library of Congress, 1993), 125–53, and Zinner, *Regiomontanus*, 51–2.

22 Fazlıoğlu, 'Qūshjī', Huff, *Intellectual Curiosity*, 139, F. Jamil Ragep, ''Ali Qushji and Regiomontanus: Eccentric Transformations and Copernican Revolutions', *Journal for the History of Astronomy* 36 (2005), and F. Jamil Ragep, 'Copernicus and His Islamic Predecessors: Some Historical Remarks', *History of Science* 45 (2007): 74.

23 Robert Westman, *The Copernican Question: Prognostication, Skepticism, and Celestial Order* (Berkeley: University of California Press, 2011), 76–108, and Hoskin and Gingerich, 'Medieval Latin Astronomy', 90–7.

24 Ragep, 'Copernicus and His Islamic Predecessors', 65, George Saliba, 'Revisiting the Astronomical Contact between the World of Islam and Renaissance Europe', in *The Occult Sciences in Byzantium*, eds. Paul Magdalino and Maria Mavroudi (Geneva: La Pomme d'Or, 2006), and Saliba, 'Whose Science is Arabic Science in Renaissance Europe?'.

25 North, *Fontana History of Astronomy*, 217–23, Ragep, 'Copernicus and His Islamic Predecessors', 68, Saliba, *Islamic Science*, 194–232, and Hoskin and Gingerich, 'Medieval Latin Astronomy', 97.

26 Saliba, 'Revisiting the Astronomical', Saliba, *Islamic Science*, 193–201, and Ragep, 'Copernicus and His Islamic Predecessors'.

27 B. L. van der Waerden, 'The Heliocentric System in Greek, Persian and Hindu Astronomy', *Annals of the New York Academy of Sciences* 500 (1987).

28 Ben-Zaken, *Cross-Cultural Scientific Exchanges* 24–5.

29 Ben-Zaken, *Cross-Cultural Scientific Exchanges*, 8–26, and Sayılı, *Observatory in Islam*, 289–305.

30 Ben-Zaken, *Cross-Cultural Scientific Exchanges*, 8–26, and Sayılı, *Observatory in Islam*, 289–305.

31 Ben-Zaken, *Cross-Cultural Scientific Exchanges*, 8–21.

32 Ben-Zaken, *Cross-Cultural Scientific Exchanges*, 10–21, and Ekmeleddin İhsanoğlu, 'Ottoman Science', in *Encyclopaedia of the History of Science, Technology and Medicine in Non-Western Cultures*, ed. Helaine Selin, 2nd edn (New York: Springer, 2008), 3478–81.

33 Ben-Zaken, *Cross-Cultural Scientific Exchanges*, 21–4, and Sayılı, *Observatory in Islam*, 297–8.

34 Ben-Zaken, *Cross-Cultural Scientific Exchanges*, 21–4, and Sayılı, *Observatory in Islam*, 297–8.

35 Ben-Zaken, *Cross-Cultural Scientific Exchanges*, 21–4, and Sayılı, *Observatory in Islam*, 297–8.

36 Ben-Zaken, *Cross-Cultural Scientific Exchanges*, 40–2.

37 Harun Küçük, *Science Without Leisure: Practical Naturalism in Istanbul, 1660–1732* (Pittsburgh: University of Pittsburgh Press, 2019), 25–6 and 56–63, Feza Günergun, 'Ottoman Encounters with European Science: Sixteenth- and Seventeenth-Century Translations into Turkish', in *Cultural Translation in Early Modern Europe*, eds. Peter Burke and R. Po-chia Hsia (Cambridge: Cambridge University Press, 2007), 193–206, and Ekmeleddin İhsanoğlu, 'The Ottoman Scientific-Scholarly Lit-

erature', in *History of the Ottoman State, Society & Civilisation*, ed. Ekmeleddin İhsanoğlu (Istanbul: Research Centre for Islamic History, Art and Culture, 1994), 521–66.

38 Küçük, *Science Without Leisure*, 109 and 237–40, İhsanoğlu, 'Ottoman Science', 5, Günergun, 'Ottoman Encounters', 194–5, and Ekmeleddin İhsanoğlu, 'The Introduction of Western Science to the Ottoman World: A Case Study of Modern Astronomy (1660–1860)', in *Science, Technology and Learning in the Ottoman Empire*, ed. Ekmeleddin İhsanoğlu (Aldershot: Ashgate, 2004), 1–4.

39 Küçük, *Science Without Leisure*, 1–3, and Goody, *Renaissances*, 98.

40 Existing secondary literature attributes the sighting of the comet to Mahmud al-Kati in 1583. However, recent work by Mauro Nobili has shown that al-Kati was not the author of the famous work attributed to him, the *Tarikh al-fattash*. Furthermore, no reference to the comet appears in the *Tarikh al-fattash*. It therefore seems unlikely that al-Kati recorded an observation of a comet. In light of this, I have instead drawn on the *Tarikh al-Sudan* of Abd al-Sadi which does refer to a comet, as do later works such as the anonymous *Tadhkirat al-nisyan*. I am extremely grateful to Rémi Dewière for pointing this out and bringing these sources to my attention, as well as for his general advice on the history of the Sahel. Thebe Rodney Medupe et al., 'The Timbuktu Astronomy Project: A Scientific Exploration of the Secrets of the Archives of Timbuktu', in *African Cultural Astronomy: Current Archaeoastronomy and Ethnoastronomy Research in Africa*, eds. Jarita Holbrook, Johnson Urama, and Thebe Rodney Medupe (Dordecht: Springer Netherlands, 2008), 182, Thebe Rodney Medupe, 'Astronomy as Practiced in the West African City of Timbuktu', in *Handbook of Archaeoastronomy and Ethnoastronomy*, ed. Clive Ruggles (New York: Springer, 2014), Sékéné Mody Cissoko, 'The Songhay from the 12th to the 16th Century', in *General History of Africa: Africa from the Twelfth to the Sixteenth Century*, ed. Djibril Tamsir Niane (Paris: UNESCO, 1984), Aslam Farouk-Alli, 'Timbuktu's Scientific Manuscript Heritage: The Reopening of an Ancient Vista?', *Journal for the Study of Religion* 22 (2009), Mauro Nobili, *Sultan, Caliph, and the Renewer of the Faith: Aḥmad Lobbo, the Tārīkh al-fattāsh and the Making of an Islamic State in West Africa* (Cambridge: Cambridge University Press, 2020), John Hunwick, *Timbuktu and the Songhay Empire: Al-Sa'di's Ta'rīkh al-Sūdān down to 1613, and Other Contemporary Documents* (Leiden: Brill, 1999), 155, and Abd al-Sadi, *Tarikh es-Soudan*, trans. Octave Houdas (Paris: Ernest Leroux, 1900), 341.

41 Souleymane Bachir Diagne, 'Toward an Intellectual History of West Africa: The Meaning of Timbuktu', in *The Meanings of Timbuktu*, eds. Shamil Jeppie and Souleymane Bachir Diagne (Cape Town: HSRC Press, 2008), 24.

42 Cissoko, 'The Songhay', 186–209, Toby Green, *A Fistful of Shells: West Africa from the Rise of the Slave Trade to the Age of Revolution* (London: Allen Lane, 2019), 25–62, Lalou Meltzer, Lindsay Hooper, and Gerald Klinghardt, *Timbuktu: Script and Scholarship* (Cape Town: Iziko Museums, 2008), and Douglas Thomas, 'Timbuktu, Mahmud Kati (Kuti) Ibn Mutaw', in *African Religions: Beliefs and Practices through History*, eds. Douglas Thomas and Temilola Alanamu (Santa Barbara: ABC-Clio, 2019).

43 Medupe et al., 'The Timbuktu Astronomy Project', Farouk-Alli, 'Timbuktu's Scientific Manuscript Heritage', 45, Shamil Jeppie and Souleymane Bachir Diagne, eds., *The Meanings of Timbuktu* (Cape Town: HSRC Press, 2008), and Ismaël

Diadié Haidara and Haoua Taore, 'The Private Libraries of Timbuktu', in Jeppie and Diagne, eds., *The Meanings of Timbuktu*, 274.

44 Claudia Zaslavsky, *Africa Counts: Number and Pattern in African Cultures* (Chicago: Lawrence Hill Books, 1999), 201 and 222–3, Suzanne Preston Blier, 'Cosmic References in Ancient Ife', in *African Cosmos*, ed. Christine Mullen Kreamer (Washington, DC: National Museum of African Art, 2012), Peter Alcock, 'The Stellar Knowledge of Indigenous South Africans', in *African Indigenous Knowledge and the Sciences*, eds. Gloria Emeagwali and Edward Shizha (Rotterdam: Sense Publishers, 2016), 128, and Keith Snedegar, 'Astronomy in Sub-Saharan Africa', in Selin, ed., *Encyclopaedia of the History of Science*.

45 Medupe et al., 'The Timbuktu Astronomy Project', Meltzer, Hooper, and Klinghardt, *Timbuktu*, 94, Diagne, 'Toward an Intellectual History of West Africa', 19, Cissoko, 'The Songhay', 209, Cheikh Anta Diop, *Precolonial Black Africa*, trans. Harold Salemson (Westport: Lawrence Hill and Company, 1987), 176–9, Elias Saad, *Social History of Timbuktu: The Role of Muslim Scholars and Notables 1400–1900* (Cambridge: Cambridge University Press, 1983), 74 and 80–1, and 'Knowledge of the Movement of the Stars and What It Portends in Every Year', Library of Congress, accessed 11 September 2020, http://hdl.loc.gov/loc.amed/aftmh.tam010.

46 Medupe, 'Astronomy as Practiced in the West African City of Timbuktu', 1102–4, Meltzer, Hooper, and Klinghardt, *Timbuktu*, 80, and Hunwick, *Timbuktu and the Songhay Empire*, 62–5.

47 Green, *A Fistful of Shells*, 57, Salisu Bala, 'Arabic Manuscripts in the Arewa House (Kaduna, Nigeria)', *History in Africa* 39 (2012), 334, WAAMD ID #2579, #3955, and #15480, West African Arabic Manuscript Database, accessed 11 September 2020, https://waamd.lib.berkeley.edu, and Ulrich Seetzen, 'Nouveaux renseignements sur le royaume ou empire de Bornou', *Annales des voyages, de la géographie et de l'histoire* 19 (1812), 176–7. (Translation my own. Although I am grateful to Rémi Dewière for this last reference.)

48 Mervyn Hiskett, 'The Arab Star-Calendar and Planetary System in Hausa Verse', *Bulletin of the School of Oriental and African Studies* 30 (1967), and Keith Snedegar, 'Astronomical Practices in Africa South of the Sahara', in *Astronomy Across Cultures: The History of Non-Western Astronomy*, ed. Helaine Selin (Dordrecht: Springer, 2000), 470.

49 Zaslavsky, *Africa Counts*, 137–52, Adam Gacek, ed., *Catalogue of the Arabic Manuscripts in the Library of the School of Oriental and African Studies* (London: School of Oriental and African Studies, 1981), 24, and Dorrit van Dalen, *Doubt, Scholarship and Society in 17th-Century Central Sudanic Africa* (Leiden: Brill, 2016).

50 Zaslavsky, *Africa Counts*, 137–52, and Musa Salih Muhammad and Sulaiman Shehu, 'Science and Mathematics in Arabic Manuscripts of Nigerian Repositories', Paper Presented at the Middle Eastern Libraries Conference, University of Cambridge, 3–6 July 2017.

51 Medupe et al., 'The Timbuktu Astronomy Project', 183, and H. R. Palmer, ed., *Sudanese Memoirs* (London: Frank Cass and Co., 1967), 90.

52 Augustín Udías, *Searching the Heavens and Earth: The History of Jesuit Observatories* (Dordrecht: Kluwer Academic, 2003), 1–40, Michela Fontana, *Matteo Ricci: A Jesuit in the Ming Court*, trans. Paul Metcalfe (Lanham: Rowman & Littlefield, 2011), 1–12 and

185–209, Benjamin Elman, *On Their Own Terms: Science in China, 1550–1900* (Cambridge, MA: Harvard University Press, 2005), 64–5, and R. Po-Chia Hsia, *A Jesuit in the Forbidden City: Matteo Ricci, 1552–1610* (Oxford: Oxford University Press, 2010), 206–7.

53 Fontana, *Matteo Ricci*, 30 and 193–209.

54 Huff, *Intellectual Curiosity*, 74, and Willard J. Peterson, 'Learning from Heaven: The Introduction of Christianity and Other Western Ideas into Late Ming China', in *China and Maritime Europe, 1500–1800: Trade, Settlement, Diplomacy and Missions*, ed. John E. Wills Jr (Cambridge: Cambridge University Press, 2011), 100.

55 Catherine Jami, Peter Engelfriet, and Gregory Blue, 'Introduction', in *Statecraft and Intellectual Renewal in Late Ming China: The Cross-Cultural Synthesis of Xu Guangqi (1562–1633)*, eds. Catherine Jami, Peter Engelfriet, and Gregory Blue (Leiden: Brill, 2001), Timothy Brook, 'Xu Guangqi in His Context', in Jami, Engelfriet, and Blue, eds., *Statecraft and Intellectual Renewal*, Keizo Hashimoto and Catherine Jami, 'From the *Elements* to Calendar Reform: Xu Guangqi's Shaping of Mathematics and Astronomy', in Jami, Engelfriet, and Blue, eds., *Statecraft and Intellectual Renewal*, Peter Engelfriet and Siu Man-Keung, 'Xu Guangqi's Attempts to Integrate Western and Chinese Mathematics', in Jami, Engelfriet, and Blue, eds., *Statecraft and Intellectual Renewal*, and Catherine Jami, *The Emperor's New Mathematics: Western Learning and Imperial Authority during the Kangxi Reign* (Oxford: Oxford University Press, 2011), 25–6.

56 Han Qi, 'Astronomy, Chinese and Western: The Influence of Xu Guangqi's Views in the Early and Mid-Qing', in Jami, Engelfriet, and Blue, eds., *Statecraft and Intellectual Renewal*, 362.

57 Engelfriet and Siu, 'Xu Guangqi's Attempts to Integrate Western and Chinese Mathematics', 279–99.

58 Jami, *The Emperor's New Mathematics*, 15 and 45, Engelfriet and Siu, 'Xu Guangqi's Attempts to Integrate Western and Chinese Mathematics', 279–99, and Goody, *Renaissances*, 198–240.

59 Jami, *The Emperor's New Mathematics*, 31, Joseph Needham, *Science and Civilisation in China* (Cambridge: Cambridge University Press, 1959), 3:171–6 and 3:367, and Elman, *On Their Own Terms*, 63–6.

60 Huff, *Intellectual Curiosity*, 90–8, and Elman, *On Their Own Terms*, 90.

61 Elman, *On Their Own Terms*, 84.

62 Udías, *Searching the Heavens*, 18, and Elman, *On Their Own Terms*, 64.

63 Jami, *The Emperor's New Mathematics*, 33, Needham, *Science and Civilisation*, 3:170–370, and Elman, *On Their Own Terms*, 65–8.

64 Udías, *Searching the Heavens*, 41–3.

65 Sun Xiaochun, 'On the Star Catalogue and Atlas of *Chongzhen Lishu*', in Jami, Engelfriet, and Blue, eds., *Statecraft and Intellectual Renewal*, 311–21, and Joseph Needham, *Chinese Astronomy and the Jesuit Mission: An Encounter of Cultures* (London: China Society, 1958), 1–12.

66 Needham, *Science and Civilisation*, 3:456, Jami, *The Emperor's New Mathematics*, 92, and Han, 'Astronomy, Chinese and Western', 365.

67 Virendra Nath Sharma, *Sawai Jai Singh and His Observatories* (Delhi: Motilal Banarsidass Publishers, 1995), 1–4 and 235–312, and George Rusby Kaye, *Astronomical Observatories of Jai Singh* (Calcutta: Superintendent Government Printing, 1918), 1–3.

68 Dhruv Raina, 'Circulation and Cosmopolitanism in 18th Century Jaipur', in *Cosmopolitismes en Asie du Sud: sources, itinéraires, langues (XVIe–XVIIIe siècle)*, eds. Corinne Lefèvre, Ines G. Županov, and Jorge Flores (Paris: Éditions de l'École des hautes études en sciences sociales, 2015), 307–29, S. A. Khan Ghori, 'Development of Zīj Literature in India', in *History of Astronomy in India*, eds. S. N. Sen and K. S. Shukla (Delhi: Indian National Science Academy, 1985), K. V. Sharma, 'A Survey of Source Material', in Sen and Shukla, eds., *History of Astronomy in India*, 8, Takanori Kusuba and David Pingree, *Arabic Astronomy in Sanskrit* (Leiden: Brill, 2002), 4–5.

69 Raina, 'Circulation and Cosmopolitanism', 307–29, and Huff, *Intellectual Curiosity*, 123–6.

70 Sharma, *Sawai Jai Singh*, 41–2, and Anisha Shekhar Mukherji, *Jantar Mantar: Maharaj Sawai Jai Singh's Observatory in Delhi* (New Delhi: Ambi Knowledge Resources, 2010), 15.

71 Sharma, *Sawai Jai Singh*, 304–8, and Mukherji, *Jantar Mantar*, 15.

72 Sharma, *Sawai Jai Singh*, 254, 284–97, 312, and 329–34, and S. M. R. Ansari, 'Introduction of Modern Western Astronomy in India during 18–19 Centuries', in Sen and Shukla, eds., *History of Astronomy in India*, 372.

73 Sharma, *Sawai Jai Singh*, 3 and 235–6, and Kaye, *Astronomical Observatories*, 1–14.

74 Kaye, *Astronomical Observatories*, 4–14, Mukherji, *Jantar Mantar*, 13–16, and Sharma, *Sawai Jai Singh*, 235–43.

75 Kaye, *Astronomical Observatories*, 4–14, Mukherji, *Jantar Mantar*, 13–16, and Sharma, *Sawai Jai Singh*, 235–43.

76 Kaye, *Astronomical Observatories*, 11–13.

Part Two: Empire and Enlightenment, c.1650–1800

3. Newton's Slaves

1 Simon Schaffer, 'Newton on the Beach: The Information Order of *Principia Mathematica*', *History of Science* 47 (2009): 250, Andrew Odlyzko, 'Newton's Financial Misadventures in the South Sea Bubble', *Notes and Records of the Royal Society* 73 (2019), and Helen Paul, *The South Sea Bubble: An Economic History of Its Origins and Consequences* (London: Routledge, 2011), 62.

2 Paul Lovejoy, 'The Volume of the Atlantic Slave Trade: A Synthesis', *The Journal of African History* 4 (1982): 478, John Craig, *Newton at the Mint* (Cambridge: Cambridge University Press, 1946), 106–9, Schaffer, 'Newton on the Beach', Odlyzko, 'Newton's Financial Misadventures', and MINT 19/2/261r, National Archives, London, UK, via 'MINT00256', The Newton Papers, accessed 15 November 2020, http://www.newtonproject.ox.ac.uk/view/texts/normalized/MINT00256.

3 Roy Porter, 'Introduction', in *The Cambridge History of Science: Eighteenth-Century Science*, ed. Roy Porter (Cambridge: Cambridge University Press, 2003), Gerd Buchdahl, *The Image of Newton and Locke in the Age of Reason* (London: Sheed and Ward, 1961), Thomas Hankins, *Science and the Enlightenment* (Cambridge: Cambridge University Press, 1985), and Dorinda Outram, *The Enlightenment* (Cambridge: Cambridge University Press, 1995).

4 Lovejoy, 'The Volume of the Atlantic Slave Trade', 485, John Darwin, *After Tamerlane: The Global History of Empire since 1405* (London: Allen Lane, 2007), 157–218, and Felicity Nussbaum, 'Introduction', in *The Global Eighteenth Century*, ed. Felicity Nussbaum (Baltimore: Johns Hopkins University Press, 2003).

5 Richard Drayton, 'Knowledge and Empire', in *The Oxford History of the British Empire: The Eighteenth Century*, ed. Peter Marshall (Oxford: Oxford University Press, 1998), Charles Withers and David Livingstone, 'Introduction: On Geography and Enlightenment', in *Geography and Enlightenment*, eds. Charles Withers and David Livingstone (Chicago: University of Chicago Press, 1999), Larry Stewart, 'Global Pillage: Science, Commerce, and Empire', in Porter, ed., *The Cambridge History of Science: Eighteenth-Century Science*, Mark Govier, 'The Royal Society, Slavery and the Island of Jamaica, 1660–1700', *Notes and Records of the Royal Society* 53 (1999), and Sarah Irving, *Natural Science and the Origins of the British Empire* (London: Pickering & Chatto, 2008), 1.

6 Anthony Grafton with April Shelford and Nancy Siraisi, *New Worlds, Ancient Texts: The Power of Tradition and the Shock of Discovery* (Cambridge, MA: The Belknap Press, 1992), 198, Irving, *Natural Science*, 1–44, and Jorge Cañizares-Esguerra, *Nature, Empire, and Nation: Explorations of the History of Science in the Iberian World* (Stanford: Stanford University Press, 2006), 15–18.

7 Steven Harris, 'Long-Distance Corporations, Big Sciences, and the Geography of Knowledge', *Configurations* 6 (1998), and Rob Iliffe, 'Science and Voyages of Discovery', in Porter, ed., *The Cambridge History of Science: Eighteenth-Century Science*.

8 Schaffer, 'Newton on the Beach'.

9 Isaac Newton, *The Principia: The Authoritative Translation and Guide*, trans. I. Bernard Cohen and Anne Whitman (Berkeley: The University of California Press, 2016), 829–32, John Olmsted, 'The Scientific Expedition of Jean Richer to Cayenne (1672–1673)', *Isis* 34 (1942), Nicholas Dew, 'Scientific Travel in the Atlantic World: The French Expedition to Gorée and the Antilles, 1681–1683', *The British Journal for the History of Science* 43 (2010), and Nicholas Dew, '*Vers la ligne*: Circulating Measurements around the French Atlantic', in *Science and Empire in the Atlantic World*, eds. James Delbourgo and Nicholas Dew (New York: Routledge, 2008).

10 Olmsted, 'The Scientific Expedition of Jean Richer', 118–22, and Jean Richer, *Observations astronomiques et physiques faites en l'Isle de Caienne* (Paris: De l'Imprimerie Royale, 1679).

11 Dew, 'Scientific Travel in the Atlantic World', 8–17.

12 Schaffer, 'Newton on the Beach', 261.

13 Newton, *Principia*, 832.

14 Schaffer, 'Newton on the Beach', 250–7, and David Cartwright, 'The Tonkin Tides Revisited', *Notes and Records of the Royal Society* 57 (2003).

15 Michael Hoskin, 'Newton and Newtonianism', in *The Cambridge Illustrated History of Astronomy*, ed. Michael Hoskin (Cambridge: Cambridge University Press, 1997), Larrie Ferreiro, *Measure of the Earth: The Enlightenment Expedition That Reshaped Our World* (New York: Basic Books, 2011), 7–8, and Henry Alexander ed., *The Leibniz–Clarke Correspondence: Together with Extracts from Newton's Principia and Opticks* (Manchester: Manchester University Press, 1956), 184.

16 Hoskin, 'Newton and Newtonianism', and Rob Iliffe and George Smith, 'Introduction', in *The Cambridge Companion to Newton*, eds. Rob Iliffe and George Smith (Cambridge: Cambridge University Press, 2016).
17 Iliffe, 'Science and Voyages of Discovery', and John Shank, *The Newton Wars and the Beginning of the French Enlightenment* (Chicago: University of Chicago Press, 2008).
18 Ferreiro, *Measure of the Earth*, 132–6.
19 Ferreiro, *Measure of the Earth*, xiv–xvii, Neil Safier, *Measuring the New World: Enlightenment Science and South America* (Chicago: University of Chicago Press, 2008), 2–7, Michael Hoare, *The Quest for the True Figure of the Earth: Ideas and Expeditions in Four Centuries of Geodesy* (Aldershot: Ashgate, 2005), 81–141, Mary Terrall, *The Man Who Flattened the Earth: Maupertuis and the Sciences in the Enlightenment* (Chicago: University of Chicago Press, 2002), and Rob Iliffe, '"Aplatisseur du Monde et de Cassini": Maupertuis, Precision Measurement, and the Shape of the Earth in the 1730s', *History of Science* 31 (1993).
20 Safier, *Measuring the New World*, 7, and Ferreiro, *Measure of the Earth*, 31–8.
21 Ferreiro, *Measure of the Earth*, 62–89.
22 Hoare, *The Quest for the True Figure of the Earth*, 12–13, and Ferreiro, *Measure of the Earth*, 133–4.
23 Ferreiro, *Measure of the Earth*, 105–8 and 114.
24 Ferreiro, *Measure of the Earth*, 108, Iván Ghezzi and Clive Ruggles, 'Chankillo', in *Handbook of Archaeoastronomy and Ethnoastronomy*, ed. Clive Ruggles (New York: Springer Reference, 2015), 808–13, Clive Ruggles, 'Geoglyphs of the Peruvian Coast', in Ruggles, ed., *Handbook of Archaeoastronomy and Ethnoastronomy*, 821–2.
25 Brian Bauer and David Dearborn, *Astronomy and Empire in the Ancient Andes: The Cultural Origins of Inca Sky Watching* (Austin: University of Texas Press, 1995), 14–16, Brian Bauer, *The Sacred Landscape of the Inca: The Cusco Ceque System* (Austin: University of Texas Press, 1998), 4–9, and Reiner Tom Zuidema, 'The Inca Calendar', in *Native American Astronomy*, ed. Anthony Aveni (Austin: University of Texas Press, 1977), 220–33.
26 Zuidema, 'Inca Calendar', 250, and Bauer, *Sacred Landscape*, 8.
27 Ferreiro, *Measure of the Earth*, 26 and 107–11, Bauer and Dearborn, *Astronomy and Empire*, 27, and Safier, *Measuring the New World*, 87–8.
28 Ferreiro, *Measure of the Earth*, 108.
29 Ferreiro, *Measure of the Earth*, 221–2.
30 Teuira Henry, 'Tahitian Astronomy', *Journal of the Polynesian Society* 16 (1907): 101–4, and William Frame and Laura Walker, *James Cook: The Voyages* (Montreal: McGill-Queen's University Press, 2018), 40.
31 Henry, 'Tahitian Astronomy', 101–2, Frame and Walker, *James Cook*, 40, Andrea Wulf, *Chasing Venus: The Race to Measure the Heavens* (London: William Heinemann, 2012), xix–xxvi, and Harry Woolf, *The Transits of Venus: A Study of Eighteenth-Century Science* (Princeton: Princeton University Press, 1959), 3–22.
32 Iliffe, 'Science and Voyages of Discovery', 624–8, Wulf, *Chasing Venus*, 128, and Anne Salmond, *The Trial of the Cannibal Dog: Captain Cook and the South Seas* (London: Penguin Books, 2004), 31–2.

33 Newton, *Principia*, 810–15, and Woolf, *Transits of Venus*, 3.
34 Wulf, *Chasing Venus*, xix–xxiv, and Woolf, *Transits of Venus*, 3–16.
35 Wulf, *Chasing Venus*, 185, and Woolf, *Transits of Venus*, 182–7.
36 Rebekah Higgitt and Richard Dunn, 'Introduction', in *Navigational Empires in Europe and Its Empires, 1730–1850*, eds. Rebekah Higgitt and Richard Dunn (Basingstoke: Palgrave Macmillan, 2016), Wayne Orchiston, 'From the South Seas to the Sun', in *Science and Exploration in the Pacific: European Voyages to the Southern Oceans in the Eighteenth Century*, ed. Margarette Lincoln (Woodbridge: Boydell & Brewer, 1998), 55–6, and Iliffe, 'Science and Voyages of Discovery', 635.
37 Salmond, *Trial*, 51.
38 Salmond, *Trial*, 64–7, Wulf, *Chasing Venus*, 168, and Simon Schaffer, 'In Transit: European Cosmologies in the Pacific', in *The Atlantic World in the Antipodes: Effects and Transformations since the Eighteenth Century*, ed. Kate Fullagar (Newcastle: Cambridge Scholars Publishing, 2012), 70.
39 Salmond, *Trial*, 79, Orchiston, 'From the South Seas', 58–9, Charles Green, 'Observations Made, by Appointment of the Royal Society, at King George's Island in the South Seas', *Philosophical Transactions* 61 (1771): 397 and 411.
40 Wulf, *Chasing Venus*, 192–3, Orchiston, 'From the South Seas', 59, and Vladimir Shiltsev, 'The 1761 Discovery of Venus' Atmosphere: Lomonosov and Others', *Journal of Astronomical History and Heritage* 17 (2014): 85–8.
41 Wulf, *Chasing Venus*, 201.
42 Salmond, *Trial*, 95, and David Lewis, *We, the Navigators: The Ancient Art of Landfinding in the Pacific* (Honolulu: University of Hawaii Press, 1994).
43 Salmond, *Trial*, 38–9, Lewis, *We, the Navigators*, 7–8, Joan Druett, *Tupaia: Captain Cook's Polynesian Navigator* (Auckland: Random House, 2011), 1–11, and Lars Eckstein and Anja Schwarz, 'The Making of Tupaia's Map: A Story of the Extent and Mastery of Polynesian Navigation, Competing Systems of Wayfinding on James Cook's *Endeavour*, and the Invention of an Ingenious Cartographic System', *The Journal of Pacific History* 54 (2019): 4.
44 Lewis, *We, the Navigators*, 82–101, and Ben Finney, 'Nautical Cartography and Traditional Navigation in Oceania', in *The History of Cartography: Cartography in the Traditional African, American, Arctic, Australian, and Pacific Societies*, eds. David Woodward and G. Malcolm Lewis (Chicago: University of Chicago Press, 1998), 2:443.
45 Finney, 'Nautical Cartography', 443 and 455–79, and Lewis, *We, the Navigators*, 218–48.
46 Druett, *Tupaia*, 2, Salmond, *Trial*, 38–9, and Eckstein and Schwarz, 'Tupaia's Map', 4.
47 Salmond, *Trial*, 37–40, and Eckstein and Schwarz, 'Tupaia's Map', 4.
48 Salmond, *Trial*, 112, Eckstein and Schwarz, 'Tupaia's Map', 93–4, and Finney, 'Nautical Cartography', 446.
49 Salmond, *Trial*, 99–101, and Eckstein and Schwarz, 'Tupaia's Map', 5.
50 Eckstein and Schwarz, 'Tupaia's Map'.
51 Eckstein and Schwarz, 'Tupaia's Map', 29–52.
52 Eckstein and Schwarz, 'Tupaia's Map', 32–52.
53 Salmond, *Trial*, 110–13, and Eckstein and Schwarz, 'Tupaia's Map', 5.
54 Eckstein and Schwarz, 'Tupaia's Map', 6–13.

55 Valentin Boss, *Newton and Russia: The Early Influence, 1698–1796* (Cambridge, MA: Harvard University Press, 1972), 2–5, Loren Graham, *Science in Russia and the Soviet Union: A Short History* (Cambridge: Cambridge University Press, 1993), 17, and Alexander Vucinich, *Science in Russian Culture: A History to 1860* (London: P. Owen, 1965), 1:51.

56 Boss, *Newton and Russia*, 5–14, Vucinich, *Science in Russian Culture*, 1:43–4, Arthur MacGregor, 'The Tsar in England: Peter the Great's Visit to London in 1698', *The Seventeenth Century* 19 (2004): 129–31, and Papers Connected with the *Principia*, MS Add. 3965.12, ff.357–358, Cambridge University Library, Cambridge, UK, via 'NATP00057', The Newton Papers, accessed 15 November 2020, http://www.newtonproject.ox.ac.uk/view/texts/normalized/NATP00057.

57 Boss, *Newton and Russia*, 9, and Vucinich, *Science in Russian Culture*, 1:51–4 and 1:74.

58 Boss, *Newton and Russia*, 116 and 235, Vucinich, *Science in Russian Culture*, 1:45 and 1:75–6, Wulf, *Chasing Venus*, 97, and Simon Werrett, 'Better Than a Samoyed: Newton's Reception in Russia', in *Reception of Isaac Newton in Europe*, eds. Helmut Pulte and Scott Mandelbrote (London: Bloomsbury, 2019), 1:217–23.

59 Boss, *Newton and Russia*, 94–5, and John Appleby, 'Mapping Russia: Farquharson, Delisle and the Royal Society', *Notes and Records of the Royal Society* 55 (2001): 192.

60 Andreĭ Grinëv, *Russian Colonization of Alaska: Preconditions, Discovery, and Initial Development, 1741–1799*, trans. Richard Bland (Lincoln, NE: University of Nebraska Press, 2018), 73, Alexey Postnikov and Marvin Falk, *Exploring and Mapping Alaska: The Russian America Era, 1741–1867*, trans. Lydia Black (Fairbanks: University of Alaska Press, 2015), 2–6, and Orcutt Frost, *Bering: The Russian Discovery of America* (New Haven: Yale University Press, 2003), xiii–xiv.

61 Frost, *Bering*, xiii and 34.

62 Robin Inglis, *Historical Dictionary of the Discovery and Exploration of the Northwest Coast of America* (Lanham: Scarecrow Press, 2008), xxxi–xxxii.

63 Frost, *Bering*, 40–63.

64 Frost, *Bering*, 65–158, Postnikov and Falk, *Exploring and Mapping*, 32 and 46, and Carol Urness, 'Russian Mapping of the North Pacific to 1792', in *Enlightenment and Exploration in the North Pacific, 1741–1805*, eds. Stephen Haycox, James Barnett, and Caedmon Liburd (Seattle: University of Washington Press, 1997), 132–7.

65 Frost, *Bering*, 144–58, Frank Golder, ed., *Bering's Voyages: An Account of the Efforts of the Russians to Determine the Relation of Asia and America* (New York: American Geographical Society, 1922), 1:91–9, and Dean Littlepage, *Steller's Island: Adventures of a Pioneer Naturalist in Alaska* (Seattle: Mountaineers Books, 2006), 61–2.

66 Inglis, *Historical Dictionary*, xlix and 39, Urness, 'Russian Mapping', 139–42, Postnikov and Falk, *Exploring and Mapping*, 78–174, and Simon Werrett, 'Russian Responses to the Voyages of Captain Cook', in *Captain Cook: Explorations and Reassessments*, ed. Glyndwr Williams (Woodbridge: Boydell & Brewer, 2004), 184–7.

67 Postnikov and Falk, *Exploring and Mapping*, 159–61, Werrett, 'Better Than a Samoyed', 226, and Alekseĭ Postnikov, 'Learning from Each Other: A History of Russian–Native Contacts in Late Eighteenth–Early Nineteenth Century Exploration and Mapping of Alaska and the Aleutian Islands', *International Hydrographic Review* 6 (2005): 10.

68 Postnikov and Falk, *Exploring and Mapping*, 99.

69 John MacDonald, *The Arctic Sky: Inuit Astronomy, Star Lore, and Legend* (Toronto: Royal Ontario Museum, 1998), 5–15, 101, and 164–7, and Ülo Siimets, 'The Sun, the Moon and Firmament in Chukchi Mythology and on the Relations of Celestial Bodies and Sacrifices', *Folklore* 32 (2006): 133–48.

70 MacDonald, *Arctic Sky*, 9, 44–5, and Siimets, 'Sun, Moon and Firmament', 148–50.

71 MacDonald, *Arctic Sky*, 173–8, and David Lewis and Mimi George, 'Hunters and Herders: Chukchi and Siberian Eskimo Navigation across Snow and Frozen Sea', *The Journal of Navigation* 44 (1991): 1–5.

72 Postnikov and Falk, *Exploring and Mapping*, 99–100, Inglis, *Historical Dictionary*, 96, and John Bockstoce, *Fur and Frontiers in the Far North: The Contest among Native and Foreign Nations for the Bering Fur Trade* (New Haven: Yale University Press, 2009), 75–6.

73 Postnikov and Falk, *Exploring and Mapping*, 161–74.

74 Dew, '*Vers la ligne*', 53.

75 Shino Konishi, Maria Nugent, and Tiffany Shellam, 'Exploration Archives and Indigenous Histories', in *Indigenous Intermediaries: New Perspectives on Exploration Archives*, eds. Shino Konishi, Maria Nugent, and Tiffany Shellam (Acton: Australian National University Press, 2015), Simon Schaffer, Lissa Roberts, Kapil Raj, and James Delbourgo, 'Introduction', in *The Brokered World: Go-Betweens and Global Intelligence, 1770–1820*, eds. Simon Schaffer, Lissa Roberts, Kapil Raj, and James Delbourgo (Sagamore Beach: Science History Publications, 2009), and Schaffer, 'Newton on the Beach', 267.

76 Vincent Carretta, 'Who was Francis Williams?', *Early American Literature* 38 (2003), and Gretchen Gerzina, *Black London: Life before Emancipation* (New Brunswick: Rutgers University Press, 1995), 6 and 40–1.

4. *Economy of Nature*

1 Natural History Museum [hereafter NHM], 'Slavery and the Natural World, Chapter 2: People and Slavery', accessed 15 October 2019, https://www.nhm.ac.uk/content/dam/nhmwww/discover/slavery-natural-world/chapter-2-people-and-slavery.pdf, and Susan Scott Parrish, *American Curiosity: Cultures of Natural History in the Colonial British Atlantic World* (Chapel Hill: University of North Carolina Press, 2006), 1–10.

2 NHM, 'Slavery and the Natural World, Chapter 2: People and Slavery', Parrish, *American Curiosity*, 1–10, and Londa Schiebinger, *Plants and Empire: Colonial Bioprospecting in the Atlantic World* (Cambridge, MA: Harvard University Press, 2009), 8.

3 Parrish, *American Curiosity*, 1–10, Schiebinger, *Plants and Empire*, 209–19, and Lisbet Koerner, 'Carl Linnaeus in His Time and Place', in *Cultures of Natural History*, eds. Nicholas Jardine, James Secord, and Emma Spary (Cambridge: Cambridge University Press, 1996), 145–9.

4 NHM, 'Slavery and the Natural World, Chapter 2: People and Slavery', and Parrish, *American Curiosity*, 1–10.

5 Richard Drayton, *Nature's Government: Science, Imperial Britain, and the 'Improvement' of the World* (New Haven: Yale University Press, 2000), Harold Cook, *Matters of Exchange: Commerce, Medicine, and Science in the Dutch Golden Age* (New Haven: Yale University Press, 2007), Dániel Margócsy, *Commercial Visions: Science, Trade, and Visual Culture in the Dutch Golden Age* (Chicago: University of Chicago Press, 2014), Londa Schiebinger and Claudia Swan, eds., *Colonial Botany: Science, Commerce, and Politics in the Early Modern World* (Philadelphia: University of Pennsylvania Press, 2005), Kris Lane, 'Gone Platinum: Contraband and Chemistry in Eighteenth-Century Colombia', *Colonial Latin American Review* 20 (2011), and Schiebinger, *Plants and Empire*, 194.

6 Schiebinger, *Plants and Empire*, 7–8, and Lisbet Koerner, *Linnaeus: Nature and Nation* (Cambridge, MA: Harvard University Press, 1999), 1–2.

7 Drayton, *Nature's Government*, Schiebinger, *Plants and Empire*, Miles Ogborn, 'Vegetable Empire', in *Worlds of Natural History*, eds. Helen Curry, Nicholas Jardine, James Secord, and Emma Spary (Cambridge: Cambridge University Press, 2018), and James McClellan III, *Colonialism and Science: Saint Domingue and the Old Regime* (Chicago: University of Chicago Press, 2010), 148–59.

8 Schiebinger, *Plants and Empire*, 25–30, James Delbourgo, 'Sir Hans Sloane's Milk Chocolate and the Whole History of the Cacao', *Social Text* 29 (2011), James Delbourgo, *Collecting the World: The Life and Curiosity of Hans Sloane* (London: Allen Lane, 2015), 35–59, and Edwin Rose, 'Natural History Collections and the Book: Hans Sloane's *A Voyage to Jamaica* (1707–1725) and His Jamaican Plants', *Journal of the History of Collections* 30 (2018).

9 NHM, 'Slavery and the Natural World, Chapter 2: People and Slavery', Schiebinger, *Plants and Empire*, 28, Delbourgo, *Collecting the World*, 35–59, and Hans Sloane, *A Voyage to the Islands Madera, Barbados, Nieves, S. Christophers and Jamaica* (London: B.M. for the Author, 1707).

10 Miles Ogborn, 'Talking Plants: Botany and Speech in Eighteenth-Century Jamaica', *History of Science* 51 (2013): 264, Judith Carney and Richard Rosomoff, *In the Shadow of Slavery: Africa's Botanical Legacy in the Atlantic World* (Berkeley: University of California Press, 2011), 71 and 124, and Bertram Osuagwu, *The Igbos and Their Traditions*, trans. Frances W. Pritchett (Lagos: Macmillan Nigeria, 1978), 1–22.

11 Carney and Rosomoff, *In the Shadow of Slavery*, 123–4.

12 Londa Schiebinger, *Secret Cures of Slaves: People, Plants, and Medicine in the Eighteenth-Century Atlantic World* (Stanford: Stanford University Press, 2017), 1–9 and 45–59.

13 Ogborn, 'Talking Plants', 255–71, Kathleen Murphy, 'Collecting Slave Traders: James Petiver, Natural History, and the British Slave Trade', *William and Mary Quarterly* 70 (2013), and NHM, 'Slavery and the Natural World, Chapter 7: Fevers', accessed 15 October 2019, https://www.nhm.ac.uk/content/dam/nhmwww/discover/slavery-natural-world/chapter-7-fevers.pdf.

14 Schiebinger, *Secret Cures of Slaves*, 90, Ogborn, 'Talking Plants', 275, and Kwasi Konadu, *Indigenous Medicine and Knowledge in African Society* (London: Routledge, 2007), 85–9.

15 Schiebinger, *Plants and Empire*, 1–35, NHM, 'Slavery and the Natural World, Chapter 2: People and Slavery', and Julie Hochstrasser, 'The Butterfly Effect: Embodied Cognition and Perceptual Knowledge in Maria Sibylla Merian's *Metamorphosis*

Insectorum Surinamensium', in *The Dutch Trading Companies as Knowledge Networks*, eds. Siegfried Huigen, Jan de Jong, and Elmer Kolfin (Leiden: Brill, 2010), 59–60.

16 Schiebinger, *Secret Cures of Slaves*, 12, NHM, 'Slavery and the Natural World, Chapter 6: Resistance', accessed 15 October 2019, https://www.nhm.ac.uk/content/dam/nhmwww/discover/slavery-natural-world/chapter-6-resistance.pdf, and Susan Scott Parrish, 'Diasporic African Sources of Enlightenment Knowledge', in *Science and Empire in the Atlantic World*, eds. James Delbourgo and Nicholas Dew (New York: Routledge, 2008), 294.

17 Richard Grove, 'Indigenous Knowledge and the Significance of South-West India for Portuguese and Dutch Constructions of Tropical Nature', *Modern Asian Studies* 30 (1996), K. S. Manilal, ed., *Botany and History of Hortus Malabaricus* (Rotterdam: A. A. Balkema, 1980), 1–3, J. Heniger, *Hendrik Adriaan van Reede tot Drakenstein (1636–1691) and Hortus Malabaricus* (Rotterdam: A. A. Balkema, 1986), vii–xii and 3–95, Kapil Raj, *Relocating Modern Science: Circulation and the Construction of Knowledge in South Asia and Europe, 1650–1900* (Basingstoke: Palgrave Macmillan, 2007), 44–5, and Hendrik van Rheede, *Hortus Indicus Malabaricus* (Amsterdam: Johannis van Someren, 1678), vol. 1, pl. 9.

18 Grove, 'Indigenous Knowledge', 134–5, and Heniger, *Hendrik Adriaan van Reede*, 3–33.

19 Grove, 'Indigenous Knowledge', 136–9, Heniger, *Hendrik Adriaan van Reede*, 41–64, 144–8, and H. Y. Mohan Ram, 'On the English Edition of van Rheede's *Hortus Malabaricus*', *Current Science* 89 (2005).

20 Heniger, *Hendrik Adriaan van Reede*, 147–8, and Rajiv Kamal, *Economy of Plants in the Vedas* (New Delhi: Commonwealth Publishers, 1988), 1–23.

21 Heniger, *Hendrik Adriaan van Reede*, 43 and 143–8, and Grove, 'Indigenous Knowledge', 139.

22 E. M. Beekman, 'Introduction: Rumphius' Life and Work', in Georg Eberhard Rumphius, *The Ambonese Curiosity Cabinet*, trans. E. M. Beekman (New Haven: Yale University Press, 1999), xxxv–lxvii, and Genie Yoo, 'Wars and Wonders: The Inter-Island Information Networks of Georg Everhard Rumphius', *The British Journal for the History of Science* 51 (2018): 561.

23 Beekman, 'Introduction', xxxv–xcviii, and George Sarton, 'Rumphius, Plinius Indicus (1628–1702)', *Isis* 27 (1937).

24 Matthew Sargent, 'Global Trade and Local Knowledge: Gathering Natural Knowledge in Seventeenth-Century Indonesia', in *Intercultural Exchange in Southeast Asia: History and Society in the Early Modern World*, eds. Tara Alberts and David Irving (London: I. B. Taurus, 2013), 155–6.

25 Beekman, 'Introduction', lxvii, Sargent, 'Global Trade', 156, Jeyamalar Kathirithamby-Wells, 'Unlikely Partners: Malay-Indonesian Medicine and European Plant Science', in *The East India Company and the Natural World*, eds. Vinita Damodaran, Anna Winterbottom, and Alan Lester (Basingstoke: Palgrave Macmillan, 2014), 195–203, and Benjamin Schmidt, *Inventing Exoticism: Geography, Globalism, and Europe's Early Modern World* (Philadelphia: University of Pennsylvania Press, 2015), 136–8.

26 Yoo, 'Wars and Wonders', 567–9.

27 Georg Eberhard Rumphius, *The Ambonese Curiosity Cabinet*, trans. E. M. Beekman (New Haven: Yale University Press, 1999), 93–4, Georg Eberhard Rumphius, *Rumphius'*

Orchids: Orchid Texts from The Ambonese Herbal, trans. E. M. Beekman (New Haven: Yale University Press, 2003), 87, and Maria-Theresia Leuker, 'Knowledge Transfer and Cultural Appropriation: Georg Everhard Rumphius's *D'Amboinsche Rariteitkamer* (1705)', in Huigen, de Jong, and Kolfin, eds. *The Dutch Trading Companies*.

28 Beekman, 'Introduction', lxii–lxiii.

29 Ray Desmond, *The European Discovery of the Indian Flora* (Oxford: Oxford University Press, 1992), 57–9, and Tim Robinson, *William Roxburgh: The Founding Father of Indian Botany* (Chichester: Phillimore, 2008), 41–3.

30 Desmond, *European Discovery*, 59, and Robinson, *William Roxburgh*, 41.

31 Robinson, *William Roxburgh*, 5–10, Pratik Chakrabarti, *Materials and Medicine: Trade, Conquest and Therapeutics in the Eighteenth Century* (Manchester: Manchester University Press, 2010), 41, Minakshi Menon, 'Medicine, Money, and the Making of the East India Company State: William Roxburgh in Madras, c. 1790', in *Histories of Medicine and Healing in the Indian Ocean World*, eds. Anna Winterbottom and Facil Tesfaye (Basingstoke: Palgrave, 2016), 2:152–9, and Arthur MacGregor, 'European Enlightenment in India: An Episode of Anglo-German Collaboration in the Natural Sciences on the Coromandel Coast, Late 1700s–Early 1800s', in *Naturalists in the Field: Collecting, Recording and Preserving the Natural World from the Fifteenth to the Twenty-First Century* (Leiden: Brill, 2018), 383.

32 Prakash Kumar, *Indigo Plantations and Science in Colonial India* (Cambridge: Cambridge University Press, 2012), 68–75, and Menon, 'Medicine, Money, and the Making of the East India Company State', 160.

33 Robinson, *William Roxburgh*, 43–56.

34 Robinson, *William Roxburgh*, 95, Chakrabarti, *Materials and Medicine*, 126, Beth Tobin, *Picturing Imperial Power: Colonial Subjects in Eighteenth-Century British Painting* (Durham, NC: Duke University Press, 1999), 194–201, M. Lazarus and H. Pardoe, eds., *Catalogue of Botanical Prints and Drawings: The National Museums & Galleries of Wales* (Cardiff: National Museums & Galleries of Wales, 2003), 35, I. G. Khan, 'The Study of Natural History in 16th–17th Century Indo-Persian Literature', *Proceedings of the Indian History Congress* 67 (2002), and Versha Gupta, *Botanical Culture of Mughal India* (Bloomington: Partridge India, 2018).

35 Markman Ellis, Richard Coulton, and Matthew Mauger, *Empire of Tea: The Asian Leaf That Conquered the World* (London: Reaktion Books, 2015), 32–5 and 105, and Erika Rappaport, *A Thirst for Empire: How Tea Shaped the Modern World* (Princeton: Princeton University Press, 2017), 23.

36 Ellis, Coulton, and Mauger, *Empire of Tea*, 9 and 22–57, Rappaport, *A Thirst for Empire*, 41, Linda Barnes, *Needles, Herbs, Gods, and Ghosts: China, Healing, and the West to 1848* (Cambridge, MA: Harvard University Press, 2005), 93–116 and 181–5, and Jane Kilpatrick, *Gifts from the Gardens of China* (London: Frances Lincoln, 2007), 9–16.

37 Markman Ellis, 'The British Way of Tea: Tea as an Object of Knowledge between Britain and China, 1690–1730', in *Curious Encounters: Voyaging, Collecting, and Making Knowledge in the Long Eighteenth Century*, eds. Adriana Craciun and Mary Terrall (Toronto: University of Toronto Press, 2019), 27–33.

38 Ellis, Coulton, and Mauger, *Empire of Tea*, 66–7 and 109–10.
39 Ellis, 'The British Way of Tea', 23–8, and James Ovington, *An Essay upon the Nature and Qualities of Tea* (London: R. Roberts, 1699), 7–14.
40 Ellis, 'The British Way of Tea', 29–32, Kilpatrick, *Gifts from the Gardens of China*, 34–48, and Charles Jarvis and Philip Oswald, 'The Collecting Activities of James Cuninghame FRS on the Voyage of *Tuscan* to China (Amoy) between 1697 and 1699', *Notes and Records of the Royal Society* 69 (2015).
41 Ellis, 'The British Way of Tea', 29–32, and James Cuninghame, 'Part of Two Letters to the Publisher from Mr James Cunningham, F. R.S.', *Philosophical Transactions of the Royal Society* 23 (1703): 1205–6.
42 Ellis, Coulton, and Mauger, *Empire of Tea*, 15–19, Huang Hsing-Tsung, *Science and Civilisation in China: Biology and Biological Technology, Fermentations and Food Science* (Cambridge: Cambridge University Press, 2000), vol. 6, part 5, 506–15, and James A. Benn, *Tea in China: A Religious and Cultural History* (Hong Kong: Hong Kong University Press, 2015), 117–44.
43 Carla Nappi, *The Monkey and the Inkpot: Natural History and Its Transformations in Early Modern China* (Cambridge, MA: Harvard University Press, 2009), 10–33 and 141–2, and Federico Marcon, *The Knowledge of Nature and the Nature of Knowledge in Early Modern Japan* (Chicago: University of Chicago Press, 2015), 25–50.
44 Nappi, *The Monkey and the Inkpot*, 10–33, Marcon, *The Knowledge of Nature*, 25–50, Ellis, 'The British Way of Tea', 27, Georges Métailié, *Science and Civilisation in China: Biology and Biological Technology, Traditional Botany: An Ethnobotanical Approach* (Cambridge: Cambridge University Press, 2015), vol. 6, part 4, 77–8, and Joseph Needham, *Science and Civilisation in China: Biology and Biological Technology, Botany* (Cambridge: Cambridge University Press, 1986), vol. 6, part 1, 308–21.
45 Nappi, *The Monkey and the Inkpot*, 155–8, Needham, *Science and Civilisation*, vol. 6, part 1, 308–21, Métailié, *Science and Civilisation in China*, vol. 6, part 4, 36 and 77, and Marcon, *The Knowledge of Nature*, 25–50.
46 Nappi, *The Monkey and the Inkpot*, 19, Métailié, *Science and Civilisation in China*, vol. 6, part 4, 620–5.
47 Nappi, *The Monkey and the Inkpot*, 19, Métailié, *Science and Civilisation in China*, vol. 6, part 4, 620–5, and Jordan Goodman and Charles Jarvis, 'The John Bradby Blake Drawings in the Natural History Museum, London: Joseph Banks Puts Them to Work', *Curtis's Botanical Magazine* 34 (2017): 264.
48 Marcon, *The Knowledge of Nature*, 128–31 and 161–3, and Ishiyama Hiroshi, 'The Herbal of Dodonaeus', in *Bridging the Divide: 400 Years, The Netherlands–Japan*, eds. Leonard Blussé, Willem Remmelink, and Ivo Smits (Leiden: Hotei, 2000), 100–1.
49 Marcon, *The Knowledge of Nature*, 128–31, 161–3, and 171–203.
50 Marcon, *The Knowledge of Nature*, x and 3–6, Iioka Naoko, 'Wei Zhiyan and the Subversion of the *Sakoku*', in *Offshore Asia: Maritime Interactions in Eastern Asia before Steamships*, eds. Fujita Kayoko, Shiro Momoki, and Anthony Reid (Singapore: Institute of Southeast Asian Studies, 2013), and Ronald Toby, 'Reopening the Question of *Sakoku*: Diplomacy in the Legitimation of the Tokugawa Bakufu', *Journal of Japanese Studies* 3 (1977): 358.

51 Marcon, *The Knowledge of Nature*, 113–28 and 141–6, and Marie-Christine Skuncke, *Carl Peter Thunberg: Botanist and Physician* (Uppsala: Swedish Collegium for Advanced Study, 2014), 113.

52 Marcon, *The Knowledge of Nature*, 128–31 and 161–3, and Harmen Beukers, 'Dodonaeus in Japanese: Deshima Surgeons as Mediators in the Early Introduction of Western Natural History', in *Dodonaeus in Japan: Translation and the Scientific Mind in the Tokugawa Period*, eds. W. F. Vande Walle and Kazuhiko Kasaya (Leuven: Leuven University Press, 2002), 291.

53 Marcon, *The Knowledge of Nature*, 55–73.

54 Marcon, *The Knowledge of Nature*, 6 and 87–102.

55 Marcon, *The Knowledge of Nature*, 90–6.

56 Marcon, *The Knowledge of Nature*, 91 (emphasis added).

57 Timon Screech, 'The Visual Legacy of Dodonaeus in Botanical and Human Categorisation', in Vande Walle and Kasaya, eds., *Dodonaeus in Japan*, 221–3, T. Yoshida, ' "Dutch Studies" and Natural Sciences', in Blussé, Remmelink, and Smits, eds., *Bridging the Divide*, Kenkichiro Koizumi, 'The Emergence of Japan's First Physicists: 1868–1900', *History and Philosophy of the Physical Sciences* 6 (1975): 7–13, James Bartholomew, *The Formation of Science in Japan: Building a Research Tradition* (New Haven: Yale University Press, 1989), 10–15, Marcon, *The Knowledge of Nature*, 128–30, Hiroshi, 'The Herbal of Dodonaeus', 100–1, and Tôru Haga, 'Dodonaeus and Tokugawa Culture: Hiraga Gennai and Natural History in Eighteenth-Century Japan', in Vande Walle and Kasaya, eds., *Dodonaeus in Japan*, 242–51.

58 Marcon, *The Knowledge of Nature*, 135–7, and Skuncke, *Carl Peter Thunberg*, 93–9 and 101–4.

59 Skuncke, *Carl Peter Thunberg*, 120–6.

60 Skuncke, *Carl Peter Thunberg*, 122–6.

61 Skuncke, *Carl Peter Thunberg*, 105 and 128–35, and Marcon, *The Knowledge of Nature*, 135–7.

62 Skuncke, *Carl Peter Thunberg*, 130 and 206, and Richard Rudolph, 'Thunberg in Japan and His *Flora Japonica* in Japanese', *Monumenta Nipponica* 29 (1974): 168.

63 Carl Thunberg, *Flora Japonica* (Leipzig: I. G. Mülleriano, 1784), 229.

Part Three: Capitalism and Conflict, c.1790–1914

5. Struggle for Existence

1 Justin Smith, 'The Ibis and the Crocodile: Napoleon's Egyptian Campaign and Evolutionary Theory in France, 1801–1835', *Republic of Letters* 6 (2018), Paul Nicholson, 'The Sacred Animal Necropolis at North Saqqara: The Cults and Their Catacombs', in *Divine Creatures: Animal Mummies in Ancient Egypt*, ed. Salima Ikram (Cairo: American University in Cairo Press, 2005), and Caitlin Curtis, Craig Millar, and David Lambert, 'The Sacred Ibis Debate: The First Test of Evolution', *PLOS Biology* 16 (2018).

2 Jean Herold, *Bonaparte in Egypt* (London: Hamish Hamilton, 1962), 164–200, Charles Gillispie, 'Scientific Aspects of the French Egyptian Expedition 1798–1801', *Proceedings of the American Philosophical Society* 133 (1989), Nina Burleigh, *Mirage: Napoleon's Scientists and the Unveiling of Egypt* (New York: Harper, 2007), vi–x, and Jane Murphy, 'Locating the Sciences in Eighteenth-Century Egypt', *The British Journal for the History of Science* 43 (2010).

3 Toby Appel, *The Cuvier–Geoffroy Debate: French Biology in the Decades before Darwin* (Oxford: Oxford University Press, 1987), 1–10 and 69–97, Burleigh, *Mirage*, 195–207, Curtis, Millar, and Lambert, 'The Sacred Ibis Debate', Smith, 'The Ibis and the Crocodile', and Murphy, 'Locating the Sciences', 558–65.

4 Appel, *The Cuvier–Geoffroy Debate*, 72–7, and Nicholson, 'The Sacred Animal Necropolis', 44–52.

5 Curtis, Millar, and Lambert, 'The Sacred Ibis Debate', 2–5, Smith, 'The Ibis and the Crocodile', 5–9, and Martin Rudwick, *Bursting the Limits of Time: The Reconstruction of Geohistory in the Age of Revolution* (Chicago: University of Chicago Press, 2007), 394–6.

6 Curtis, Millar, and Lambert, 'The Sacred Ibis Debate', 2–5, Smith, 'The Ibis and the Crocodile', 5–9, Rudwick, *Bursting the Limits of Time*, 394–6, Appel, *The Cuvier–Geoffroy Debate*, 82, and Martin Rudwick, *Georges Cuvier, Fossil Bones, and Geological Catastrophes: New Translations and Interpretations of the Primary Texts* (Chicago: University of Chicago Press, 2008), 229.

7 Smith, 'The Ibis and the Crocodile', 4, Robert Young, *Darwin's Metaphor: Nature's Place in Victorian Culture* (Cambridge: Cambridge University Press, 1985), 40–1, Marwa Elshakry, 'Spencer's Arabic Readers', in *Global Spencerism: The Communication and Appropriation of a British Evolutionist*, ed. Bernard Lightman (Leiden: Brill, 2016), and G. Clinton Godart, 'Spencerism in Japan: Boom and Bust of a Theory', in *Global Spencerism*, ed. Lightman.

8 Janet Browne, *Charles Darwin: Voyaging* (London: Jonathan Cape, 1995), and Ana Sevilla, '*On the Origin of Species* and the Galapagos Islands', in *Darwin, Darwinism and Conservation in the Galapagos Islands*, eds. Diego Quiroga and Ana Sevilla (Cham: Springer International, 2017).

9 James Secord, 'Global Darwin', in *Darwin*, eds. William Brown and Andrew Fabian (Cambridge: Cambridge University Press, 2010), Alexander Vucinich, *Darwin in Russian Thought* (Berkeley: University of California Press, 1989), 12, and G. Clinton Godart, *Darwin, Dharma, and the Divine: Evolutionary Theory and Religion in Modern Japan* (Honolulu: University of Hawaii Press, 2017), 19–20.

10 Alex Levine and Adriana Novoa, *¡Darwinistas! The Construction of Evolutionary Thought in Nineteenth Century Argentina* (Leiden: Brill, 2012), x–xii, 85, and 91–5, and Adriana Novoa and Alex Levine, *From Man to Ape: Darwinism in Argentina, 1870–1920* (Chicago: University of Chicago Press, 2010), 17.

11 Levine and Novoa, *¡Darwinistas!*, 91–5, and Novoa and Levine, *From Man to Ape*, 33–7.

12 Levine and Novoa, *¡Darwinistas!*, 85–95, Novoa and Levine, *From Man to Ape*, 33–7, Charles Darwin to Francisco Muñiz, 26 February 1847, Darwin Correspondence Project, Letter no. 1063, accessed 14 August 2020, https://www.darwinproject.ac.uk/letter/DCP-LETT-1063.xml, Charles Darwin to Richard Owen, 12 February

[1847], Darwin Correspondence Project, Letter no. 1061, accessed 14 August 2020, https://www.darwinproject.ac.uk/letter/DCP-LETT-1061.xml, and Charles Darwin to Richard Owen, [4 February 1842], Darwin Correspondence Project, Letter no. 617G, accessed 14 August 2020, https://www.darwinproject.ac.uk/letter/DCP-LETT-617G.xml.

13 Levine and Novoa, *¡Darwinistas!*, 85, Novoa and Levine, *From Man to Ape*, 31, Arturo Argueta Villamar, 'Darwinism in Latin America: Reception and Introduction', in Quiroga and Sevilla, eds., *Darwin, Darwinism and Conservation*, and Thomas Glick, Miguel Ángel Puig-Samper, and Rosaura Ruiz, eds., *The Reception of Darwinism in the Iberian World: Spain, Spanish America, and Brazil* (Dordrecht: Springer Netherlands, 2001).

14 Novoa and Levine, *From Man to Ape*, 18–19, 30, and 78–81, Maria Margaret Lopes and Irina Podgorny, 'The Shaping of Latin American Museums of Natural History, 1850–1990', *Osiris* 15 (2000): 108–18, and Carolyne Larson, '"Noble and Delicate Sentiments": Museum Natural Scientists as an Emotional Community in Argentina, 1862–1920', *Historical Studies in the Natural Sciences* 47 (2017): 43–50.

15 Levine and Novoa, *¡Darwinistas!*, 113–6, Novoa and Levine, *From Man to Ape*, 83–7, Larson, ' "Noble and Delicate Sentiments" ', 53, Marcelo Montserrat, 'The Evolutionist Mentality in Argentina: An Ideology of Progress', in Glick, Puig-Samper, and Ruiz, eds., *The Reception of Darwinism*, 6, and Francisco Moreno, *Viaje a la patagonia austral* (Buenos Aires: Sociedad de Abogados Editores), 28 and 199.

16 Levine and Novoa, *¡Darwinistas!*, 113–6, Novoa and Levine, *From Man to Ape*, 83–7, Carolyne Larson, *Our Indigenous Ancestors: A Cultural History of Museums, Science, and Identity in Argentina, 1877–1943* (University Park: Penn State University Press, 2015), 17–20, and Frederico Freitas, 'The Journeys of Francisco Moreno', accessed 5 June 2020, https://fredericofreitas.org/2009/08/18/the-journeys-of-francisco-moreno/.

17 Levine and Novoa, *¡Darwinistas!*, 113–23, and Novoa and Levine, *From Man to Ape*, 83–7 and 148–50.

18 Levine and Novoa, *¡Darwinistas!*, 116, Larson, *Our Indigenous Ancestors*, 35–42, and Sadiah Qureshi, 'Looking to Our Ancestors', in *Time Travelers: Victorian Encounters with Time and History*, eds. Adelene Buckland and Sadiah Qureshi (Chicago: University of Chicago Press, 2020).

19 Larson, *Our Indigenous Ancestors*, 35–42, Novoa and Levine, *From Man to Ape*, 125, and Carlos Gigoux, ' "Condemned to Disappear": Indigenous Genocide in Tierra del Fuego', *Journal of Genocide Research* (2020).

20 Levine and Novoa, *¡Darwinistas!*, 113–5, and Novoa and Levine, *From Man to Ape*, 149–53.

21 Levine and Novoa, *¡Darwinistas!*, 195–9, Novoa and Levine, *From Man to Ape*, 145, Montserrat, 'The Evolutionist Mentality in Argentina', 6, Larson, '"Noble and Delicate Sentiments"', 57–66, and Irina Podgorny, 'Bones and Devices in the Constitution of Paleontology in Argentina at the End of the Nineteenth Century', *Science in Context* 18 (2005).

22 Levine and Novoa, *¡Darwinistas!*, 200–2.

23 Levine and Novoa, *¡Darwinistas!*, 200–2.

24 Thomas Glick, 'The Reception of Darwinism in Uruguay', in Glick, Puig-Samper, and Ruiz, eds., *The Reception of Darwinism*, Pedro M. Pruna Goodgall, 'Biological Evolutionism in Cuba at the End of the Nineteenth Century', in Glick, Puig-Samper, and Ruiz, eds., *The Reception of Darwinism*, Roberto Moreno, 'Mexico', in *The Comparative Reception of Darwinism*, ed. Thomas Glick (Chicago: University of Chicago Press, 1988).
25 Levine and Novoa, *¡Darwinistas!*, 138, and Podgorny, 'Bones and Devices', 261.
26 Vucinich, *Darwin in Russian Thought*, 217–8, Daniel Todes, *Darwin Without Malthus: The Struggle for Existence in Russian Evolutionary Thought* (Oxford: Oxford University Press, 1989), 143–6, Nikolai Severtzov, 'The Mammals of Turkestan', *Annals and Magazine of Natural History* 36 (1876), and Nikolai Severtzov to Charles Darwin, 26 September [1875], Darwin Correspondence Project, Letter no. 10172, accessed 14 August 2020, https://www.darwinproject.ac.uk/letter/DCP-LETT-10172.xml.
27 Todes, *Darwin Without Malthus*, 144–7.
28 Todes, *Darwin Without Malthus*, 146–51, and Severtzov, 'The Mammals of Turkestan', 41–5, 172–217, and 330–3.
29 Todes, *Darwin Without Malthus*, 148–51.
30 Vucinich, *Darwin in Russian Thought*, 12–32, and James Rogers, 'The Reception of Darwin's *Origin of Species* by Russian Scientists', *Isis* 64 (1973).
31 Alexander Vucinich, *Science in Russian Culture: A History to 1860* (London: Peter Owen, 1965), 247–384, and Alexander Vucinich, *Science in Russian Culture, 1861–1917* (Stanford: Stanford University Press, 1970), 3–86.
32 Vucinich, *Darwin in Russian Thought*, 18–19 and 84, Michael Katz, 'Dostoevsky and Natural Science', *Dostoevsky Studies* 9 (1988), George Kline, 'Darwinism and the Russian Orthodox Church', in *Continuity and Change in Russian and Soviet Thought*, ed. Ernest Simmons (Cambridge, MA: Harvard University Press, 1955), Anna Berman, 'Darwin in the Novels: Tolstoy's Evolving Literary Response', *The Russian Review* 76 (2017), and Leo Tolstoy, *Anna Karenina*, trans. Constance Garnett (New York: The Modern Library, 2000), 533.
33 Todes, *Darwin Without Malthus*, 3–29.
34 Todes, *Darwin Without Malthus*, 82–102, Vucinich, *Darwin in Russian Thought*, 278–81, Kirill Rossiianov, 'Taming the Primitive: Elie Metchnikov and His Discovery of Immune Cells', *Osiris* 23 (2008), and Ilya Mechnikov, 'Nobel Lecture: On the Present State of the Question of Immunity in Infectious Diseases', The Nobel Prize, accessed 14 August 2020, https://www.nobelprize.org/prizes/medicine/1908/mechnikov/lecture/
35 Todes, *Darwin Without Malthus*, 82–5 and 91.
36 Todes, *Darwin Without Malthus*, 82–102, and Vucinich, *Darwin in Russian Thought*, 278–81.
37 Rossiianov, 'Taming the Primitive', 223, and Vucinich, *Darwin in Russian Thought*, 281.
38 Rossiianov, 'Taming the Primitive', 214.
39 Ann Koblitz, 'Science, Women, and the Russian Intelligentsia: The Generation of the 1860s', *Isis* 79 (1988), Mary Creese, *Ladies in the Laboratory IV: Imperial Russia's Women in Science, 1800–1900: A Survey of Their Contributions to Research* (Lanham: Rowman & Littlefield, 2015), xi–xii and 76–8, and Marilyn Ogilvie and Joy

Harvey, 'Sofia Pereiaslavtseva', in *The Biographical Dictionary of Women in Science: Pioneering Lives from Ancient Times to the Mid-20th Century*, eds. Marilyn Ogilvie and Joy Harvey (London: Routledge, 2000).

40 Creese, *Ladies in the Laboratory IV*, 76–8.

41 Creese, *Ladies in the Laboratory IV*, 76–8.

42 Todes, *Darwin Without Malthus*, 123–34, and Jerry Bergman, *The Darwin Effect: Its Influence on Nazism, Eugenics, Racism, Communism, Capitalism, and Sexism* (Master Books: Green Forest, 2014), 288–9.

43 Todes, *Darwin Without Malthus*, 45–7.

44 Todes, *Darwin Without Malthus*, 51–9.

45 Todes, *Darwin Without Malthus*, 51–9.

46 Vucinich, *Darwin in Russian Thought*, 87.

47 Godart, *Darwin, Dharma, and the Divine*, 2–3 and 26–30, Masao Watanabe, *The Japanese and Western Science* (Philadelphia: University of Pennsylvania Press, 1990), 41–67, Kuang-chi Hung, 'Alien Science, Indigenous Thought and Foreign Religion: Reconsidering the Reception of Darwinism in Japan', *Intellectual History Review* 19 (2009), and Ian Miller, *The Nature of the Beasts: Empire and Exhibition at the Tokyo Imperial Zoo* (Berkeley: University of California Press, 2013), 51.

48 Miller, *The Nature of the Beasts*, 51–2.

49 Miller, *The Nature of the Beasts*, 49–50, Taku Komai, 'Genetics of Japan, Past and Present', *Science* 123 (1956): 823, and James Bartholomew, *The Formation of Science in Japan: Building a Research Tradition* (New Haven: Yale University Press, 1989), 59.

50 Bartholomew, *The Formation of Science in Japan*, 49–100, and Watanabe, *The Japanese and Western Science*, 41–67.

51 Hung, 'Alien Science', 231, Godart, *Darwin, Dharma, and the Divine*, 28, Watanabe, *The Japanese and Western Science*, 39–50, Eikoh Shimao, 'Darwinism in Japan, 1877–1927', *Annals of Science* 38 (1981): 93, and Isono Naohide, 'Contributions of Edward S. Morse to Developing Young Japan', in *Foreign Employees in Nineteenth-Century Japan*, eds. Edward Beauchamp and Akira Iriye (Boulder: Westview, 1990).

52 Komai, 'Genetics of Japan', 823, Bartholomew, *The Formation of Science in Japan*, 68–70, and Frederick Churchill, *August Weismann: Development, Heredity, and Evolution* (Cambridge, MA: Harvard University Press, 2015), 354–6.

53 Churchill, *August Weismann*, 354–6 and 644–5, and Komai, 'Genetics of Japan', 823.

54 Watanabe, *The Japanese and Western Science*, 71–3.

55 Godart, *Darwin, Dharma, and the Divine*, 2–21.

56 Godart, *Darwin, Dharma, and the Divine*, 103–12, Watanabe, *The Japanese and Western Science*, 84–95, Shimao, 'Darwinism in Japan', 95, Gregory Sullivan, 'Tricks of Transference: Oka Asajirō (1868–1944) on Laissez-Faire Capitalism', *Science in Context* 23 (2010): 370–85, and Gregory Sullivan, *Regenerating Japan: Organicism, Modernism and National Destiny in Oka Asajirō's Evolution and Human Life* (Budapest: Central European University Press, 2018), 1–3.

57 Godart, *Darwin, Dharma, and the Divine*, 103–12, Watanabe, *The Japanese and Western Science*, 84–95, and Sullivan, 'Tricks of Transference', 373–85.

58 Godart, *Darwin, Dharma, and the Divine*, 103, Watanabe, *The Japanese and Western Science*, 84–95, Sullivan, 'Tricks of Transference', 370–85, and Ernest Lee and Stef-

anos Kales, 'Chemical Weapons', in *War and Public Health*, eds. Barry Levy and Victor Sidel (Oxford: Oxford University Press, 2008), 128.

59 Bartholomew, *The Formation of Science in Japan*, 69–70, and Watanabe, *The Japanese and Western Science*, 95.

60 Xiaoxing Jin, 'The Evolution of Evolutionism in China, 1870–1930', *Isis* 111 (2020): 50–1.

61 Jin, 'The Evolution of Evolutionism in China', 50–2, Xiaoxing Jin, 'Translation and Transmutation: *The Origin of Species* in China', *The British Journal for the History of Science* 52 (2019): 122–3, and Yang Haiyan, 'Knowledge Across Borders: The Early Communication of Evolution in China', in *The Circulation of Knowledge between Britain, India, and China*, eds. Bernard Lightman, Gordon McOuat, and Larry Stewart (Leiden: Brill, 2013).

62 Jin, 'The Evolution of Evolutionism in China', 48–50, and James Pusey, *China and Charles Darwin* (Cambridge, MA: Harvard University Press, 1983), 16 and 58–60.

63 Jin, 'The Evolution of Evolutionism in China', 50–2, Yang Haiyan, 'Encountering Darwin and Creating Darwinism in China', in *The Cambridge Encyclopedia of Darwin and Evolutionary Thought*, ed. Michael Ruse (Cambridge: Cambridge University Press, 2013), 253, Frank Dikötter, *The Discourse of Race in Modern China* (Oxford: Oxford University Press, 2015), 140, and Ke Zunke and Li Bin, 'Spencer and Science Education in China', in Lightman, ed., *Global Spencerism*.

64 Pusey, *China and Charles Darwin*, 92–117 and 317–8.

65 Pusey, *China and Charles Darwin*, 58–9, Joseph Needham, *Science and Civilisation in China: The History of Scientific Thought* (Cambridge: Cambridge University Press, 1956), vol. 2, 74–81 and 317–8, and Joseph Needham and Donald Leslie, 'Ancient and Mediaeval Chinese Thought on Evolution', in *Theories and Philosophies of Medicine* (New Delhi: Institute of History of Medicine and Medical Research, 1973).

66 Jixing Pan, 'Charles Darwin's Chinese Sources', *Isis* 75 (1984).

67 Benjamin Elman, *A Cultural History of Modern Science in China* (Cambridge, MA: Harvard University Press, 2009), 198–220, Peter Lavelle, 'Agricultural Improvement at China's First Agricultural Experiment Stations', in *New Perspectives on the History of Life Sciences and Agriculture*, eds. Denise Phillips and Sharon Kingsland (Cham: Springer International, 2015), 323–41, and Joseph Lawson, 'The Chinese State and Agriculture in an Age of Global Empires, 1880–1949', in *Eco-Cultural Networks and the British Empire: New Views on Environmental History*, eds. James Beattie, Edward Melillo, and Emily O'Gorman (London: Bloomsbury, 2015).

68 Elman, *A Cultural History of Modern Science in China*, 198 and 220.

69 Jin, 'Translation and Transmutation', 125–40, and Yang, 'Encountering Darwin and Creating Darwinism in China', 254–5.

70 Jin, 'Translation and Transmutation', 125–40, Jin, 'The Evolution of Evolutionism in China', 52–4, and Yang, 'Encountering Darwin and Creating Darwinism in China', 254–5.

71 Jin, 'Translation and Transmutation', 125–40, Jin, 'The Evolution of Evolutionism in China', 52–4, Yang, 'Encountering Darwin and Creating Darwinism in China', 254–5, Pusey, *China and Charles Darwin*, 318, and Zhou Rong, *The Revolutionary Army: A Chinese Nationalist Tract of 1903*, trans. John Lust (Paris: Mouton, 1968), 58.

72 Yang, 'Encountering Darwin and Creating Darwinism in China', 254–5.

73 Pusey, *China and Charles Darwin*, 321–2, and Dikötter, *The Discourse of Race in Modern China*, 140.

74 Secord, 'Global Darwin', 51, and Todes, *Darwin Without Malthus*, 11.

6. Industrial Experiments

1 Richard Staley, *Einstein's Generation: The Origins of the Relativity Revolution* (Chicago: University of Chicago Press, 2008), 169–70, Paul Greenhalgh, *Ephemeral Vistas: The Expositions Universelles, Great Exhibitions and World's Fairs, 1851–1939* (Manchester: Manchester University Press, 1988), and 'Liste de membres du Congrès international de physique', in *Rapports présentés au Congrès international de physique réuni à Paris en 1900*, eds. Charles-Édouard Guillaume and Lucien Poincaré (Paris: Gauthier-Villars, 1901), 4:129–69.

2 Staley, *Einstein's Generation*, 138–63, Charles-Édouard Guillaume, 'The International Physical Congress', *Nature* 62 (1900), and Richard Mandell, *Paris 1900: The Great World's Fair* (Toronto: Toronto University Press, 1967), 62–88.

3 Staley, *Einstein's Generation*, 137, and Charles-Édouard Guillaume and Lucien Poincaré, 'Avertissement', in Guillaume and Poincaré, eds., *Rapports présentés*, 1:v. (Final two quotes in this paragraph are my own translation. The first quote is from Staley.)

4 Iwan Rhys Morus, *When Physics Became King* (Chicago: University of Chicago Press, 2005), 77–81, and James Clerk Maxwell, 'A Dynamical Theory of the Electromagnetic Field', *Philosophical Transactions of the Royal Society* 155 (1865): 460 and 466.

5 Morus, *When Physics Became King*, 170–2 and 188–91, and 'Liste de membres du Congrès international de physique'.

6 Peter Lebedev, 'Les forces de Maxwell-Bartoli dues à la pression de la lumière', in Guillaume and Poincaré, eds., *Rapports présentés*, 2:133–40, and Alexander Vucinich, *Science in Russian Culture, 1861–1917* (Stanford: Stanford University Press, 1963), 2:367–8.

7 Hantaro Nagaoka, 'La magnetostriction', in Guillaume and Poincaré, eds., *Rapports présentés*, 2:536–56, Subrata Dasgupta, *Jagadis Chandra Bose and the Indian Response to Western Science* (New Delhi: Oxford University Press, 1999), 109–10, and Jagadish Chandra Bose, 'De la généralité des phénomènes moléculaires produits par l'électricité sur la matière inorganique et sur la matière vivante', in Guillaume and Poincaré, eds., *Rapports présentés*, 3:581–7 (translation my own).

8 Morus, *When Physics Became King*, and Daniel Headrick, *The Tentacles of Progress: Technology Transfer in the Age of Imperialism, 1850–1940* (Oxford: Oxford University Press, 1988), 97–144.

9 Aaron Ihde, *The Development of Modern Chemistry* (New York: Harper & Row, 1964 [1984]), 94, 231–58, 443–74, and 747–9, and V. N. Pitchkov, 'The Discovery of Ruthenium', *Platinum Metals Review* 40 (1996): 184.

10 Ihde, *The Development of Modern Chemistry*, 249 and 488.

11 Charles Édouard Guillaume, 'The International Physical Congress', *Nature* 62 (1900): 428.

12 Moisei Radovsky, *Alexander Popov: Inventor of the Radio*, trans. G. Yankovsky (Moscow: Foreign Languages Publishing House, 1957), 23–61.
13 Sungook Hong, *Wireless: From Marconi's Black-Box to the Audion* (Cambridge, MA: The MIT Press, 2001), 4, and Radovsky, *Alexander Popov*, 54–61.
14 Radovsky, *Alexander Popov*, 5–23.
15 Radovsky, *Alexander Popov*, 23–38, 69–73, and 79.
16 Radovsky, *Alexander Popov*, 69–73 and 79, Daniel Headrick, *The Invisible Weapon: Telecommunications and International Politics, 1851–1945* (Oxford: Oxford University Press, 1991), 123, and Robert Lochte, 'Invention and Innovation of Early Radio Technology', *Journal of Radio Studies* 7 (2000).
17 Vucinich, *Science in Russian Culture*, 2:1–78, Paul Josephson, *Physics and Politics in Revolutionary Russia* (Berkeley: University of California Press, 1991), 9–39, and Natalia Nikiforova, 'Electricity at Court: Technology in Representation of Imperial Power', in *Electric Worlds: Creations, Circulations, Tensions, Transitions*, eds. Alain Beltran, Léonard Laborie, Pierre Lanthier, and Stéphanie Le Gallic (Brussels: Peter Lang, 2016), 66–8.
18 Joseph Bradley, *Voluntary Associations in Tsarist Russia: Science, Patriotism, and Civil Society* (Cambridge, MA: Harvard University Press, 2009), 171–2, and Radovsky, *Alexander Popov*, 18.
19 Vucinich, *Science in Russian Culture*, 2:366–8.
20 Vucinich, *Science in Russian Culture*, 2:151–63, Loren Graham, *Science in Russia and the Soviet Union: A Short History* (Cambridge: Cambridge University Press, 1993), 45–53, and Michael Gordin, *A Well-Ordered Thing: Dmitrii Mendeleev and the Shadow of the Periodic Table* (New York: Basic Books, 2004).
21 Vucinich, *Science in Russian Culture*, 2:163, and Gordin, *A Well-Ordered Thing*, 8–9.
22 Michael Gordin, 'A Modernization of "Peerless Homogeneity": The Creation of Russian Smokeless Gunpowder', *Technology and Culture* 44 (2003): 682–93, and Michael Gordin, 'No Smoking Gun: D. I. Mendeleev and Pyrocollodion Gunpowder', in *Troisièmes journées scientifiques Paul Vieille* (Paris: A3P, 2000).
23 Gordin, 'The Creation of Russian Smokeless Gunpowder', 678–82.
24 Gordin, 'The Creation of Russian Smokeless Gunpowder', 680–2.
25 Gordin, 'The Creation of Russian Smokeless Gunpowder', 682–90, and Gordin, 'No Smoking Gun', 73–4.
26 Francis Michael Stackenwalt, 'Dmitrii Ivanovich Mendeleev and the Emergence of the Modern Russian Petroleum Industry, 1863–1877', *Ambix* 45 (1998), and Zack Pelta-Hella, 'Braving the Elements: Why Mendeleev Left Russian Soil for American Oil', Science History Institute, accessed 9 August 2020, https://www.sciencehistory.org/distillations/braving-the-elements-why-mendeleev-left-russian-soil-for-american-oil.
27 Mary Creese, *Ladies in the Laboratory IV: Imperial Russia's Women in Science, 1800–1900* (Lanham: Rowman & Littlefield, 2015), 54–61.
28 Creese, *Ladies in the Laboratory IV*, 52–5.
29 Creese, *Ladies in the Laboratory IV*, 55–6, and Ann Koblitz, *Science, Women and Revolution in Russia* (London: Routledge, 2014), 129.
30 Creese, *Ladies in the Laboratory IV*, 55–6, and Gisela Boeck, 'Ordering the Platinum Metals – The Contribution of Julia V. Lermontova (1846/47–1919)', in *Women in*

Their Element: Selected Women's Contributions to the Periodic System, eds. Annette Lykknes and Brigitte Van Tiggelen (New Jersey: World Scientific, 2019), 112–23.

31 Creese, *Ladies in the Laboratory IV*, 57–8.

32 Gordin, *A Well-Ordered Thing*, 63–4, and 'Liste de membres du Congrès international de physique', 159.

33 Josephson, *Physics and Politics*, 16–18, Alexei Kojevnikov, *Stalin's Great Science: The Times and Adventures of Soviet Physicists* (London: Imperial College Press, 2004), 1–22, and Nathan Brooks, 'Chemistry in War, Revolution, and Upheaval: Russia and the Soviet Union, 1900–1929', *Centaurus* 39 (1997): 353–8.

34 Yakup Bektas, 'The Sultan's Messenger: Cultural Constructions of Ottoman Telegraphy, 1847–1880', *Technology and Culture* 41 (2000): 671–2, Yakup Bektas, 'Displaying the American Genius: The Electromagnetic Telegraph in the Wider World', *The British Journal for the History of Science* 34 (2001): 199–214, and John Porter Brown, 'An Exhibition of Professor Morse's Magnetic Telegraph before the Sultan', *Journal of the American Oriental Society* 1 (1849): liv–lvii.

35 Bektas, 'Displaying the American Genius', 199–216, Bektas, 'The Sultan's Messenger', 672, and Brown, 'An Exhibition', lv.

36 Roderic Davison, *Essays in Ottoman and Turkish History, 1774–1923: The Impact of the West* (Austin: University of Texas Press, 2013), 133–54, and Bektas, 'The Sultan's Messenger', 669–94.

37 Ekmeleddin İhsanoğlu, *The House of Sciences: The First Modern University in the Muslim World* (Oxford: Oxford University Press, 2019), 1–5, Meltem Akbaş, 'The March of Military Physics – I: Physics and Mechanical Sciences in the Curricula of the 19th Century Ottoman Military Schools', *Studies in Ottoman Science* 13 (2012), Meltem Akbaş, 'The March of Military Physics – II: Teachers and Textbooks of Physics and Mechanical Sciences of the 19th Century Ottoman Military Schools', *Studies in Ottoman Science* 14 (2012), and Mustafa Kaçar, 'The Development in the Attitude of the Ottoman State towards Science and Education and the Establishment of the Engineering Schools (Mühendishanes)', in *Science, Technology and Industry in the Ottoman World*, eds. Ekmeleddin İhsanoğlu, Ahmed Djebbar, and Feza Günergun (Turnhout: Brepols Publishers, 2000).

38 Feza Günergun, 'Chemical Laboratories in Nineteenth-Century Istanbul: A Case-Study on the Laboratory of the Hamidiye Etfal Children's Hospital', *Spaces and Collections in the History of Science*, eds. Marta Lourenço and Ana Carneiro (Lisbon: Museum of Science of the University of Lisbon, 2009), 91, Ekmeleddin İhsanoğlu, 'Ottoman Educational and Scholarly Scientific Institutions', in *History of the Ottoman State, Society, and Civilization*, ed. Ekmeleddin İhsanoğlu (Istanbul: Research Center for Islamic History, Art and Culture, 2001), 2:484–5, and İhsanoğlu, *The House of Sciences*, 1–5.

39 İhsanoğlu, *The House of Sciences*, xii, 2, and 77.

40 Akbaş, 'The March of Military Physics – II', 91–2, Feza Günergun, 'Derviş Mehmed Emin pacha (1817–1879), serviteur de la science et de l'État ottoman', in *Médecins et ingénieurs ottomans a l'âge des nationalismes*, ed. Méropi Anastassiadou-Dumont (Paris: L'Institut français d'études anatoliennes, 2003), 174–6 (translation from the French my own), and George Vlahakis, Isabel Maria Malaquias, Nathan Brooks, François Regourd, Feza Günergun, and David Wright, *Imperialism and Science: Social Impact and Interaction* (Santa Barbara: ABC-CLIO, 2006), 103–4.

41 Vlahakis et al., *Imperialism and Science*, 104–5, M. Alper Yalçinkaya, *Learned Patriots: Debating Science, State, and Society in the Nineteenth-Century Ottoman Empire* (Chicago: University of Chicago Press, 2015), 65, and Emre Dölen, 'Ottoman Scientific Literature during the 18th and 19th Centuries', 168–71.
42 Günergun, 'Derviş Mehmed Emin', İhsanoğlu, *The House of Sciences*, 23–6, Alper Yalçinkaya, *Learned Patriots*, 73–5, and Murat Şiviloğlu, *The Emergence of Public Opinion: State and Society in the Late Ottoman Empire* (Cambridge: Cambridge University Press, 2018), 148–9.
43 İhsanoğlu, *The House of Sciences*, 28, Alper Yalçinkaya, *Learned Patriots*, 76, and Marwa Elshakry, 'When Science Became Western: Historiographical Reflections', *Isis* 101 (2010).
44 Daniel Stolz, *The Lighthouse and the Observatory: Islam, Science, and Empire in Late Ottoman Egypt* (Cambridge: Cambridge University Press, 2018), 207–42, Vanessa Ogle, *The Global Transformation of Time, 1870–1950* (Cambridge, MA: Harvard University Press, 2015), 149–76, and James Gelvin and Nile Green, eds., *Global Muslims in the Age of Steam and Print* (Berkeley: University of California Press, 2014).
45 Ferhat Ozcep, 'Physical Earth and Its Sciences in Istanbul: A Journey from Pre-Modern (Islamic) to Modern Times', *History of Geo- and Space Sciences* 11 (2020): 189.
46 Amit Bein, 'The Istanbul Earthquake of 1894 and Science in the Late Ottoman Empire', *Middle Eastern Studies* 44 (2008): 916, and Ozcep, 'Physical Earth', 186.
47 Bein, 'The Istanbul Earthquake of 1894', and Ozcep, 'Physical Earth'.
48 Ozcep, 'Physical Earth', 189–93.
49 Bein, 'The Istanbul Earthquake of 1894', 920, Ozcep, 'Physical Earth', 186, and Demetrios Eginitis, 'Le tremblement de terre de Constantinople du 10 juillet 1894', *Annales de géographie* 15 (1895): 165 (translation my own).
50 İhsanoğlu, *The House of Sciences*, 86–93 and 218–22, and Lâle Aka Burk, 'Fritz Arndt and His Chemistry Books in the Turkish Language', *Bulletin of the History of Chemistry* 28 (2003).
51 Jagadish Chandra Bose, 'Electro-Magnetic Radiation and the Polarisation of the Electric Ray', in *Collected Physical Pages of Sir Jagadis Chunder Bose* (London: Longmans, Green and Co., 1927), and Dasgupta, *Jagadis Chandra Bose*, 1–3.
52 Bose, 'Electro-Magnetic Radiation', 77–101.
53 Bose, 'Electro-Magnetic Radiation', 100–1.
54 Dasgupta, *Jagadis Chandra Bose*, 16–28, John Lourdusamy, *Science and National Consciousness in Bengal: 1870–1930* (New Delhi: Orient Blackswan, 2004), 100–1, and Deepak Kumar, 'Science in Higher Education: A Study in Victorian India', *Indian Journal of History of Science* 19 (1984): 253–5.
55 Lourdusamy, *Science and National Consciousness*, 56–95, and Pratik Chakrabarti, *Western Science in Modern India: Metropolitan Methods, Colonial Practices* (New Delhi: Orient Blackswan, 2004), 157.
56 Lourdusamy, *Science and National Consciousness*, 101, and Dasgupta, *Jagadis Chandra Bose*, 32–4.
57 Lourdusamy, *Science and National Consciousness*, 101, and Dasgupta, *Jagadis Chandra Bose*, 43.

58 Dasgupta, *Jagadis Chandra Bose*, 51–5 and 72–3, and Jagadish Chandra Bose, 'On the Rotation of Plane of Polarisation of Electric Waves by a Twisted Structure', *Proceedings of the Royal Society of London* 63 (1898): 150–2.
59 Dasgupta, *Jagadis Chandra Bose*, 48–9 and 82, Viśvapriya Mukherji, 'Some Historical Aspects of Jagadis Chandra Bose's Microwave Research during 1895–1900', *Indian Journal of History of Science* 14 (1979): 97, and Jagadish Chandra Bose, 'On a Self-Recovering Coherer and the Study of the Cohering Action of Different Metals', *Proceedings of the Royal Society of London* 65 (1900).
60 Dasgupta, *Jagadis Chandra Bose*, 56.
61 Dasgupta, *Jagadis Chandra Bose*, 109, and Lourdusamy, *Science and National Consciousness*, 115.
62 David Arnold, *Science, Technology and Medicine in Colonial India* (Cambridge: Cambridge University Press, 2000), 129–34 and 191, Deepak Kumar, *Science and the Raj, 1857–1905* (New Delhi: Oxford University Press, 1995), 74–179, and Aparajito Basu, 'Chemical Research in India (1876–1918)', *Annals of Science* 52 (1995): 592.
63 Suvobrata Sarkar, *Let There be Light: Engineering, Entrepreneurship, and Electricity in Colonial Bengal, 1880–1945* (Cambridge: Cambridge University Press, 2020), 119, and Aparajita Basu, 'The Conflict and Change-Over in Indian Chemistry', *Indian Journal of History of Science* 39 (2004): 337–46.
64 Arnold, *Science, Technology and Medicine*, 138–40 and 166, and Kumar, 'Science in Higher Education', 253–5.
65 Chakrabarti, *Western Science*, 157–62, and Lourdusamy, *Science and National Consciousness*, 56–95.
66 Lourdusamy, *Science and National Consciousness*, 144–5, David Arnold, *Toxic Histories: Poison and Pollution in Modern India* (Cambridge: Cambridge University Press, 2016), 114, Priyadaranjan Ray, 'Prafulla Chandra Ray: 1861–1944', *Biographical Memoirs of Fellows of the Indian National Science Academy* 1 (1944), and Prafulla Chandra Ray, *Life and Experiences of a Bengali Chemist* (London: Kegan Paul, French, Trübner, 1923), 1–47.
67 Lourdusamy, *Science and National Consciousness*, 144–5, and Ray, *Life and Experiences*, 50–76.
68 Ray, *Life and Experiences*, 112–3, and Madhumita Mazumdar, 'The Making of an Indian School of Chemistry, Calcutta, 1889–1924', in *Science and Modern India: An Institutional History, c.1784–1947*, ed. Uma Das Gupta (New Delhi: Pearson Longman, 2011), 806–12.
69 Ray, *Life and Experiences*, 113–5, Mazumdar, 'The Making of an Indian School of Chemistry', 807, and Dhruv Raina, *Images and Contexts: The Historiography of Science and Modernity in India* (New Delhi: Oxford University Press, 2010), 75.
70 Mazumdar, 'The Making of an Indian School of Chemistry', 807, Ray, *Life and Experiences*, 113–4, Arnab Rai Choudhuri and Rajinder Singh, 'The FRS Nomination of Sir Prafulla C. Ray and the Correspondence of N. R. Dhar', *Notes and Records* 721 (2018): 58–61, and Prafulla Chandra Ray, 'On Mercurous Nitrite', *Journal of the Asiatic Society of Bengal* 65 (1896): 2–9.
71 Lourdusamy, *Science and National Consciousness*, 143–52 and 170–2, Ray, *Life and Experiences*, 92–111, and Pratik Chakrabarti, 'Science and Swadeshi: The Establish-

ment and Growth of the Bengal Chemical and Pharmaceutical Works, 1893–1947', in Gupta, ed., *Science and Modern India*, 117–8.

72 Lourdusamy, *Science and National Consciousness*, 154.

73 Ray, *Life and Experiences*, 104–14, Lourdusamy, *Science and National Consciousness*, 154, Raina, *Images and Contexts*, 61–72, Projit Bihari Mukharji, 'Parachemistries: Colonial Chemopolitics in a Zone of Contest', *History of Science* 54 (2016): 362–5, Prafulla Chandra Ray, 'Antiquity of Hindu Chemistry', in *Essays and Discourses*, ed. Prafulla Chandra Ray (Madras: G. A. Natesan & Co., 1918), 102, Prafulla Chandra Ray, 'The Bengali Brain and Its Misuse', in Ray, ed., *Essays and Discourses*, 207, and Prafulla Chandra Ray, *A History of Hindu Chemistry* (Calcutta: Bengal Chemical and Pharmaceutical Works, 1902–4), 2 vols.

74 Mukharji, 'Parachemistries', 362–5, Raina, *Images and Contexts*, 61–72, Ray, *Life and Experiences*, 115–8, and Prafulla Chandra Ray, *The Rasārṇavam, or The Ocean of Mercury and Other Metals and Minerals* (Calcutta: Satya Press, 1910), 1–2.

75 Basu, 'Conflict and Change-Over', 337–44, and Arnold, *Science, Technology and Medicine*, 191.

76 Arnold, *Science, Technology and Medicine*, 165, and Mazumdar, 'The Making of an Indian School of Chemistry', 23.

77 Greg Clancey, *Earthquake Nation: The Cultural Politics of Japanese Seismicity, 1868–1930* (Berkeley: University of California Press, 2006), 128–50.

78 Haruyo Yoshida, 'Aikitu Tanakadate and the Controversy over Vertical Electrical Currents in Geomagnetic Research', *Earth Sciences History* 20 (2001): 156–60.

79 Kenkichiro Koizumi, 'The Emergence of Japan's First Physicists: 1868–1900', *Historical Studies in the Physical Sciences* 6 (1975): 72–81.

80 Koizumi, 'The Emergence of Japan's First Physicists', 72–81, James Bartholomew, *The Formation of Science in Japan: Building a Research Tradition* (New Haven: Yale University Press, 1989), 62–75, and Aikitsu Tanakadate, 'Mean Intensity of Magnetization of Soft Iron Bars of Various Lengths in a Uniform Magnetic Field', *The Philosophical Magazine* 26 (1888).

81 Yoshida, 'Aikitu Tanakadate', 159–72.

82 John Cawood, 'The Magnetic Crusade: Science and Politics in Early Victorian Britain', *Isis* 70 (1979), Yoshida, 'Aikitu Tanakadate', 159–72, and Cargill Knott and Aikitsu Tanakadate, 'A Magnetic Survey of All Japan', *The Journal of the College of Science, Imperial University, Japan* 2 (1889): 168 and 216.

83 Yoshida, 'Aikitu Tanakadate', 159–72, and Aikitsu Tanakadate and Hantaro Nagaoka, 'The Disturbance of Isomagnetics Attending the Mino-Owari Earthquake of 1891', *The Journal of the College of Science, Imperial University, Japan* 5 (1893): 150 and 175.

84 Koizumi, 'The Emergence of Japan's First Physicists', 4–16, Bartholomew, *The Formation of Science in Japan*, 49–50, and William Brock, 'The Japanese Connexion: Engineering in Tokyo, London, and Glasgow at the End of the Nineteenth Century', *The British Journal for the History of Science* 14 (1981): 229.

85 Bartholomew, *The Formation of Science in Japan*, 52, Koizumi, 'The Emergence of Japan's First Physicists', 77, and Yoshiyuki Kikuchi, *Anglo-American Connections in Japanese Chemistry: The Lab as Contact Zone* (Basingstoke: Palgrave Macmillan, 2013), 97–8.

86 Kikuchi, *Anglo-American Connections*, 45–6 and 90, and Togo Tsukahara, *Affinity and Shinwa Ryoku: Introduction of Western Chemical Concepts in Early Nineteenth-Century Japan* (Amsterdam: J. C. Gieben, 1993), 1–3 and 149–50.

87 Tetsumori Yamashima, 'Jokichi Takamine (1854–1922), the Samurai Chemist, and His Work on Adrenalin', *Journal of Medical Biography* 11 (2003), and William Shurtleff and Akiko Aoyagi, *Jokichi Takamine (1854–1922) and Caroline Hitch Takamine (1866–1954): Biography and Bibliography* (Lafayette: Soyinfo Center, 2012), 5–14.

88 Yamashima, 'Jokichi Takamine (1854–1922)', and Shurtleff and Aoyagi, *Jokichi Takamine*, 224.

89 Bartholomew, *The Formation of Science in Japan*, 63, and Koizumi, 'The Emergence of Japan's First Physicists', 82–4.

90 Koizumi, 'The Emergence of Japan's First Physicists', 84–7.

91 Koizumi, 'The Emergence of Japan's First Physicists', 90–2, Eri Yagi, 'On Nagaoka's Saturnian Atom (1903)', *Japanese Studies in the History of Science* 3 (1964), and Hantaro Nagaoka, 'Motion of Particles in an Ideal Atom Illustrating the Line and Band Spectra and the Phenomena of Radioactivity', *Journal of the Tokyo Mathematico-Physical Society* 2 (1904).

92 'Liste de membres du Congrès international de physique', 156, Koizumi, 'The Emergence of Japan's First Physicists', 89, and Tanakadate and Nagaoka, 'The Disturbance of Isomagnetics'.

93 Eri Yagi, 'The Development of Nagaoka's Saturnian Atomic Model, I – Dispersion of Light', *Japanese Studies in the History of Science* 6 (1967): 25, and Eri Yagi, 'The Development of Nagaoka's Saturnian Atomic Model, II – Nagaoka's Theory of the Structure of Matter', *Japanese Studies in the History of Science* 11 (1972): 76–8.

94 Yagi, 'On Nagaoka's Saturnian Atom', 29–47, Lawrence Badash, 'Nagaoka to Rutherford, 22 February 1911', *Physics Today* 20 (1967), and Ernest Rutherford, 'The Scattering of α and β Particles by Matter and the Structure of the Atom', *Philosophical Magazine* 21 (1911): 688.

95 Koizumi, 'The Emergence of Japan's First Physicists', 65.

96 Bartholomew, *The Formation of Science in Japan*, 199–201.

97 Koizumi, 'The Emergence of Japan's First Physicists', 96.

98 'In Memory of Pyotr Nikolaevich Lebedev', *Physics-Uspekhi* 55 (2012).

99 Morus, *When Physics Became King*, 167.

100 Koizumi, 'The Emergence of Japan's First Physicists', 18.

Part Four: Ideology and Aftermath, c.1914–2000

7. Faster Than Light

1 Josef Eisinger, *Einstein on the Road* (Amherst: Prometheus Books, 2011), 32–4, Danian Hu, *China and Albert Einstein: The Reception of the Physicist and His Theory in China, 1917–1979* (Cambridge, MA: Harvard University Press, 2009), 66–74, Albert Einstein, *The Travel Diaries of Albert Einstein: The Far East, Palestine, and*

Spain, 1922–1923, ed. Ze'ev Rosenkranz (Princeton: Princeton University Press, 2018), 135, and Alice Calaprice, ed., *The Ultimate Quotable Einstein* (Princeton: Princeton University Press, 2011), 419.

2 Eisinger, *Einstein on the Road*, 34–51, and Einstein, *Travel Diaries*, 143.

3 Eisinger, *Einstein on the Road*, 36–46, and Seiya Abiko, 'Einstein's Kyoto Address: "How I Created the Theory of Relativity"', *Historical Studies in the Physical and Biological Sciences* 31 (2000): 1–6.

4 Eisinger, *Einstein on the Road*, 58–63, David Rowe and Robert Schulmann, eds., *Einstein on Politics: His Private Thoughts and Public Stands on Nationalism, Zionism, War, Peace, and the Bomb* (Princeton: Princeton University Press, 2007), 95–105 and 125–6, and Richard Crockatt, *Einstein and Twentieth-Century Politics* (Oxford: Oxford University Press, 2016), 77–106.

5 Eisinger, *Einstein on the Road*, 58–63, Calaprice, ed., *Quotable Einstein*, 194 and 202, and Rowe and Schulmann, *Einstein on Politics*, 156–9.

6 Calaprice, ed., *Quotable Einstein*, 165.

7 Calaprice, ed., *Quotable Einstein*, 292, Crockatt, *Einstein and Twentieth-Century Politics*, 29, Rowe and Schulmann, *Einstein on Politics*, 189–97, and Kenkichiro Koizumi, 'The Emergence of Japan's First Physicists: 1868–1900', *Historical Studies in the Physical Sciences* 6 (1975): 80.

8 Ashish Lahiri, 'The Creative Mind: A Mirror or a Component of Reality?', in *Tagore, Einstein and the Nature of Reality: Literary and Philosophical Reflections*, ed. Partha Ghose (London: Routledge, 2019), 215–7.

9 Abraham Pais, 'Paul Dirac: Aspects of His Life and Work', in *Paul Dirac: The Man and His Work*, ed. Peter Goddard (Cambridge: Cambridge University Press, 1998), 14–16, Kenji Ito, 'Making Sense of Ryôshiron (Quantum Theory): Introduction of Quantum Physics into Japan, 1920–1940' (PhD diss., Harvard University, 2002), 260–1, and Yan Kangnian, 'Niels Bohr in China', in *Chinese Studies in the History and Philosophy of Science and Technology*, eds. Fan Dainian and Robert Cohen (Dordrecht: Springer Netherlands, 1996), 433–7.

10 Alexei Kojevnikov, *Stalin's Great Science: The Times and Adventures of Soviet Physicists* (London: Imperial College Press, 2004), 103–6, and Istvan Hargittai, *Buried Glory: Portraits of Soviet Scientists* (Oxford: Oxford University Press, 2013), 98–102.

11 Kojevnikov, *Stalin's Great Science*, 107–8, and Hargittai, *Buried Glory*, 103.

12 Hargittai, *Buried Glory*, 104–5, and Jack Boag, David Shoenberg, and P. Rubinin, eds., *Kapitza in Cambridge and Moscow: Life and Letters of a Russian Physicist* (Amsterdam: North-Holland, 1990), 235.

13 Kojevnikov, *Stalin's Great Science*, 107–9, and Hargittai, *Buried Glory*, 104–5.

14 Kojevnikov, *Stalin's Great Science*, 116–7, Peter Kapitza, 'Viscosity of Liquid Helium below the λ-Point', *Nature* 74 (1938): 74, and Sébastien Balibar, 'Superfluidity: How Quantum Mechanics Became Visible', in *History of Artificial Cold, Scientific, Technological and Cultural Issues*, ed. Kostas Gavroglu (Dordrecht: Springer, 2014).

15 Kojevnikov, *Stalin's Great Science*, 1–28, Valerii Ragulsky, 'About People with the Same Life Attitude: 100th Anniversary of Lebedev's Lecture on the Pressure of Light', *Physics-Uspekhi* 54 (2011): 294, Paul Josephson, *Physics and Politics in Revolutionary Russia* (Berkeley: University of California Press, 1991), 1–6 and 62, Loren Graham, *Science in*

Russia and the Soviet Union: A Short History (Cambridge: Cambridge University Press, 1993), 79–98, and R. W. Davies, 'Soviet Military Expenditure and the Armaments Industry, 1929–33: A Reconsideration', *Europe–Asia Studies* 45 (1993): 578.

16 Kojevnikov, *Stalin's Great Science*, 41, and Josephson, *Physics and Politics*, 1–6, 106, and 134–5.

17 Josephson, *Physics and Politics*, 6 and 23, Loren Graham, *Science, Philosophy, and Human Behavior in the Soviet Union* (New York: Columbia University Press, 1987), 322–3, and Clemens Dutt, ed., *V. I. Lenin: Collected Works*, trans. Abraham Fineberg (Moscow: Progress Publishers, 1962), 14:252–7 and 33:227–36.

18 Alexander Vucinich, *Einstein and Soviet Ideology* (Stanford: Stanford University Press, 2001), 1–5, 13, and 58–68, V. P. Vizgin and G. E. Gorelik, 'The Reception of the Theory of Relativity in Russia and the USSR', in *The Comparative Reception of Relativity*, ed. Thomas Glick (Dordrecht: Springer, 1987), and Ethan Pollock, *Stalin and the Soviet Science Wars* (Princeton: Princeton University Press, 2009), 78–9.

19 Kojevnikov, *Stalin's Great Science*, 49–53, and Josephson, *Physics and Politics*, 114–6.

20 Kojevnikov, *Stalin's Great Science*, 53–6, and Victor Frenkel, *Yakov Illich Frenkel*, trans. Alexander Silbergleit (Basel: Springer Basel, 1996), 28–9.

21 Kojevnikov, *Stalin's Great Science*, 48–55.

22 Kojevnikov, *Stalin's Great Science*, 48–55, and Yakov Frenkel, 'Beitrag zur Theorie der Metalle', *Zeitschrift für Physik* 29 (1924).

23 Josephson, *Physics and Politics*, 221, and M. Shpak, 'Antonina Fedorovna Prikhot'ko (On Her Sixtieth Birthday)', *Soviet Physics Uspekhi* 9 (1967): 785–6.

24 Shpak, 'Antonina Fedorovna Prikhot'ko', 785–6.

25 Kojevnikov, *Stalin's Great Science*, 74–6, Hargittai, *Buried Glory*, 119–20, Josephson, *Physics and Politics*, 224, and Karl Hall, 'The Schooling of Lev Landau: The European Context of Postrevolutionary Soviet Theoretical Physics', *Osiris* 23 (2008).

26 Kojevnikov, *Stalin's Great Science*, 85–92, Hargittai, *Buried Glory*, 121, and Nikolai Krementsov and Susan Gross Solomon, 'Giving and Taking across Borders: The Rockefeller Foundation and Russia, 1919–1928', *Minerva* 39 (2001).

27 Kojevnikov, *Stalin's Great Science*, 117, and L. Reinders, *The Life, Science and Times of Lev Vasilevich Shubnikov: A Pioneer of Soviet Cryogenics* (Cham: Springer, 2018), 23–32.

28 Reinders, *Lev Vasilevich Shubnikov*, 171–92.

29 Kojevnikov, *Stalin's Great Science*, 85–8, Hargittai, *Buried Glory*, 109–10 and 125, and Josephson, *Physics and Politics*, 312.

30 Hargittai, *Buried Glory*, 128.

31 Hargittai, *Buried Glory*, 112 and 122.

32 Hu, *China and Albert Einstein*, 58–9, Gao Pingshu, 'Cai Yuanpei's Contributions to China's Science', in Dainian and Cohen, eds., *Chinese Studies*, 399, and Dai Nianzu, 'The Development of Modern Physics in China: The 50th Anniversary of the Founding of the Chinese Physical Society', in Dainian and Cohen, eds., *Chinese Studies*, 208.

33 Hu, *China and Albert Einstein*, 89–92.

34 Hu, *China and Albert Einstein*, 92–7.

35 Hu, *China and Albert Einstein*, 58–61 and 133.

36 Hu, *China and Albert Einstein*, 66–9, and Gao, 'Cai Yuanpei's Contributions', 397–404.

37 Hu, *China and Albert Einstein*, 127, and Dai, 'Development of Modern Physics', 209–10.
38 Danian Hu, 'American Influence on Chinese Physics Study in the Early Twentieth Century', *Physics in Perspective* 17 (2016): 277.
39 Hu, *China and Albert Einstein*, 44–6.
40 Hu, *China and Albert Einstein*, 116–7, and Mary Bullock, 'American Science and Chinese Nationalism: Reflections on the Career of Zhou Peiyuan', in *Remapping China: Fissures in Historical Terrain*, eds. Gail Hershatter, Emily Honig, Jonathan Lipman, and Randall Stross (Stanford: Stanford University Press, 1996), 214–5.
41 Hu, *China and Albert Einstein*, 116–7, and Bullock, 'American Science and Chinese Nationalism', 214–6.
42 Hu, *China and Albert Einstein*, 116–9, and P'ei-yuan Chou, 'The Gravitational Field of a Body with Rotational Symmetry in Einstein's Theory of Gravitation', *American Journal of Mathematics* 53 (1931).
43 Hu, *China and Albert Einstein*, 119–20, and Bullock, 'American Science and Chinese Nationalism', 217.
44 Hu, *China and Albert Einstein*, 119–20, and Dai, 'Development of Modern Physics', 210–13.
45 Zhang Wei, 'Millikan and China', in Dainian and Cohen, eds., *Chinese Studies*.
46 Dai, 'Development of Modern Physics', 210, Zuoyue Wang, 'Zhao Zhongyao', in *New Dictionary of Scientific Biography*, ed. Noretta Koertge (Detroit: Charles Scribner's Sons, 2008), 8:397–402, and William Duane, H. H. Palmer, and Chi-Sun Yeh, 'A Remeasurement of the Radiation Constant, *h*, by Means of X-Rays', *Proceedings of the National Academy of Sciences of the United States of America* 7 (1921).
47 Zhang, 'Millikan and China', 441–2, Dai, 'Development of Modern Physics', 210, Zuoyue, 'Zhao Zhongyao', 397–402, and C. Y. Chao, 'The Absorption Coefficient of Hard γ-Rays', *Proceedings of the National Academy of Sciences of the United States of America* 16 (1930).
48 Jagdish Mehra and Helmut Rechenberg, *The Historical Development of Quantum Theory* (New York: Springer, 1982), 6:804, and Cong Cao, 'Chinese Science and the "Nobel Prize Complex" ', *Minerva* 42 (2004): 154.
49 Gao, 'Cai Yuanpei's Contributions', 398.
50 Ito, 'Making Sense of Ryôshiron', 20–1, 91–2, and 165–6.
51 Ito, 'Making Sense of Ryôshiron', 56–7 and 87–8, Tsutomu Kaneko, 'Einstein's Impact on Japanese Intellectuals', in Glick, ed., *The Comparative Reception of Relativity*, 354, Morris Low, *Science and the Building of a New Japan* (Basingstoke: Palgrave Macmillan, 2005), 1–16, and Dong-Won Kim, 'The Emergence of Theoretical Physics in Japan: Japanese Physics Community between the Two World Wars', *Annals of Science* 52 (1995).
52 Ito, 'Making Sense of Ryôshiron', 171, Kaneko, 'Einstein's Impact on Japanese Intellectuals', 354, Low, *Science and the Building of a New Japan*, 9, and Kim, 'Emergence of Theoretical Physics', 386.
53 Low, *Science and the Building of a New Japan*, 10, Kim, 'Emergence of Theoretical Physics', 386–7, and L. M. Brown et al., 'Cosmic Ray Research in Japan before World War II', *Progress of Theoretical Physics Supplement* 105 (1991): 25.

54 Ito, 'Making Sense of Ryôshiron', 173–206, Low, *Science and the Building of a New Japan*, 18–20, and Dong-Won Kim, *Yoshio Nishina: Father of Modern Physics in Japan* (London: Taylor and Francis, 2007), 1–15.
55 Kim, *Yoshio Nishina*, 15–46, Ito, 'Making Sense of Ryôshiron', 206–8, and Low, *Science and the Building of a New Japan*, 20.
56 Kim, *Yoshio Nishina*, 15–46, Low, *Science and the Building of a New Japan*, 20–2, and *A Century of Discovery: The History of RIKEN* (Wako: Riken, 2019), 22.
57 Ito, 'Making Sense of Ryôshiron', 208–9 and 239–45, Kim, *Yoshio Nishina*, 26–39, and Low, *Science and the Building of a New Japan*, 20–2.
58 Kim, *Yoshio Nishina*, 26–39, and Yuji Yazaki, 'How the Klein–Nishina Formula was Derived: Based on the Sangokan Nishina Source Materials', *Proceedings of the Japan Academy. Series B, Physical and Biological Sciences* 93 (2017).
59 Ito, 'Making Sense of Ryôshiron', 110–16 and 260, Low, *Science and the Building of a New Japan*, 22, and Kim, *Yoshio Nishina*, 55.
60 Ito, 'Making Sense of Ryôshiron', 261, Low, *Science and the Building of a New Japan*, 22, and Kim, *Yoshio Nishina*, 64.
61 Ito, 'Making Sense of Ryôshiron', 1, Low, *Science and the Building of a New Japan*, 106–7, Nicholas Kemmer, 'Hideki Yukawa, 23 January 1907–8 September 1981', *Biographical Memoirs of Fellows of the Royal Society* 29 (1983), L. M. Brown et al., 'Yukawa's Prediction of the Mesons', *Progress of Theoretical Physics Supplement* 105 (1991): 10, and Hideki Yukawa, *Tabibito (The Traveler)*, trans. L. Brown and R. Yoshida (Singapore: World Scientific, 1982), 10–11 and 36–7.
62 Ito, 'Making Sense of Ryôshiron', 280, Kim, 'Emergence of Theoretical Physics', 395, Low, *Science and the Building of a New Japan*, 106–7 and 119–21, Yukawa, *Tabibito*, 12, and Hideki Yukawa, *Creativity and Intuition: A Physicist Looks at East and West*, trans. John Bester (Tokyo: Kodansha International, 1973), 31–5.
63 Kim, 'Emergence of Theoretical Physics', 395–9, Low, *Science and the Building of a New Japan*, 106–7, and Yukawa, *Tabibito*, 170.
64 Ito, 'Making Sense of Ryôshiron', 280–1, Kim, 'Emergence of Theoretical Physics', 395, Low, *Science and the Building of a New Japan*, 108, Brown et al., 'Yukawa's Prediction of the Mesons', 14, and L. M. Brown et al., 'Particle Physics in Japan in the 1940s Including Meson Physics in Japan after the First Meson Paper', *Progress of Theoretical Physics Supplement* 105 (1991): 35–40.
65 Low, *Science and the Building of a New Japan*, 120, Yukawa, *Tabibito*, 24, Brown et al., 'Particle Physics in Japan', 35, and Hideki Yukawa, 'On the Interaction of Elementary Particles', *Proceedings of the Physico-Mathematical Society of Japan* 17 (1935).
66 Brown et al., 'Cosmic Ray Research in Japan', 31, Kim, 'Emergence of Theoretical Physics', 387, and Low, *Science and the Building of a New Japan*, 77–9.
67 Robert Anderson, *Nucleus and Nation: Scientists, International Networks, and Power in India* (Chicago: University of Chicago Press, 2010), 24–6, Pramod Naik, *Meghnad Saha: His Life in Science and Politics* (Cham: Springer, 2017), 32–3, and D. S. Kothari, 'Meghnad Saha, 1893–1956', *Biographical Memoirs of Fellows of the Royal Society* 5 (1960): 217–8.
68 Anderson, *Nucleus and Nation*, 24–6, Naik, *Meghnad Saha*, 32–3, and Kothari, 'Meghnad Saha', 217–9.

69 Anderson, *Nucleus and Nation*, 26–31, Naik, *Meghnad Saha*, 33–47, and Kothari, 'Meghnad Saha', 218–9.
70 Anderson, *Nucleus and Nation*, 26–31, Naik, *Meghnad Saha*, 33–47, and Kothari, 'Meghnad Saha', 218–9.
71 Anderson, *Nucleus and Nation*, 1–15 and 57, David Arnold, 'Nehruvian Science and Postcolonial India', *Isis* 104 (2013): 262–5, David Arnold, *Science, Technology and Medicine in Colonial India* (Cambridge: Cambridge University Press, 2000), 169–210, G. Venkataraman, *Journey into Light: Life and Science of C. V. Raman* (Bangalore: Indian Academy of Sciences, 1988), 457, and Benjamin Zachariah, *Developing India: An Intellectual and Social History, c. 1930–50* (New Delhi: Oxford University Press, 2005), 236–8.
72 Anderson, *Nucleus and Nation*, 23–35, Naik, *Meghnad Saha*, 48–65, Kothari, 'Meghnad Saha', 223–4, and Purabi Mukherji and Atri Mukhopadhyay, *History of the Calcutta School of Physical Sciences* (Singapore: Springer, 2018), 14–15.
73 Kothari, 'Meghnad Saha', 220–1, and Meghnad Saha, 'Ionization in the Solar Chromosphere', *Philosophical Magazine* 40 (1920).
74 Naik, *Meghnad Saha*, 94–123, Kothari, 'Meghnad Saha', 229, and Abha Sur, 'Scientism and Social Justice: Meghnad Saha's Critique of the State of Science in India', *Historical Studies in the Physical and Biological Sciences* 33 (2002).
75 Mukherji and Mukhopadhyay, *History of the Calcutta School*, 111–5, and Jagdish Mehra, 'Satyendra Nath Bose, 1 January 1894–4 February 1974', *Biographical Memoirs of Fellows of the Royal Society* 21 (1975): 118–20.
76 Anderson, *Nucleus and Nation*, 26–7, and Mehra, 'Satyendra Nath Bose', 118–20.
77 Anderson, *Nucleus and Nation*, 28, Mehra, 'Satyendra Nath Bose', 122, and Meghnad Saha and Satyendra Nath Bose, *The Principle of Relativity* (Calcutta: University of Calcutta, 1920).
78 Anderson, *Nucleus and Nation*, 41, Mehra, 'Satyendra Nath Bose', 123–9, and Rajinder Singh, *Einstein Rediscovered: Interactions with Indian Academics* (Düren: Shaker Verlag, 2019), 23.
79 Mehra, 'Satyendra Nath Bose', 123–9.
80 Mehra, 'Satyendra Nath Bose', 130–42, Singh, *Einstein Rediscovered*, 23, Wali Kameshwar, ed., *Satyendra Nath Bose, His Life and Times: Selected Works* (Hackensack: World Scientific Publishing, 2009), xxix, and Satyendra Nath Bose, 'Plancks Gesetz und Lichtquantenhypothese', *Zeitschrift für Physik* 26 (1924).
81 Singh, *Einstein Rediscovered*, 10, and Rasoul Sorkhabi, 'Einstein and the Indian Minds: Tagore, Gandhi and Nehru', *Current Science* 88 (2005): 1187–90.
82 Venkataraman, *Journey into Light*, 186–91 and 267, Mukherji and Mukhopadhyay, *History of the Calcutta School*, 53–5, S. Bhagavantam, 'Chandrasekhara Venkata Raman. 1888–1970', *Biographical Memoirs of Fellows of the Royal Society* 17 (1971): 569, and Chandrasekhara Venkata Raman, 'The Colour of the Sea', *Nature* 108 (1921): 367.
83 Raman, 'The Colour of the Sea', 367, Venkataraman, *Journey into Light*, 195–6, and Bhagavantam, 'Chandrasekhara Venkata Raman', 568–9.
84 Arnold, *Science, Technology and Medicine*, 169, and Chandrasekhara Venkata Raman, 'A New Radiation', *Indian Journal of Physics* 2 (1928).
85 Anderson, *Nucleus and Nation*, 65–7, and Venkataraman, *Journey into Light*, 255–66.
86 Venkataraman, *Journey into Light*, 389.

87 Venkataraman, *Journey into Light*, 318–9, Abha Sur, 'Dispersed Radiance: Women Scientists in C. V. Raman's Laboratory', *Meridians* 1 (2001), and Arvind Gupta, *Bright Sparks: Inspiring Indian Scientists from the Past* (Delhi: Indian National Academy of Sciences, 2012), 123–6.

88 Venkataraman, *Journey into Light*, 318–9, Sur, 'Dispersed Radiance', and Gupta, *Bright Sparks*, 115–8.

89 Venkataraman, *Journey into Light*, 459, Arnold, *Science, Technology and Medicine*, 210, and Anderson, *Nucleus and Nation*, 42.

90 David Holloway, *Stalin and the Bomb: The Soviet Union and Atomic Energy, 1939–1956* (New Haven: Yale University Press, 1994), 294, and Lawrence Sullivan and Nancy Liu-Sullivan, *Historical Dictionary of Science and Technology in Modern China* (Lanham: Rowman & Littlefield, 2015), 424.

8. Genetic States

1 Masao Tsuzuki, 'Report on the Medical Studies of the Effects of the Atomic Bomb', in *General Report Atomic Bomb Casualty Commission* (Washington, DC: National Research Council, 1947), 68–74, Susan Lindee, *Suffering Made Real: American Science and the Survivors at Hiroshima* (Chicago: University of Chicago Press, 1994), 24–5, Frank Putnam, 'The Atomic Bomb Casualty Commission in Retrospect', *Proceedings of the National Academy of Sciences* 95 (1998): 5246–7, and 'Damage Surveys in the Post-War Turmoil', Hiroshima Peace Memorial Museum, accessed 25 August 2020, http://www.pcf.city.hiroshima.jp/virtual/VirtualMuseum_e/exhibit_e/exh0307_e/exh03075_e.html.

2 'Japanese Material: Organization for Study of Atomic Bomb Casualties, Monthly Progress Reports', in *General Report Atomic Bomb Casualty Commission*, 16, John Beatty, 'Genetics in the Atomic Age: The Atomic Bomb Casualty Commission, 1947–1956', in *The Expansion of American Biology*, eds. Keith Benson, Janes Maienschein, and Ronald Rainger (New Brunswick: Rutgers University Press, 1991), 285 and 297, and Susan Lindee, 'What is a Mutation? Identifying Heritable Change in the Offspring of Survivors at Hiroshima and Nagasaki', *Journal of the History of Biology* 25 (1992).

3 Lindee, *Suffering Made Real*, 24–5 and 73–4, Lindee, 'What is a Mutation?', 232–3, Beatty, 'Genetics in the Atomic Age', 285–7, and Putnam, 'The Atomic Bomb Casualty Commission', 5426.

4 Lindee, *Suffering Made Real*, 178–84, and Lindee, 'What is a Mutation?', 234–45.

5 Lindee, 'What is a Mutation?', 250, and Vassiliki Smocovitis, 'Genetics behind Barbed Wire: Masuo Kodani, Émigré Geneticists, and Wartime Genetics Research at Manzanar Relocation Center', *Genetics* 187 (2011).

6 Smocovitis, 'Genetics behind Barbed Wire', Soraya de Chadarevian, *Heredity under the Microscope: Chromosomes and the Study of the Human Genome* (Chicago: University of Chicago Press, 2020), 5–6, and Masuo Kodani, 'The Supernumerary Chromosome of Man', *American Journal of Human Genetics* 10 (1958).

7 Lindee, 'What is a Mutation?', 232–3, Beatty, 'Genetics in the Atomic Age', 287–93, Lisa Onaga, 'Measuring the Particular: The Meanings of Low-Dose Radiation

Experiments in Post-1954 Japan', *Positions: Asia Critique* 26 (2018), Aya Homei, 'Fall-out from Bikini: The Explosion of Japanese Medicine', *Endeavour* 31 (2007), and Kaori Iida, 'Peaceful Atoms in Japan: Radioisotopes as Shared Technical and Socio-political Resources for the Atomic Bomb Casualty Commission and the Japanese Scientific Community in the 1950s', *Studies in History and Philosophy of Science Part C: Studies in History and Philosophy of Biological and Biomedical Sciences* 80 (2020).

8 Lindee, *Suffering Made Real*, 59–60, Iida, 'Peaceful Atoms in Japan', 2, and Onaga, 'Measuring the Particular', 271.

9 Beatty, 'Genetics in the Atomic Age', 312, and 'The Fourth Geneva Conference', *IAEA Bulletin* 13 (1971): 2–18.

10 James Watson, *The Double Helix: A Personal Account of the Discovery of the Structure of DNA* (London: Weidenfeld & Nicolson, 1968), Soraya de Chadarevian, *Designs for Life: Molecular Biology after World War II* (Cambridge: Cambridge University Press, 2002), and Francis Crick, 'On Protein Synthesis', *Symposia of the Society for Experimental Biology* 12 (1958): 161.

11 Susan Lindee, 'Scaling Up: Human Genetics as a Cold War Network', *Studies in History and Philosophy of Science Part C: Studies in History and Philosophy of Biological and Biomedical Sciences* 47 (2014), and Susan Lindee, 'Human Genetics after the Bomb: Archives, Clinics, Proving Grounds and Board Rooms', *Studies in History and Philosophy of Science Part C: Studies in History and Philosophy of Biological and Biomedical Sciences* 55 (2016).

12 Robin Pistorius, *Scientists, Plants and Politics: A History of the Plant Genetic Resources Movement* (Rome: International Plant Genetic Resources Institute, 1997), 55–7, Helen Curry, 'From Working Collections to the World Germplasm Project: Agricultural Modernization and Genetic Conservation at the Rockefeller Foundation', *History and Philosophy of the Life Sciences* 39 (2017), John Perkins, *Geopolitics and the Green Revolution: Wheat, Genes, and the Cold War* (Oxford: Oxford University Press, 1997), R. Douglas Hurt, *The Green Revolution in the Global South: Science, Politics, and Unintended Consequences* (Tuscaloosa: University of Alabama Press, 2020), Alison Bashford, *Global Population: History, Geopolitics, and Life on Earth* (New York: Columbia University Press, 2014), and David Grigg, 'The World's Hunger: A Review, 1930–1990', *Geography* 82 (1997): 201.

13 Perkins, *Geopolitics and the Green Revolution*, Joseph Cotter, *Troubled Harvest: Agronomy and Revolution in Mexico, 1880–2002* (Westport: Praeger, 2003), 249–50, and Bruce Jennings, *Foundations of International Agricultural Research: Science and Politics in Mexican Agriculture* (Boulder: CRC Press, 1988), 145.

14 Lindee, 'Human Genetics after the Bomb', de Chadarevian, *Designs for Life*, 50 and 74–5, Michelle Brattain, 'Race, Racism, and Antiracism: UNESCO and the Politics of Presenting Science to the Postwar Public', *American Historical Review* 112 (2007): 1387, and Elise Burton, *Genetic Crossroads: The Middle East and the Science of Human Heredity* (Stanford: Stanford University Press, 2021).

15 Naomi Oreskes and John Krige, eds., *Science and Technology in the Global Cold War* (Cambridge, MA: The MIT Press, 2014), Ana Barahona, 'Transnational Knowledge during the Cold War: The Case of the Life and Medical Sciences', *História, Ciências, Saúde-Manguinhos* 26 (2019), Heike Petermann, Peter Harper, and Susanne Doetz,

eds., *History of Human Genetics: Aspects of Its Development and Global Perspectives* (Cham: Springer, 2017), and Patrick Manning and Mat Savelli, eds., *Global Transformations in the Life Sciences, 1945–1980* (Pittsburgh: University of Pittsburgh Press, 2018).

16 Efraím Hernández Xolocotzi, 'Experiences in the Collection of Maize Germplasm', in *Recent Advances in the Conservation and Utilization of Genetic Resources*, ed. Nathan Russel (Mexico City: CIMMYT, 1988), and Elvin Stakman, Richard Bradfield, and Paul Christoph Mangelsdorf, *Campaigns Against Hunger* (Cambridge, MA: The Belknap Press, 1967), 61.

17 Cotter, *Troubled Harvest*, 11–12, and Curry, 'From Working Collections', 3–6.

18 Cotter, *Troubled Harvest*, 1–12, and Jennings, *Foundations of International Agricultural Research*, 1–37, 145, and 162.

19 Artemio Cruz León, Marcelino Ramírez Castro, Francisco Collazo-Reyes, Xóchitl Flores Vargas, 'La obra escrita de Efraím Hernández Xolocotzi, patrimonio y legado', *Revista de Geografía Agrícola* 50 (2013), 'Efraim Hernandez Xolocotzi', Instituto de Biología, Universidad Nacional Autónama de México, accessed 24 April 2020, http://www.ibiologia.unam.mx/jardin/gela/page4.html, 'Efraim Hernández Xolocotzi', Biodiversidad Mexicana, accessed 6 May 2020, https://www.biodiversidad.gob.mx/biodiversidad/curiosos/sXX/EfrainHdezX.php, and Edwin Wellhausen, Louis Roberts, Efraím Hernández Xolocotzi, and Paul Mangelsdorf, *Races of Maize in Mexico* (Cambridge, MA: The Bussey Institution, 1952), 9. I am very grateful to Ricardo Aguilar-González for sharing his knowledge of Mexican history and Nahuatl names so that I could better understand Efraím Hernández Xolocotzi's background. After completing this chapter, I was also introduced to the following PhD dissertation: Matthew Caire-Pérez, 'A Different Shade of Green: Efraím Hernández Xolocotzi, Chapingo, and Mexico's Green Revolution, 1950–1967' (PhD diss., University of Oklahoma, 2016), which gives a detailed account of Hernández's biography, particularly at 73–81, as well as his broader role in the Green Revolution.

20 Hernández, 'Experiences', 1–6, Edwin Wellhausen, 'The Indigenous Maize Germplasm Complexes of Mexico', in Russel, ed., *Recent Advances*, 18, Paul Mangelsdorf, *Corn: Its Origin, Evolution, and Improvement* (Cambridge, MA: Harvard University Press, 1974), 101–5, and Garrison Wilkes, 'Teosinte and the Other Wild Relatives of Maize', in Russel, ed., *Recent Advances*, 72.

21 Helen Curry, 'Breeding Uniformity and Banking Diversity: The Genescapes of Industrial Agriculture, 1935–1970', *Global Environment* 10 (2017), Mangelsdorf, *Corn*, 24 and 106, and Wellhausen, Roberts, Hernández, and Mangelsdorf, *Races of Maize*, 22.

22 Cotter, *Troubled Harvest*, 232, Mangesldorf, *Corn*, 101, Wellhausen, Roberts, Hernández, and Mangelsdorf, *Races of Maize*, 34, and Hernández, 'Experiences', 6.

23 Hernández, 'Experiences', 1, Cotter, *Troubled Harvest*, 192 and 234, Curry, 'From Working Collections', 6, and Jonathan Harwood, 'Peasant Friendly Plant Breeding and the Early Years of the Green Revolution in Mexico', *Agricultural History* 83 (2009).

24 Gisela Mateos and Edna Suárez Díaz, 'Mexican Science during the Cold War: An Agenda for Physics and the Life Sciences', *Ludus Vitalis* 20 (2012): 48–59, Ana Barahona, 'Medical Genetics in Mexico: The Origins of Cytogenetics and the Health Care System', *Historical Studies in the Natural Sciences* 45 (2015), José Alonso-Pavon and

Ana Barahona, 'Genetics, Radiobiology and the Circulation of Knowledge in Cold War Mexico, 1960–1980', in *The Scientific Dialogue Linking America, Asia and Europe between the 12th and the 20th Century*, ed. Fabio D'Angelo (Naples: Associazione culturale Viaggiatori, 2018), Thomas Glick, 'Science in Twentieth-Century Latin America', in *Ideas and Ideologies in Twentieth-Century Latin America*, ed. Leslie Bethel (Cambridge: Cambridge University Press, 1996), 309, Larissa Lomnitz, 'Hierarchy and Peripherality: The Organisation of a Mexican Research Institute', *Minerva* 17 (1979), and *Biomedical Research Policies in Latin America: Structures and Processes* (Washington, DC: Pan American Health Organization, 1965), 165–7.

25 Ana Barahona, Susana Pinar, and Francisco Ayala, 'Introduction and Institutionalization of Genetics in Mexico', *Journal of the History of Biology* 38 (2005): 287–9.

26 Barahona, Pinar, and Ayala, 'Introduction and Institutionalization', 287–9, Ana Barahona, 'Transnational Science and Collaborative Networks: The Case of Genetics and Radiobiology in Mexico, 1950–1970', *Dynamis* 35 (2015): 347–8, and Eucario López-Ochoterena, '*In Memoriam*: Rodolfo Félix Estrada (1924–1990)', Ciencias UNAM, accessed 3 July 2020, http://repositorio.fciencias.unam.mx:8080/xmlui/bitstream/handle/11154/143333/41VMemoriamRodolfo.pdf.

27 Alfonso León de Garay, Louis Levine, and J. E. Lindsay Carter, *Genetic and Anthropological Studies of Olympic Athletes* (New York: Academic Press, 1974), ix–xvi, 1–23, and 30.

28 Barahona, Pinar, and Ayala, 'Introduction and Institutionalization', 289, James Rupert, 'Genitals to Genes: The History and Biology of Gender Verification in the Olympics', *Canadian Bulletin of Medical History* 28 (2011), and De Garay, Levine, and Carter, *Genetic and Anthropological Studies*, ix–xvi, 1–23, and 30.

29 De Garay, Levine, and Carter, *Genetic and Anthropological Studies*, 43, 147, and 230, James Meade and Alan Parkes, eds., *Genetic and Environmental Factors in Human Ability* (London: Eugenics Society, 1966), Angela Saini, *Superior: The Return of Race Science* (London: Fourth Estate, 2019), and Alison Bashford, 'Epilogue: Where Did Eugenics Go?', in *The Oxford Handbook of the History of Eugenics*, eds. Alison Bashford and Philippa Levine (Oxford: Oxford University Press, 2010).

30 Ana Barahona and Francisco Ayala, 'The Emergence and Development of Genetics in Mexico', *Nature Reviews Genetics* 6 (2005): 860, Glick, 'Science in Twentieth-Century Latin America', 297, and Francisco Salzano, 'The Evolution of Science in a Latin-American Country: Genetics and Genomics in Brazil', *Genetics* 208 (2018).

31 Gita Gopalkrishnan, *M. S. Swaminathan: One Man's Quest for a Hunger-Free World* (Chennai: Sri Venkatesa Printing House, 2002), 8–24, and Hurt, *The Green Revolution in the Global South*, 45–6.

32 Gopalkrishnan, *M. S. Swaminathan*, 24–5.

33 Gopalkrishnan, *M. S. Swaminathan*, 28–9, Debi Prosad Burma and Maharani Chakravorty, 'Biochemistry: A Hybrid Science Giving Birth to Molecular Biology', in *History of Science, Philosophy, and Culture in Indian Civilization: From Physiology and Chemistry to Biochemistry*, eds. Debi Prosad Burma and Maharani Chakravorty (Delhi: Longman, 2011), vol. 13, part 2, 157, and David Arnold, 'Nehruvian Science and Postcolonial India', *Isis* 104 (2013): 366.

34 Gopalkrishnan, *M. S. Swaminathan*, 35–42.

35 Gopalkrishnan, *M. S. Swaminathan*, 43–4, Cotter, *Troubled Harvest*, 252, Curry, 'From Working Collections', 7–9, Hurt, *The Green Revolution in the Global South*, 46, and Srabani Sen, '1960–1999: Four Decades of Biochemistry in India', *Indian Journal of History of Science* 46 (2011): 175–9.

36 Gopalkrishnan, *M. S. Swaminathan*, 45, Hurt, *The Green Revolution in the Global South*, 46, and 'Dilbagh Athwal, Geneticist and "Father of the Wheat Revolution" – Obituary', The Telegraph, accessed 2 September 2020, https://www.telegraph.co.uk/obituaries/2017/05/22/dilbagh-athwal-geneticist-father-wheat-revolution-obituary/.

37 Arnold, 'Nehruvian Science', 362 and 368, Sen, 'Four Decades of Biochemistry', 175, Sigrid Schmalzer, *Red Revolution, Green Revolution: Scientific Farming in Socialist China* (Chicago: University of Chicago Press, 2016), 5.

38 Jawaharlal Nehru, *Jawaharlal Nehru on Science and Society: A Collection of His Writings and Speeches* (New Delhi: Nehru Memorial Museum and Library, 1988), 137–8, and Robert Anderson, *Nucleus and Nation: Scientists, International Networks, and Power in India* (Chicago: University of Chicago Press, 2010), 4 and 237.

39 Indira Chowdhury, *Growing the Tree of Science: Homi Bhabha and the Tata Institute of Fundamental Research* (New Delhi: Oxford University Press, 2016), 175, Krishnaswamy VijayRaghavan, 'Obaid Siddiqi: Celebrating His Life in Science and the Cultural Transmission of Its Values', *Journal of Neurogenetics* 26 (2012), Zinnia Ray Chaudhuri, 'Her Father's Voice: A Photographer Pays Tribute to Her Celebrated Scientist-Father', Scroll.in, accessed 5 May 2020, https://scroll.in/roving/802600/her-fathers-voice-a-photographer-pays-tribute-to-her-celebrated-scientist-father, and 'India Mourns Loss of "Aristocratic" & Gutsy Molecular Biology Guru', Nature India, accessed 4 May 2020, https://www.natureasia.com/en/nindia/article/10.1038/nindia.2013.102.

40 'India Mourns', VijayRaghavan, 'Obaid Siddiqi', 257–9, and Chowdhury, *Growing the Tree of Science*, 175.

41 VijayRaghavan, 'Obaid Siddiqi', 257–9, Chowdhury, *Growing the Tree of Science*, 175, and Alan Garen and Obaid Siddiqi, 'Suppression of Mutations in the Alkaline Phosphatase Structural Cistron of *E. coli*', *Proceedings of the National Academy of Sciences of the United States of America* 48 (1962).

42 Chowdhury, *Growing the Tree of Science*, 175–8.

43 Chowdhury, *Growing the Tree of Science*, 181–2, VijayRaghavan, 'Obaid Siddiqi', 259, and Obaid Siddiqi and Seymour Benzer, 'Neurophysiological Defects in Temperature-Sensitive Paralytic Mutants of Drosophila Melanogaster', *Proceedings of the National Academy of Sciences of the United States of America* 73 (1976).

44 Chowdhury, *Growing the Tree of Science* 183, Krishnaswamy VijayRaghavan and Michael Bate, 'Veronica Rodrigues (1953–2010)', *Science* 330 (2010), Namrata Gupta and A. K. Sharma, 'Triple Burden on Women Academic Scientists', in *Women and Science in India: A Reader*, ed. Neelam Kumar (Delhi: Oxford University Press, 2009), 236, and Malathy Duraisamy and P. Duraisamy, 'Women's Participation in Scientific and Technical Education and Labour Markets in India', in Kumar, ed., *Women and Science in India*, 293.

45 Chowdhury, *Growing the Tree of Science*, 183, and VijayRaghavan and Bate, 'Veronica Rodrigues', 1493–4.

46 Chowdhury, *Growing the Tree of Science* 183, and VijayRaghavan and Bate, 'Veronica Rodrigues', 1493–4.
47 Chowdhury, *Growing the Tree of Science* 183, VijayRaghavan and Bate, 'Veronica Rodrigues', 1493–4, and Veronica Rodrigues and Obaid Siddiqi, 'Genetic Analysis of Chemosensory Path', *Proceedings of the Indian Academy of Sciences* 87 (1978).
48 Arnold, 'Nehruvian Science', 368, and 'Teaching', Indian Agricultural Research Institute, accessed 2 September 2020, https://www.iari.res.in/index.php?option=com_content&view=article&id=284&Itemid=889.
49 VijayRaghavan and Bate, 'Veronica Rodrigues', 1493.
50 Laurence Schneider, *Biology and Revolution in Twentieth-Century China* (Lanham: Rowman & Littlefield, 2005), 123, Eliot Spiess, 'Ching Chun Li, Courageous Scholar of Population Genetics, Human Genetics, and Biostatistics: A Living History Essay', *American Journal of Medical Genetics* 16 (1983): 610–11, and Aravinda Chakravarti, 'Ching Chun Li (1912–2003): A Personal Remembrance of a Hero of Genetics', *The American Journal of Human Genetics* 74 (2004): 790.
51 Schneider, *Biology and Revolution*, 122, and Spiess, 'Ching Chun Li', 604–5.
52 Schneider, *Biology and Revolution*, 117–44, Li Peishan, 'Genetics in China: The Qingdao Symposium of 1956', *Isis* 79 (1988), and Trofim Lysenko, 'Concluding Remarks on the Report on the Situation in the Biological Sciences, in *Death of a Science in Russia: The Fate of Genetics as Described in Pravda and Elsewhere*, ed. Conway Zirkle (Philadelphia: University of Pennsylvania Press, 1949), 257.
53 Schneider, *Biology and Revolution*, 117–44, Li, 'Genetics in China', 228, and Mao Zedong, 'On the Correct Handling of Contradictions among the People', in *Selected Readings from the Works of Mao Tsetung* (Peking: Foreign Languages Press, 1971), 477–8.
54 Li Jingzhun, 'Genetics Dies in China', *Journal of Heredity* 41 (1950).
55 Spiess, 'Ching Chun Li', 613.
56 Schmalzer, *Red Revolution*, 27, Sigrid Schmalzer, 'On the Appropriate Use of Rose-Colored Glasses: Reflections on Science in Socialist China', *Isis* 98 (2007), and Chunjuan Nancy Wei and Darryl E. Brock, eds., *Mr. Science and Chairman Mao's Cultural Revolution: Science and Technology in Modern China* (Lanham: Lexington Books, 2013).
57 Schmalzer, *Red Revolution*, 4, Schneider, *Biology and Revolution*, 3 and 196, Jack Harlan, 'Plant Breeding and Genetics', in *Science in Contemporary China*, ed. Leo Orleans (Stanford: Stanford University Press, 1988), 296–7, John Lewis and Litai Xue, *China Builds the Bomb* (Stanford: Stanford University Press, 1991), and Mao Zedong, *Speech at the Chinese Communist Party's National Conference on Propaganda Work* (Beijing: Foreign Languages Press, 1966), 3.
58 Schneider, *Biology and Revolution*, 169–77, Li, 'Genetics in China', 230–5, Yu Guangyuan, 'Speeches at the Qingdao Genetics Conference of 1956', in *Chinese Studies in the History and Philosophy of Science and Technology*, eds. Fan Dainian and Robert Cohen (Dordrecht: Kluwer, 1996), 27–34, and Karl Marx, *The Collected Works of Karl Marx and Frederick Engels*, trans. Victor Schnittke and Yuri Sdobnikov (London: Lawrence & Wishart, 1987), 29:263.
59 Schmalzer, *Red Revolution*, 38–9.

60 Schmalzer, *Red Revolution*, 73, Deng Xiangzi and Deng Yingru, *The Man Who Puts an End to Hunger: Yuan Longping, 'Father of Hybrid Rice'* (Beijing: Foreign Languages Press, 2007), 29–37, and Yuan Longping, *Oral Autobiography of Yuan Longping*, trans. Zhao Baohua and Zhao Kuangli (Nottingham: Aurora Publishing, 2014), Kindle Edition, loc. 492 and 736.

61 Schneider, *Biology and Revolution*, 13, Schmalzer, *Red Revolution*, 4, 40–1, and 73, Deng and Deng, *Yuan Longping*, 30, and Yuan, *Oral Autobiography*, loc. 626 and 756.

62 Schmalzer, *Red Revolution*, 75, Deng and Deng, *Yuan Longping*, 42 and 60–1, and Yuan, *Oral Autobiography*, loc. 797.

63 Schmalzer, *Red Revolution*, 75.

64 Schmalzer, *Red Revolution*, 75, and Deng and Deng, *Yuan Longping*, 60–1.

65 Schmalzer, *Red Revolution*, 86, Deng and Deng, *Yuan Longping*, 88–98, and Yuan, *Oral Autobiography*, loc. 1337 and 1463.

66 Schmalzer, *Red Revolution*, 75, and Yuan, *Oral Autobiography*, loc. 1337 and 1463.

67 Schmalzer, *Red Revolution*, 4, and 'Breeding Program Management', International Rice Research Institute, accessed 2 September 2020, http://www.knowledgebank.irri.org/ricebreedingcourse/Hybrid_Rice_Breeding_&_Seed_Production.htm.

68 Nadia Abu El-Haj, *The Genealogical Science: The Search for Jewish Origins and the Politics of Epistemology* (Chicago: University of Chicago Press, 2012), 86–98, Nurit Kirsh, 'Population Genetics in Israel in the 1950s: The Unconscious Internalization of Ideology', *Isis* 94 (2003), Nurit Kirsh, 'Genetic Studies of Ethnic Communities in Israel: A Case of Values-Motivated Research', in *Jews and Sciences in German Contexts*, eds. Ulrich Charpa and Ute Deichmann (Tübingen: Mohr Sibeck, 2007), 182, and Burton, *Genetic Crossroads*, 114.

69 Burton, *Genetic Crossroads*, 114, and El-Haj, *The Genealogical Science*, 87.

70 Burton, *Genetic Crossroads*, 104–5 and 114–5.

71 El-Haj, *The Genealogical Science*, 87–97, and Joseph Gurevitch and E. Margolis, 'Blood Groups in Jews from Iraq', *Annals of Human Genetics* 19 (1955).

72 *Facts and Figures* (New York: Israel Office of Information, 1955), 56–9, Moshe Prywes, ed., *Medical and Biomedical Research in Israel* (Jerusalem: Hebrew University of Jerusalem, 1960), xiii, 12–18, and 33–9, and Yakov Rabkin, 'Middle East', in *The Cambridge History of Science: Modern Science in National, Transnational, and Global Context*, eds. Hugh Slotten, Ronald Numbers, and David Livingstone (Cambridge: Cambridge University Press, 2020), 424, 434–5, and 438–43.

73 Rabkin, 'Middle East', 424–43, Arnold Reisman, 'Comparative Technology Transfer: A Tale of Development in Neighboring Countries, Israel and Turkey', *Comparative Technology Transfer and Society* 3 (2005): 331, Burton, *Genetic Crossroads*, 107–13, 138–50, and 232–9, and Murat Ergin, *'Is the Turk a White Man?': Race and Modernity in the Making of Turkish Identity* (Leiden: Brill, 2017).

74 Kirsh, 'Population Genetics', 641, Shifra Shvarts, Nadav Davidovitch, Rhona Seidelman, and Avishay Goldberg, 'Medical Selection and the Debate over Mass Immigration in the New State of Israel (1948–1951)', *Canadian Bulletin of Medical History* 22 (2005), and Roselle Tekiner, 'Race and the Issue of National Identity in Israel', *International Journal of Middle East Studies* 23 (1991).

75 Burton, *Genetic Crossroads*, 108 and 146, El-Haj, *The Genealogical Science*, 63, Kirsh, 'Population Genetics', 635, and Joyce Donegani, Karima Ibrahim, Elizabeth Ikin, and Arthur Mourant, 'The Blood Groups of the People of Egypt', *Heredity* 4 (1950).
76 Nurit Kirsh, 'Geneticist Elisabeth Goldschmidt: A Two-Fold Pioneering Story', *Israel Studies* 9 (2004).
77 Burton, *Genetic Crossroads*, 157–9, Batsheva Bonné, 'Chaim Sheba (1908–1971)', *American Journal of Physical Anthropology* 36 (1972), Raphael Falk, *Zionism and the Biology of Jews* (Cham: Springer, 2017), 145–8, and Elisabeth Goldschmidt, ed., *The Genetics of Migrant and Isolate Populations* (New York: The Williams and Wilkins Company, 1973), v.
78 Goldschmidt, *The Genetics of Migrant and Isolate Populations*, Burton, *Genetic Crossroads*, 161–3, El-Haj, *The Genealogical Science*, 63–5 and 99, Kirsh, 'Population Genetics', 653, and Kirsh, 'Geneticist Elisabeth Goldschmidt', 90.
79 Burton, *Genetic Crossroads*, 161–3, El-Haj, *The Genealogical Science*, 63–5 and 99, Kirsh, 'Population Genetics', 653, Kirsh, 'Geneticist Elisabeth Goldschmidt', 90, Newton Freire-Maia, 'The Effect of the Load of Mutations on the Mortality Rate in Brazilian Populations', in *The Genetics of Migrant and Isolate Populations*, ed. Elisabeth Goldschmidt (New York: The Williams and Wilkins Company, 1973), 221–2, and Katumi Tanaka, 'Differences between Caucasians and Japanese in the Incidence of Certain Abnormalities', in Goldschmidt, ed., *The Genetics of Migrant and Isolate Populations*.
80 El-Haj, *The Genealogical Science*, 86, Arthur Mourant, *The Distribution of the Human Blood Groups* (Oxford: Blackwell Scientific Publishing, 1954), 1, Michelle Brattain, 'Race, Racism, and Antiracism: UNESCO and the Politics of Presenting Science to the Postwar Public', *American Historical Review* 112 (2007), and *Four Statements on Race* (Paris: UNESCO, 1969), 18.
81 Burton, *Genetic Crossroads*, 96 and 103, El-Haj, *The Genealogical Science*, 1–8, and Arthur Mourant, Ada Kopeć, and Kazimiera Domaniewska-Sobczak, *The Distribution of the Human Blood Groups and Other Polymorphisms*, 2nd edn (London: Oxford University Press, 1976), 79–83.
82 Aaron Rottenberg, 'Daniel Zohary (1926–2016)', *Genetic Resources and Crop Evolution* 64 (2017).
83 Rottenberg, 'Daniel Zohary', 1102–3, and Jack Harlan and Daniel Zohary, 'Distribution of Wild Wheats and Barley', *Science* 153 (1966): 1074.
84 Rottenberg, 'Daniel Zohary', 1104–5, Harlan and Zohary, 'Distribution of Wild Wheats and Barley', 1076, Pistorius, *Scientists, Plants and Politics*, 17, and Daniel Zohary and Maria Hopf, *Domestication of Plants in the Old World* (Oxford: Clarendon Press, 1988), 2 and 8.
85 Zohary and Hopf, *Domestication of Plants*, 8, and Prywes, *Medical and Biomedical Research*, 155.
86 Burton, *Genetic Crossroads*, 17.
87 Burton, *Genetic Crossroads*, 128–50, 167–75, and 219–41.
88 'June 2000 White House Event', National Human Genome Research Institute, accessed 1 September 2020, https://www.genome.gov/10001356/june-2000-white-house-event.

89 'June 2000 White House Event'.

90 'June 2000 White House Event' and 'Fiscal Year 2001 President's Budget Request for the National Human Genome Research Institute', National Human Genome Research Institute, accessed 1 September 2020, https://www.genome.gov/10002083/2000-release-fy-2001-budget-request.

91 Nancy Stepan, 'Science and Race: Before and after the Human Genome Project', *Socialist Register* 39 (2003), Sarah Zhang, '300 Million Letters of DNA are Missing from the Human Genome', The Atlantic, accessed 1 September 2020, https://www.theatlantic.com/science/archive/2018/11/human-genome-300-million-missing-letters-dna/576481/, Elise Burton, 'Narrating Ethnicity and Diversity in Middle Eastern National Genome Projects', *Social Studies of Science* 48 (2018), Projit Bihari Mukharji, 'The Bengali Pharaoh: Upper-Caste Aryanism, Pan-Egyptianism, and the Contested History of Biometric Nationalism in Twentieth-Century Bengal', *Comparative Studies in Society and History* 59 (2017): 452, 'The Indian Genome Variation database (IGVdb): A Project Overview', *Human Genetics* 119 (2005), 'Mission', Genome Russia Project, accessed 1 September 2020, http://genomerussia.spbu.ru, and 'Summary', Han Chinese Genomes, accessed 1 September 2020, https://www.hanchinesegenomes.org/HCGD/data/summary.

92 David Cyranoski, 'China Expands DNA Data Grab in Troubled Western Region', *Nature News* 545 (2017), Sui-Lee Wee, 'China Uses DNA to Track Its People, with the Help of American Expertise', The New York Times, accessed 1 September 2020, https://www.nytimes.com/2019/02/21/business/china-xinjiang-uighur-dna-thermo-fisher.html, 'Ethnical Non Russian Groups', Genome Russian Project, accessed 1 September 2020, http://genomerussia.spbu.ru/?page_id=862&lang=en, and 'Trump Administration to Expand DNA Collection at Border and Give Data to FBI', The Guardian, accessed 20 February 2021, https://www.theguardian.com/us-news/2019/oct/02/us-immigration-border-dna-trump-administration.

Epilogue: The Future of Science

1 'Harvard University Professor and Two Chinese Nationals Charged in Three Separate China Related Cases', Department of Justice, accessed 20 September 2020, https://www.justice.gov/opa/pr/harvard-university-professor-and-two-chinese-nationals-charged-three-separate-china-related, 'Affidavit in Support of Application for Criminal Complaint', Department of Justice, accessed 20 September 2020, https://www.justice.gov/opa/press-release/file/1239796/download, and 'Harvard Chemistry Chief's Arrest over China Links Shocks Researchers', Nature, accessed 4 April 2020, https://www.nature.com/articles/d41586-020-00291-2.

2 'Harvard University Professor and Two Chinese Nationals Charged', 'Affidavit in Support of Application for Criminal Complaint', and 'Harvard Chemistry Chief's Arrest'.

3 'Remarks Delivered by FBI Boston Division Special Agent in Charge Joseph R. Bonavolonta Announcing Charges against Harvard University Professor and Two Chinese Nationals', Federal Bureau of Investigation, accessed 20 September 2020,

https://www.fbi.gov/contact-us/field-offices/boston/news/press-releases/remarks-delivered-by-fbi-boston-special-agent-in-charge-joseph-r-bonavolonta-announcing-charges-against-harvard-university-professor-and-two-chinese-nationals, Elizabeth Gibney, 'UC Berkeley Bans New Research Funding from Huawei', *Nature* 566 (2019), Andrew Silver, Jeff Tollefson, and Elizabeth Gibney, 'How US–China Political Tensions are Affecting Science', *Nature* 568 (2019), Mihir Zaveri, 'Wary of Chinese Espionage, Houston Cancer Center Chose to Fire 3 Scientists', The New York Times, accessed 7 December 2020, https://www.nytimes.com/2019/04/22/health/md-anderson-chinese-scientists.html, and 'Meng Wanzhou: Questions over Huawei Executive's Arrest as Legal Battle Continues', BBC News, accessed 16 December 2020, https://www.bbc.co.uk/news/world-us-canada-54756044.

4 World Bank National Accounts Data, and OECD National Accounts Data Files, accessed 16 February 2021, https://data.worldbank.org. See comparative data for China and the United States, 1982–2019, for 'GDP growth (annual %)', 'GDP (current US$)', and 'GDP, PPP (current international $)'. 'China Overtakes Japan as World's Second-Biggest Economy', BBC News, accessed 20 February 2021, https://www.bbc.co.uk/news/business-12427321. See also Thomas Piketty, *Capital in the Twenty-First Century* (Cambridge, MA: Harvard University Press, 2014), 78 and 585, and Jude Woodward, *The US vs China: Asia's New Cold War?* (Manchester: Manchester University Press, 2017) for a general account of both the geopolitics and the economics.

5 Piketty, *Capital in the Twenty-First Century*, 31 and 412.

6 'Notice of the State Council: New Generation of Artificial Intelligence Development Plan', Foundation for Law and International Affairs, accessed 12 December 2020, https://flia.org/wp-content/uploads/2017/07/A-New-Generation-of-Artificial-Intelligence-Development-Plan-1.pdf (translation by Flora Sapio, Weiming Chen, and Adrian Lo), 'Home', Beijing Academy of Artificial Intelligence, accessed 13 December 2020, https://www.baai.ac.cn/en, and Sarah O'Meara, 'China's Ambitious Quest to Lead the World in AI by 2030', *Nature* 572 (2019).

7 'New Generation of Artificial Intelligence Development Plan', and Kai-Fu Lee, *AI Superpowers: China, Silicon Valley, and the New World Order* (New York: Houghton Mifflin Harcourt, 2018), 227.

8 Huiying Liang et al., 'Evaluation and Accurate Diagnoses of Pediatric Diseases Using Artificial Intelligence', *Nature Medicine* 25 (2019), and Tanveer Syeda-Mahmood, 'IBM AI Algorithms Can Read Chest X-Rays at Resident Radiologist Levels', IBM Research Blog, accessed 16 December 2020, https://www.ibm.com/blogs/research/2020/11/ai-x-rays-for-radiologists/.

9 Lee, *AI Superpowers*, 14–17, and Drew Harwell and Eva Dou, 'Huawei Tested AI Software That Could Recognize Uighur Minorities and Alert Police, Report Says', Washington Post, accessed 16 December 2020, https://www.washingtonpost.com/technology/2020/12/08/huawei-tested-ai-software-that-could-recognize-uighur-minorities-alert-police-report-says/.

10 Karen Hao, 'The Future of AI Research is in Africa', MIT Technology Review, accessed 16 December 2020, https://www.technologyreview.com/2019/06/21/134820/ai-africa-machine-learning-ibm-google/, and 'Moustapha Cissé', African Institute

for Mathematical Sciences, accessed 13 December 2020, https://nexteinstein.org/person/moustapha-cisse/.

11 Shan Jie, 'China Exports Facial ID Technology to Zimbabwe', Global Times, accessed 14 December 2020, https://www.globaltimes.cn/content/1097747.shtml, and Amy Hawkins, 'Beijing's Big Brother Tech Needs African Faces', Foreign Policy, accessed 14 December 2020, https://foreignpolicy.com/2018/07/24/beijings-big-brother-tech-needs-african-faces/.

12 Elizabeth Gibney, 'Israel–Arab Peace Accord Fuels Hope for Surge in Scientific Research', *Nature* 585 (2020).

13 Eliran Rubin, 'Tiny IDF Unit is Brains behind Israel Army Artificial Intelligence', Haaretz, accessed 12 December 2020, https://www.haaretz.com/israel-news/tiny-idf-unit-is-brains-behind-israeli-army-artificial-intelligence-1.5442911, and Jon Gambrell, 'Virus Projects Renew Questions about UAE's Mass Surveillance', Washington Post, accessed 12 December 2020, https://www.washingtonpost.com/world/the_americas/virus-projects-renew-questions-about-uaes-mass-surveillance/2020/07/09/4c9a0f42-c1ab-11ea-8908-68a2b9eae9e0_story.html.

14 Agence France-Presse, 'UAE Successfully Launches Hope Probe', The Guardian, accessed 20 November 2020, http://www.theguardian.com/science/2020/jul/20/uae-mission-mars-al-amal-hope-space, and Elizabeth Gibney, 'How a Small Arab Nation Built a Mars Mission from Scratch in Six Years', Nature, accessed 9 July 2020, https://www.nature.com/immersive/d41586-020-01862-z/index.html.

15 Gibney, 'How a Small Arab Nation', and Sarwat Nasir, 'UAE to Sign Agreement with Virgin Galactic for Spaceport in Al Ain Airport', Khaleej Times, accessed 16 December 2020, https://www.khaleejtimes.com/technology/uae-to-sign-agreement-with-virgin-galactic-for-spaceport-in-al-ain-airport.

16 'UAE Successfully Launches Hope Probe' and Jonathan Amos, 'UAE Hope Mission Returns First Image of Mars', BBC News, accessed 16 February 2021, https://www.bbc.co.uk/news/science-environment-56060890.

17 Smriti Mallapaty, 'How China is Planning to Go to Mars amid the Coronavirus Outbreak', *Nature* 579 (2020), 'China Becomes Second Nation to Plant Flag on the Moon', BBC News, accessed 4 December 2020, https://www.bbc.com/news/world-asia-china-55192692, and Jonathan Amos, 'China Mars Mission: Tianwen-1 Spacecraft Enters into Orbit', BBC News, accessed 16 February 2021, https://www.bbc.co.uk/news/science-environment-56013041.

18 Çağrı Mert Bakırcı-Taylor, 'Turkey Creates Its First Space Agency', *Nature* 566 (2019), Sanjeev Miglani and Krishna Das, 'Modi Hails India as Military Space Power after Anti-Satellite Missile Test', Reuters, accessed 16 December 2020, https://uk.reuters.com/article/us-india-satellite/modi-hails-india-as-military-space-power-after-anti-satellite-missile-test-idUKKCN1R80IA, and Umar Farooq, 'The Second Drone Age: How Turkey Defied the U.S. and Became a Killer Drone Power', The Intercept, accessed 16 February 2021, https://theintercept.com/2019/05/14/turkey-second-drone-age/.

19 John Houghton, Geoffrey Jenkins, and J. J. Ephraums, eds., *Climate Change: The IPCC Scientific Assessment* (Cambridge: Cambridge University Press, 1990), xi–xii and 343–58.

20 Matt McGrath, 'Climate Change: China Aims for "Carbon Neutrality" by 2060', BBC News, accessed 13 December 2020, https://www.bbc.com/news/science-environment-54256826, 'China's Top Scientists Unveil Road Map to 2060 Goal', The Japan Times, accessed 13 December 2020, https://www.japantimes.co.jp/news/2020/09/29/asia-pacific/science-health-asia-pacific/china-climate-change-road-map-2060/, and 'Division of New Energy and Material Chemistry', Tsinghua University Institute of Nuclear and New Energy Technology, accessed 13 December 2020, http://www.inet.tsinghua.edu.cn/publish/ineten/5685/index.html.

21 *Digital Belt and Road Program: Science Plan* (Beijing: Digital Belt and Road Program, 2017), 1–25 and 93–4, Ehsan Masood, 'Scientists in Pakistan and Sri Lanka Bet Their Futures on China', Nature, accessed 3 May 2019, https://www.nature.com/articles/d41586-019-01125-6, and Anatol Lieven, *Climate Change and the Nation State: The Realist Case* (London: Allen Lane, 2020), xi–xxiv, 1–35, and 139–46.

22 Christoph Schumann, 'SASSCAL's Newly Appointed Executive Director – Dr Jane Olwoch', Southern African Science Service Centre for Climate Change and Adaptive Land Management, accessed 16 December 2020, https://www.sasscal.org/sasscals-newly-appointed-executive-director-dr-jane-olwoch/, and *Climate Change and Adaptive Land Management in Southern Africa* (Göttingen: Klaus Hess Publishers, 2018).

23 Carolina Vera, 'Farmers Transformed How We Investigate Climate', *Nature* 562 (2018).

24 Lee, *AI Superpowers*.

25 Shan Lu et al., 'Racial Profiling Harms Science', *Science* 363 (2019), Catherine Matacic, 'Uyghur Scientists Swept Up in China's Massive Detentions', Science, accessed 10 October 2020, https://www.sciencemag.org/news/2019/10/there-s-no-hope-rest-us-uyghur-scientists-swept-china-s-massive-detentions, Declan Butler, 'Prominent Sudanese Geneticist Freed from Prison as Dictator Ousted', Nature, accessed 17 December 2020, https://www.nature.com/articles/d41586-019-01231-5, Alison Abbott, 'Turkish Science on the Brink', *Nature* 542 (2017), and John Pickrell, '"Landscape of Fear" Forces Brazilian Rainforest Researchers into Anonymity', Nature Index, accessed 6 December 2020, https://www.natureindex.com/news-blog/landscape-of-fear-forces-brazilian-forest-researchers-into-anonymity.

Acknowledgements

I would like to begin by expressing my thanks to the many scholars who have done so much to transform the field of the history of science in recent years. Whereas the history of science used to be dominated by European and American case studies, there is now a wealth of detailed literature on the wider world. Without this scholarship, much of which was published in the last decade or so, this book would not have been possible. I hope that by connecting this work together, I have been able to show just how central it is to the story of the origins of modern science.

Throughout the process of researching and writing this book, many people generously shared their expertise on different regions, languages, and periods of history. In this respect I would like to thank Ricardo Aguilar-González, David Arnold, Somak Biswas, Mary Brazelton, Janet Browne, Elise Burton, Michael Bycroft, Rémi Dewière, Rebecca Earle, Anne Gerritsen, Nicolás Gómez Baeza, Rob Iliffe, Nick Jardine, Guido van Meersbergen, Projit Bihari Mukharji, Edwin Rose, Simon Schaffer, Jim Secord, Katayoun Shafiee, Claire Shaw, Tom Simpson, Charu Singh, Ben Smith, Miki Sugiura, and Simon Werrett. I am also very grateful to my friends in the sciences who read various chapters, particularly Johannes Knolle at the Technical University of Munich and Michael Shaw at the University of Reading.

Over the past four years, I've had the great pleasure of being a member of the Department of History at the University of Warwick. I would like to thank all my colleagues at Warwick for being so supportive, and for providing such an intellectually stimulating place to work. Thank you also to Rebecca Earle, our head of department, for all her assistance during an exceptionally difficult period, and also for so generously sharing her expertise on the history of the Americas. The Global History and Culture Centre at Warwick has similarly been a constant source of both ideas and friendship, for which I am very grateful.

Prior to joining Warwick, I spent a decade at the University of Cambridge. I entered Cambridge to read computer science, but left as a historian.

That is partly a testament to the wonderful flexibility of the degree course, but also to the individuals who helped me along the way. In particular, I would like to thank Jim Secord, who supervised my PhD in the Department of History and Philosophy of Science at Cambridge, as well as Simon Schaffer, who supervised my MPhil work and has been a great mentor ever since. Thank you also to Janet Browne at Harvard University, who examined my PhD, and has supported me at all stages of my career. I would like to say a special thank you to Sujit Sivasundaram, my other PhD examiner. In 2010, Sujit published an exceptionally important article on the global history of science. This article totally transformed how I thought about the field. Reading it convinced me to return to Cambridge to undertake a PhD. I would like to thank Sujit for introducing me to the global history of science, and also for being such a wonderful mentor over the past ten years.

Publishing this book with Viking has been a real joy. I would like to thank my editors, Connor Brown and Daniel Crewe, for their support and enthusiasm throughout this project. They helped in all kinds of ways, particularly in encouraging me to refine my argument. Thanks also to the rest of the team at Penguin Random House – in production, marketing, publicity, sales, and rights, amongst others – as well as to Jack Ramm for his early editorial work on the book. I would like to extend my thanks to Alexander Littlefield, Olivia Bartz, and the whole team at Houghton Mifflin Harcourt for their work on the American edition.

A big thank you to my agent Ben Clark at the Soho Agency. I'm sure everyone thinks their agent is the best. But Ben really is! He has been a great friend and champion, helping me to navigate the world of publishing, talking through my ideas, and reading my work. I can't imagine a better agent.

Beyond the world of academia and publishing, I would like to thank a number of other individuals, without whom I could not have written this book. I am extremely grateful for the team at Hinchingbrooke Hospital (North West Anglia NHS Foundation Trust), particularly to David, Rowena, and Syed. Their care, over nearly a decade, has kept me going, quite literally.

This book is dedicated to my wife, Alice, and my mother, Nancy. Writing this book was especially challenging, and without their support, I simply could not have done it. Thank you, Alice. And thank you, Mum. For everything.

Index

Page references in *italics* indicate images.